Graduate Texts in Mathematics 158

Graduate Texts in Mathematics

1 TAKEUTI/ZARING. Introduction to Axiomatic Set Theory. 2nd ed.
2 OXTOBY. Measure and Category. 2nd ed.
3 SCHAEFER. Topological Vector Spaces. 2nd ed.
4 HILTON/STAMMBACH. A Course in Homological Algebra. 2nd ed.
5 MAC LANE. Categories for the Working Mathematician. 2nd ed.
6 HUGHES/PIPER. Projective Planes.
7 J.-P. SERRE. A Course in Arithmetic.
8 TAKEUTI/ZARING. Axiomatic Set Theory.
9 HUMPHREYS. Introduction to Lie Algebras and Representation Theory.
10 COHEN. A Course in Simple Homotopy Theory.
11 CONWAY. Functions of One Complex Variable I. 2nd ed.
12 BEALS. Advanced Mathematical Analysis.
13 ANDERSON/FULLER. Rings and Categories of Modules. 2nd ed.
14 GOLUBITSKY/GUILLEMIN. Stable Mappings and Their Singularities.
15 BERBERIAN. Lectures in Functional Analysis and Operator Theory.
16 WINTER. The Structure of Fields.
17 ROSENBLATT. Random Processes. 2nd ed.
18 HALMOS. Measure Theory.
19 HALMOS. A Hilbert Space Problem Book. 2nd ed.
20 HUSEMOLLER. Fibre Bundles. 3rd ed.
21 HUMPHREYS. Linear Algebraic Groups.
22 BARNES/MACK. An Algebraic Introduction to Mathematical Logic.
23 GREUB. Linear Algebra. 4th ed.
24 HOLMES. Geometric Functional Analysis and Its Applications.
25 HEWITT/STROMBERG. Real and Abstract Analysis.
26 MANES. Algebraic Theories.
27 KELLEY. General Topology.
28 ZARISKI/SAMUEL. Commutative Algebra. Vol. I.
29 ZARISKI/SAMUEL. Commutative Algebra. Vol. II.
30 JACOBSON. Lectures in Abstract Algebra I. Basic Concepts.
31 JACOBSON. Lectures in Abstract Algebra II. Linear Algebra.
32 JACOBSON. Lectures in Abstract Algebra III. Theory of Fields and Galois Theory.
33 HIRSCH. Differential Topology.
34 SPITZER. Principles of Random Walk. 2nd ed.
35 ALEXANDER/WERMER. Several Complex Variables and Banach Algebras. 3rd ed.
36 KELLEY/NAMIOKA et al. Linear Topological Spaces.
37 MONK. Mathematical Logic.
38 GRAUERT/FRITZSCHE. Several Complex Variables.
39 ARVESON. An Invitation to C^*-Algebras.
40 KEMENY/SNELL/KNAPP. Denumerable Markov Chains. 2nd ed.
41 APOSTOL. Modular Functions and Dirichlet Series in Number Theory. 2nd ed.
42 J.-P. SERRE. Linear Representations of Finite Groups.
43 GILLMAN/JERISON. Rings of Continuous Functions.
44 KENDIG. Elementary Algebraic Geometry.
45 LOÈVE. Probability Theory I. 4th ed.
46 LOÈVE. Probability Theory II. 4th ed.
47 MOISE. Geometric Topology in Dimensions 2 and 3.
48 SACHS/WU. General Relativity for Mathematicians.
49 GRUENBERG/WEIR. Linear Geometry. 2nd ed.
50 EDWARDS. Fermat's Last Theorem.
51 KLINGENBERG. A Course in Differential Geometry.
52 HARTSHORNE. Algebraic Geometry.
53 MANIN. A Course in Mathematical Logic.
54 GRAVER/WATKINS. Combinatorics with Emphasis on the Theory of Graphs.
55 BROWN/PEARCY. Introduction to Operator Theory I: Elements of Functional Analysis.
56 MASSEY. Algebraic Topology: An Introduction.
57 CROWELL/FOX. Introduction to Knot Theory.
58 KOBLITZ. p-adic Numbers, p-adic Analysis, and Zeta-Functions. 2nd ed.
59 LANG. Cyclotomic Fields.
60 ARNOLD. Mathematical Methods in Classical Mechanics. 2nd ed.
61 WHITEHEAD. Elements of Homotopy Theory.
62 KARGAPOLOV/MERIZJAKOV. Fundamentals of the Theory of Groups.
63 BOLLOBAS. Graph Theory.

(continued after index)

Steven Roman

Field Theory

Second Edition

With 18 Illustrations

Steven Roman
University of California, Irvine
Irvine, California 92697-3875
USA
sroman@romanpress.com

Mathematics Subject Classification (2000): 12-01, 11Txx

Library of Congress Control Number: 2005932166

ISBN 978-1-4419-2095-9

Printed on acid-free paper.

9 8 7 6 5 4 3 2 1

springeronline.com

To Donna

Preface

This book presents the basic theory of fields, starting more or less from the beginning. It is suitable for a graduate course in field theory, or independent study. The reader is expected to have taken an undergraduate course in abstract algebra, not so much for the material it contains but in order to gain a certain level of mathematical maturity.

The book begins with a preliminary chapter (Chapter 0), which is designed to be quickly scanned or skipped and used as a reference if needed. The remainder of the book is divided into three parts.

Part 1, entitled *Field Extensions*, begins with a chapter on polynomials. Chapter 2 is devoted to various types of field extensions, including finite, finitely generated, algebraic and normal. Chapter 3 takes a close look at the issue of separability. In my classes, I generally cover only Sections 3.1 to 3.4 (on perfect fields). Chapter 4 is devoted to algebraic independence, starting with the general notion of a dependence relation and concluding with Luroth's theorem on intermediate fields of a simple transcendental extension.

Part 2 of the book is entitled *Galois Theory*. Chapter 5 examines Galois theory from an historical perspective, discussing the contributions from Lagrange, Vandermonde, Gauss, Newton, and others that led to the development of the theory. I have also included a very brief look at the very brief life of Galois himself.

Chapter 6 begins with the notion of a Galois correspondence between two partially ordered sets, and then specializes to the Galois correspondence of a field extension, concluding with a brief discussion of the Krull topology. In Chapter 7, we discuss the Galois theory of equations. In Chapter 8, we view a field extension E of F as a vector space over F.

Chapter 9 and Chapter 10 are devoted to finite fields, although this material can be omitted in order to reach the topic of solvability by radicals more quickly. Möbius inversion is used in a few places, so an appendix has been included on this subject.

Part 3 of the book is entitled *The Theory of Binomials*. Chapter 11 covers the roots of unity and Wedderburn's theorem on finite division rings. We also briefly discuss the question of whether a given group is the Galois group of a field extension. In Chapter 12, we characterize cyclic extensions and splitting fields of binomials when the base field contains appropriate roots of unity. Chapter 13 is devoted to the question of solvability of a polynomial equation by radicals. (This chapter might make a convenient ending place in a graduate course.) In Chapter 14, we determine conditions that characterize the irreducibility of a binomial and describe the Galois group of a binomial. Chapter 15 briefly describes the theory of families of binomials—the so-called *Kummer theory*.

Sections marked with an asterisk may be skipped without loss of continuity.

Changes for the Second Edition

Let me begin by thanking the readers of the first edition for their many helpful comments and suggestions.

For the second edition, I have gone over the entire book, and rewritten most of it, including the exercises. I believe the book has benefited significantly from a class testing at the beginning graduate level and at a more advanced graduate level.

I have also rearranged the chapters on separability and algebraic independence, feeling that the former is more important when time is of the essence. In my course, I generally touch only very lightly (or skip altogether) the chapter on algebraic independence, simply because of time constraints.

As mentioned earlier, as several readers have requested, I have added a chapter on Galois theory from an historical perspective.

A few additional topics are sprinkled throughout, such as a proof of the Fundamental Theorem of Algebra, a discussion of *casus irreducibilis*, Berlekamp's algorithm for factoring polynomials over $\mathbb{Z}_p$ and natural and accessory irrationalities.

Thanks

I would like to thank my students Phong Le, Sunil Chetty, Timothy Choi and Josh Chan, who attended lectures on essentially the entire book and offered many helpful suggestions. I would also like to thank my editor, Mark Spencer, who puts up with my many requests and is most amiable.

Contents

Chapter 0
Preliminaries

The purpose of this chapter is to review some basic facts that will be needed in the book. The discussion is not intended to be complete, nor are all proofs supplied. We suggest that the reader quickly skim this chapter (or skip it altogether) and use it as a reference if needed.

0.1 Lattices

Definition *A* **partially ordered set** (*or* **poset**) *is a nonempty set P, together with a binary relation $\leq$ on P satisfying the following properties. For all α, β, $\gamma \in P$,*

1) **(reflexivity)**

$$\alpha \leq \alpha$$

2) **(antisymmetry)**

$$\alpha \leq \beta,\ \beta \leq \alpha \Rightarrow \alpha = \beta$$

3) **(transitivity)**

$$\alpha \leq \beta,\ \beta \leq \gamma \Rightarrow \alpha \leq \gamma$$

If, in addition,

$$\alpha, \beta \in P \Rightarrow \alpha \leq \beta \text{ or } \beta \leq \alpha$$

then P is said to be **totally ordered**. □

Any subset of a poset P is also a poset under the restriction of the relation defined on P. A totally ordered subset of a poset is called a **chain**. If $S \subseteq P$ and $s \leq \alpha$ for all $s \in S$ then α is called an **upper bound** for S. A **least upper bound** for S, denoted by $\text{lub}(S)$ or $\bigvee S$, is an upper bound that is less than or equal to any other upper bound. Similar statements hold for lower bounds and greatest lower bounds, the latter denoted by $\text{glb}(S)$, or $\bigwedge S$. A **maximal element** in a poset P is an element $\alpha \in P$ such that $\alpha \leq \beta$ implies $\alpha = \beta$. A **minimal element** in a poset P is an element $\gamma \in P$ such that $\beta \leq \gamma$ implies

$\beta = \gamma$. A **top element** $1 \in P$ is an element with the property that $\alpha \leq 1$ for all $\alpha \in P$. Similarly, a **bottom element** $0 \in P$ is an element with the property that $0 \leq \alpha$ for all $\alpha \in P$. **Zorn's lemma** says that if every chain in a poset P has an upper bound in P then P has a maximal element.

Definition *A* **lattice** *is a poset L in which every pair of elements $\alpha, \beta \in L$ has a least upper bound, or* **join**, *denoted by $\alpha \vee \beta$ and a greatest lower bound, or* **meet**, *denoted by $\alpha \wedge \beta$. If every nonempty subset of L has a join and a meet then L is called a* **complete lattice**. □

Note that any nonempty complete lattice has a greatest element, denoted by 1 and a smallest element, denoted by 0.

Definition *A* **sublattice** *of a lattice L is a subset S of L that is closed under the meet and join operation of L.* □

It is important to note that a subset S of a lattice L can be a lattice under the same order relation and yet not be a sublattice of L. As an example, consider the coll
S of all subgroups of a group G, ordered by inclusion. Then S is a subset of the power set $\mathcal{P}(G)$, which is a lattice under union and intersection. But S is not a sublattice of $\mathcal{P}(G)$ since the union of two subgroups need not be a subgroup. On the other hand, S is a lattice in its own right under set inclusion, where the meet $H \wedge K$ of two subgroups is their intersection and the join $H \vee K$ is the smallest subgroup of G containing H and K.

In a complete lattice L, joins can be defined in terms of meets, since $\bigvee T$ is the meet of all upper bounds of T. The fact that $1 \in L$ ensures that T has at least one upper bound, so that the meet is not an empty one. The following theorem exploits this idea to give conditions under which a subset of a complete lattice is itself a complete lattice.

Theorem 0.1.1 *Let L be a complete lattice. If $S \subseteq L$ has the properties*
1) $1 \in S$
2) **(Closed under arbitrary meets)** $T \subseteq S, T \neq \emptyset \Rightarrow \bigwedge T \in S$
then S is a complete lattice under the same meet.
Proof. Let $T \subseteq S$. Then $\bigwedge T \in S$ by assumption. Let U be the set of all upper bounds of T that lie in S. Since $1 \in S$, we have $U \neq \emptyset$. Hence, $\bigwedge U \in S$ and is $\bigvee T$. Thus, S is a complete lattice. (Note that S need not be a sublattice of L since $\bigwedge U$ need not equal the meet of all upper bounds of T in L.) □

0.2 Groups

Definition *A* **group** *is a nonempty set G, together with a binary operation on G, that is, a map $G \times G \to G$, denoted by juxtaposition, with the following properties:*

1) **(Associativity)** $(\alpha\beta)\gamma = \alpha(\beta\gamma)$ *for all* $\alpha, \beta, \gamma \in G$
2) **(Identity)** *There exists an element* $\epsilon \in G$ *for which* $\epsilon\alpha = \alpha\epsilon = \alpha$ *for all* $\alpha \in G$
3) **(Inverses)** *For each* $\alpha \in G$, *there is an element* $\alpha^{-1} \in G$ *for which* $\alpha\alpha^{-1} = \alpha^{-1}\alpha = \epsilon$.

A group G is **abelian**, *or* **commutative**, *if* $\alpha\beta = \beta\alpha$, *for all* $\alpha, \beta \in G$.□

The identity element is often denoted by 1. When G is abelian, the group operation is often denoted by $+$ and the identity by 0.

Subgroups

Definition *A* **subgroup** S *of a group* G *is a subset of* G *that is a group in its own right, using the restriction of the operation defined on* G. *We denote the fact that* S *is a subgroup of* G *by writing* $S < G$.□

If G is a group and $\alpha \in G$, then the set of all powers of α

$$\langle \alpha \rangle = \{\alpha^n \mid n \in \mathbb{Z}\}$$

is a subgroup of G, called the **cyclic subgroup generated by** α. A group G is **cyclic** if it has the form $G = \langle \alpha \rangle$, for some $\alpha \in G$. In this case, we say that α **generates** G.

Let G be a group. Since G is a subgroup of itself and since the intersection of subgroups of G is a subgroup of G, Theorem 0.1.1 implies that the set of subgroups of G forms a complete lattice, where $H \wedge J = H \cap J$ and $H \vee J$ is the smallest subgroup of G containing both H and J.

If H and K are subgroups of G, it does not follow that the **set product**

$$HK = \{hk \mid h \in H, k \in K\}$$

is a subgroup of G. It is not hard to show that HK is a subgroup of G precisely when $HK = KH$.

The **center** of G is the set

$$Z(G) = \{\beta \in G \mid \alpha\beta = \beta\alpha \text{ for all } \alpha \in G\}$$

of all elements of G that commute with every element of G.

Orders and Exponents

A group G is **finite** if it contains only a finite number of elements. The cardinality of a finite group G is called its **order** and is denoted by $|G|$ or $o(G)$. If $\alpha \in G$, and if $\alpha^k = \epsilon$ for some integer k, we say that k is an **exponent** of α. The smallest positive exponent for $\alpha \in G$ is called the **order** of α and is denoted by $o(\alpha)$. An integer m for which $\alpha^m = 1$ for all $\alpha \in G$ is called an

exponent of G. (Note: Some authors use the term exponent of G to refer to the *smallest* positive exponent of G.)

Theorem 0.2.1 *Let G be a group and let $\alpha \in G$. Then k is an exponent of α if and only if k is a multiple of $o(\alpha)$. Similarly, the exponents of G are precisely the multiples of the smallest positive exponent of G.* $\square$

We next characterize the smallest positive exponent for finite abelian groups.

Theorem 0.2.2 *Let G be a finite abelian group.*
1) **(Maximum order equals minimum exponent)** *If m is the maximum order of all elements in G then $\alpha^m = 1$ for all $\alpha \in G$. Thus, the smallest positive exponent of G is equal to the maximum order of all elements of G.*
2) *The smallest positive exponent of G is equal to $o(G)$ if and only if G is cyclic.* $\square$

Cosets and Lagrange's Theorem

Let $H < G$. We may define an equivalence relation on G by saying that $\alpha \sim \beta$ if $\beta^{-1}\alpha \in H$ (or equivalently $\alpha^{-1}\beta \in H$). The equivalence classes are the **left cosets** $\alpha H = \{\alpha h \mid h \in H\}$ of H in G. Thus, the distinct left cosets of H form a partition of G. Similarly, the distinct **right cosets** $H\alpha$ form a partition of G. It is not hard to see that all cosets of H have the same cardinality and that there is the same number of left cosets of H in G as right cosets. (This is easy when G is finite. Otherwise, consider the map $\alpha H \mapsto H\alpha^{-1}$.)

Definition *The* **index** *of H in G, denoted by $(G : H)$, is the cardinality of the set G/H of all distinct left cosets of H in G. If G is finite then $(G : H) = |G|/|H|$.* $\square$

Theorem 0.2.3 *Let G be a finite group.*
1) **(Lagrange)** *The order of any subgroup of G divides the order of G.*
2) *The order of any element of G divides the order of G.*
3) **(Converse of Lagrange's Theorem for Finite Abelian Groups)** *If A is a finite abelian group and if $k \mid o(A)$ then A has a subgroup of order k.* $\square$

Normal Subgroups

If S and T are subsets of a group G, then the **set product** ST is defined by

$$ST = \{st \mid s \in S, t \in T\}$$

Theorem 0.2.4 *Let $H < G$. The following are equivalent*
1) *The set product of any two cosets is a coset.*
2) *If $\alpha, \beta \in G$, then*

$$\alpha H \beta H = \alpha\beta H$$

3) *Any right coset of H is also a left coset, that is, for any $\alpha \in G$ there is a $\beta \in G$ for which $H\alpha = \beta H$.*
4) *If $\alpha \in G$, then*

$$\alpha H = H\alpha$$

5) $\alpha\beta \in H \Rightarrow \beta\alpha \in H$ *for all* $\alpha, \beta \in G$.□

Definition A subgroup H of G is **normal** in G, written $H \lhd G$, if any of the equivalent conditions in Theorem 0.2.4 holds. □

Definition *A group G is* **simple** *if it has no normal subgroups other than $\{1\}$ and G.* □

Here are some normal subgroups.

Theorem 0.2.5
1) *The center $Z(G)$ is a normal subgroup of G.*
2) *Any subgroup H of a group G with $(G : H) = 2$ is normal.*
3) *If G is a finite group and if p is the smallest prime dividing $o(G)$, then any subgroup of index p is normal in G*□

With respect to the last statement in the previous theorem, it makes some intuitive sense that if a subgroup H of a finite group G is extremely large, then it may be normal, since there is not much room for conjugates. This is true in the most extreme case. Namely, the largest possible proper subgroup of G has index equal to the smallest prime number dividing $o(G)$. This subgroup, if it exists, is normal.

If $H \lhd G$, then we have the set product formula

$$\alpha H \beta H = \alpha\beta H$$

It is not hard to see that this makes the quotient G/H into a group, called the **quotient group** of H in G. The order of G/H is called the **index** of H in G and is denoted by $(G : H)$.

Theorem 0.2.6 *If G is a group and $\{H_i\}$ is a collection of normal subgroups of G then $\bigcap H_i$ and $\bigvee H_i$ are normal subgroups of G. Hence, the collection of normal subgroups of G is a complete sublattice of the complete lattice of all subgroups of G.* □

If $H < G$ then there is always an intermediate subgroup $H < K < G$ for which $H \lhd K$, in fact, H is such an intermediate subgroup. The largest such subgroup is called the **normalizer** of H in G. It is

$$N_G(H) = \{g \in G \mid gHg^{-1} = H\}$$

Euler's Formula

We will denote a greatest common divisor of α and β by (α, β) or $\gcd(\alpha, \beta)$.

If $(\alpha, \beta) = 1$, then α and β are **relatively prime**. The **Euler phi function** ϕ is defined by letting $\phi(n)$ be the number of positive integers less than or equal to n that are relatively prime to n.

Two integers α and β are **congruent modulo** n, written $\alpha \equiv \beta \bmod n$, if $\alpha - \beta$ is divisible by n. Let $\mathbb{Z}_n$ denote the ring of integers $\{0, \ldots, n-1\}$ under addition and multiplication modulo n.

Theorem 0.2.7 (Properties of Euler's phi function)

1) *The Euler phi function is multiplicative, that is, if m and n are relatively prime, then*

$$\phi(mn) = \phi(m)\phi(n)$$

2) *If p is a prime and $n > 0$ then*

$$\phi(p^n) = p^{n-1}(p-1)$$

These two properties completely determine ϕ. □

Since the set $G = \{\beta \in \mathbb{Z}_n \mid (\beta, n) = 1\}$ is a group of order $\phi(n)$ under multiplication modulo n, it follows that $\phi(n)$ is an exponent for G.

Theorem 0.2.8 (Euler's Theorem) *If $\alpha, n \in \mathbb{Z}$ and $(\alpha, n) = 1$, then*

$$\alpha^{\phi(n)} \equiv 1 \bmod n \qquad \square$$

Corollary 0.2.9 (Fermat's Theorem) *If p is a prime not dividing the integer α, then*

$$\alpha^p \equiv \alpha \bmod p \qquad \square$$

Cyclic Groups

Theorem 0.2.10

1) *Every group of prime order is cyclic.*
2) *Every subgroup of a cyclic group is cyclic.*
3) *A finite abelian group G is cyclic if and only if its smallest positive exponent is equal to $o(G)$.* □

The following theorem contains some key results about finite cyclic groups.

Theorem 0.2.11 *Let $G = \langle \alpha \rangle$ be a cyclic group of order n.*

1) *For* $1 \le k < n$,

$$o(\alpha^k) = \frac{n}{(n,k)}$$

In particular, α^k *generates* $G = \langle \alpha \rangle$ *if and only if* $(n,k) = 1$.

2) *If* $d \mid n$, *then*

$$o(\alpha^k) = d \Leftrightarrow k = r\frac{n}{d}, \text{ where } (r,d) = 1$$

Thus the elements of G *of order* $d \mid n$ *are the elements of the form* $\alpha^{r(n/d)}$, *where* $0 \le r < d$ *and* r *is relatively prime to* d.

3) *For each* $d \mid n$, *the group* G *has exactly one subgroup* H_d *of order* d *and* $\phi(d)$ *elements of order* d, *all of which lie in* H_d.

4) **(Subgroup structure charactertizes property of being cyclic)** *If a finite group* G *of order* n *has the property that it has at most one subgroup of each order* $d \mid n$, *then* G *is cyclic.*□

Counting the elements in a cyclic group of order n gives the following corollary.

Corollary 0.2.12 *For any positive integer* n,

$$n = \sum_{d \mid n} \phi(d)$$ □

Homomorphisms

Definition *Let* G *and* H *be groups. A map* $\psi: G \to H$ *is called a* **group homomorphism** *if*

$$\psi(\alpha\beta) = (\psi\alpha)(\psi\beta)$$

A surjective homomorphism is an **epimorphism**, *an injective homomorphism is a* **monomorphism** *and a bijective homomorphism is an* **isomorphism**. *If* $\psi: G \to H$ *is an isomorphism, we say that* G *and* H *are* **isomorphic** *and write* $G \approx H$. □

If ψ is a homomorphism then $\psi\epsilon = \epsilon$ and $\psi\alpha^{-1} = (\psi\alpha)^{-1}$. The **kernel** of a homomorphism $\psi: G \to H$,

$$\ker(\psi) = \{\alpha \in G \mid \psi\alpha = \epsilon\}$$

is a normal subgroup of G. Conversely, any normal subgroup H of G is the kernel of a homomorphism. For we may define the **natural projection** $\pi: G \to G/H$ by $\pi\alpha = \alpha H$. This is easily seen to be an epimorphism with kernel H.

Let $f: S \to T$ be a function from a set S to a set T. Let $\mathcal{P}(S)$ and $\mathcal{P}(T)$ be the power sets of S and T, respectively. We define the **induced map** $f: \mathcal{P}(S) \to \mathcal{P}(T)$ by $f(U) = \{f(u) \mid u \in U\}$ and the **induced inverse map** by $f^{-1}(V) = \{s \in S \mid f(s) \in V\}$. (It is customary to denote the induced maps by the same notation as the original map.) Note that f is surjective if and only if its induced map is surjective, and this holds if and only if the induced inverse map is injective. A similar statement holds with the words surjective and injective reversed.

Theorem 0.2.13 *Let $\psi: G \to G'$ be a group homomorphism.*

1) a) If $H < G$ then $\psi(H) < G'$.
b) If ψ is surjective and $H \lhd G$ then $\psi(H) \lhd G'$.

2) a) If $H' < G'$ then $\psi^{-1}(H') < G$.
b) If $H' \lhd G'$ then $\psi^{-1}(H') \lhd G$. □

Theorem 0.2.14 *Let G be a group.*

1) **(First Isomorphism Theorem)** *Let $\psi: G \to G'$ be a group homomorphism with kernel K. Then $K \lhd G$ and the map $\overline{\psi}: G/K \to \operatorname{im}\psi$ defined by $\overline{\psi}(\alpha K) = \psi\alpha$ is an isomorphism. Hence $G/K \approx \operatorname{im}\psi$. In particular, ψ is injective if and only if $\ker(\psi) = \{\epsilon\}$.*

2) **(Second Isomorphism Theorem)** *If $H < G$ and $N \lhd G$ then $N \cap H \lhd H$ and*

$$\frac{H}{N \cap H} \approx \frac{NH}{N}$$

3) **(Third Isomorphism Theorem)** *If $H < K < G$ with H and K normal in G then $K/H \lhd G/H$ and*

$$\frac{G/H}{K/H} \approx \frac{G}{K}$$

Hence $(G : K) = (G/H : K/H)$. □

Theorem 0.2.15 *Let G_1 and G_2 be groups and let $H_i < G_i$. Then*

$$\frac{G_1 \times G_2}{H_1 \times H_2} \approx \frac{G_1}{H_1} \times \frac{G_2}{H_2}$$ □

Theorem 0.2.16 (The Correspondence Theorem) *Let $H \lhd G$ and let π be the natural projection $\pi: G \to G/H$. Thus, for any $H < I < G$,*

$$\pi(I) = I/H = \{iH \mid i \in I\}$$

1) The induced maps π and π^{-1} define a one-to-one correspondence between the lattice of subgroups of G containing H and the lattice of subgroups of G/H.

2) *π preserves index, that is, for any $H < I < J < G$, we have*

$$(J : I) = (\pi(J) : \pi(I))$$

3) *π preserves normality, that is, if $H < I < J < G$ then $I \lhd J$ if and only if $I/H \lhd J/H$, in which case*

$$\frac{J}{I} \approx \frac{\pi(J)}{\pi(I)}$$ □

Theorem 0.2.17

1) *An abelian group G is simple if and only if it is finite and has prime order.*
2) *If M is a maximal subgroup of G, that is, $M < G$ and if $M < N < G$ then $N = M$ or $N = G$, and if M is normal then G/M is cyclic of prime order.* □

Sylow Subgroups

Definition *If p is a prime, then a group G is called a p-**group** if every element of G has order a power of p. A* **Sylow p-subgroup** *S of G is a maximal p-subgroup of G.* □

Theorem 0.2.18 (Properties of p-groups)

1) *A finite group G is a p-group if and only if $|G| = p^n$ for some n.*
2) *If H is a finite p-group, then the center of H is nontrivial.*
3) *If $o(G) = p^2$, p prime, then G is abelian.*
4) *If H is a proper subgroup of G, then H is also a proper subgroup of its normalizer $N_G(H)$.*
5) *If H is a maximal subgroup of G then H is normal and has index p.* □

For finite groups, if $\alpha \in G$ then $o(\alpha) \mid o(G)$. The converse does not hold in general, but we do have the following.

Theorem 0.2.19 Let G be a finite group.

1) **(Cauchy's Theorem)** *If $o(G)$ is divisible by a prime p then G contains an element of order p.*
2) **(Partial converse of Lagrange's theorem)** *If p is a prime and $p^k \mid o(G)$, then for any* Sylow p-subgroup *S of G, there is a subgroup H of G, normal in S and of order p^k.* □

Here is the famous result on maximal p-subgroups of a finite group.

Theorem 0.2.20 (Sylow's Theorem) *Let G have order $p^n m$ where $p \nmid m$.*

1) *All Sylow p-subgroups of G have order p^n.*
2) *All Sylow p-subgroups are conjugate (and hence isomorphic).*
3) *The number of Sylow p-subgroups of G divides $o(G)$ and is congruent to* 1 mod p.
4) *Any p-subgroup of G is contained is a Sylow p-subgroup of G.* □

0.3 The Symmetric Group

Definition *The* **symmetric group** S_n *on the set* $I_n = \{1, \ldots, n\}$ *is the group of all permutations of* I_n, *under composition of maps. A* **transposition** *is a permutation that interchanges two distinct elements of* I_n *and leaves all other elements fixed. The* **alternating group** A_n *is the subgroup of* S_n *consisting of all even permutations, that is, all permutations that can be written as a product of an even number of transpositions.* □

Theorem 0.3.1
1) *The order of* S_n *is* $n!$.
2) *The order of* A_n *is* $n!/2$. *Thus,* $[S_n : A_n] = 2$ *and* $A_n \lhd S_n$.
3) A_n *is the only subgroup of* S_n *of index* 2.
4) A_n *is simple (no nontrivial normal subgroups) for* $n \geq 5$. □

A subgroup H of S_n is **transitive** if for any $k, j \in I_n$ there is a $\sigma \in H$ for which $\sigma k = j$.

Theorem 0.3.2 *If* H *is a transitive subgroup of* S_n *then the order* $o(H)$ *is a multiple of* n. □

0.4 Rings

Definition A **ring** is a nonempty set R, together with two binary operations on R, called **addition** (denoted by $+$), and **multiplication** (denoted by juxtaposition), satisfying the following properties.
1) *R is an abelian group under the operation* $+$.
2) **(Associativity)** $(\alpha\beta)\gamma = \alpha(\beta\gamma)$ *for all* $\alpha, \beta, \gamma \in R$.
3) **(Distributivity)** *For all* $\alpha, \beta, \gamma \in R$,

$$(\alpha + \beta)\gamma = \alpha\gamma + \alpha\beta \quad \textit{and} \quad \gamma(\alpha + \beta) = \gamma\alpha + \gamma\beta$$ □

Definition *Let R be a ring.*
1) *R is called a* **ring with identity** *if there exists an element* $1 \in R$ *for which* $\alpha 1 = 1\alpha = \alpha$, *for all* $\alpha \in R$. *In a ring R with identity, an element α is called a* **unit** *if it has a multiplicative inverse in R, that is, if there exists a* $\beta \in R$ *such that* $\alpha\beta = \beta\alpha = 1$.
2) *R is called a* **commutative ring** *if multiplication is commutative, that is, if* $\alpha\beta = \beta\alpha$ *for all* $\alpha, \beta \in R$.
3) *A* **zero divisor** *in a commutative ring R is a nonzero element* $\alpha \in R$ *such that* $\alpha\beta = 0$ *for some* $\beta \neq 0$. *A commutative ring R with identity is called an* **integral domain** *if R contains no zero divisors.*
4) *A ring R with identity* $1 \neq 0$ *is called a* **field** *if the nonzero elements of R form an abelian group under multiplication.* □

It is not hard to see that the set of all units in a ring with identity forms a group under multiplication. We shall have occasion to use the following example.

Example 0.4.1 Let $\mathbb{Z}_n = \{0, 1, \ldots, n-1\}$ be the ring of integers modulo n. Then k is a unit in $\mathbb{Z}_n$ if and only if $(k, n) = 1$. This follows from the fact that $(k, n) = 1$ if and only if there exist integers a and b such that $ak + bn = 1$, that is, if and only if $ak \equiv 1 \bmod n$. The set of units of $\mathbb{Z}_n$, denoted by $\mathbb{Z}_n^*$, is a group under multiplication. □

Definition *A* **subring** *of a ring R is a nonempty subset S of R that is a ring in its own right, using the same operations as defined on R.* □

Definition *A* **subfield** *of a field E is a nonempty subset F of E that is a field in its own right, using the same operations as defined on E. In this case, we say that E is an* **extension** *of F and write $F < E$ or $E > F$.* □

Definition *Let R and S be rings. A function $\psi: R \to S$ is a* **homomorphism** *if, for all $\alpha, \beta \in R$,*

$$\psi(\alpha + \beta) = \psi\alpha + \psi\beta \text{ and } \psi(\alpha\beta) = (\psi\alpha)(\psi\beta)$$

An injective homomorphism is a **monomorphism** *or an* **embedding**, *a surjective homomorphism is an* **epimorphism** *and a bijective homomorphism is an* **isomorphism**. *A homomorphism from R into itself is an* **endomorphism** *and an isomorphism from R onto itself is an* **automorphism**. □

Ideals

Definition *A nonempty subset $\mathcal{I}$ of a ring R is called an* **ideal** *if it satisfies*

1) $\alpha, \beta \in \mathcal{I}$ *implies* $\alpha - \beta \in \mathcal{I}$.

2) $\alpha \in R, \iota \in \mathcal{I}$ *implies* $\alpha\iota \in \mathcal{I}$ *and* $\iota\alpha \in \mathcal{I}$. □

If S is a nonempty subset of a ring R, then the **ideal generated** by S is defined to be the smallest ideal $\mathcal{I}$ of R containing S. If R is a commutative ring with identity, and if $\alpha \in R$, then the ideal generated by $\{\alpha\}$ is the set

$$\langle \alpha \rangle = R\alpha = \{\rho\alpha \mid \rho \in R\}$$

Any ideal of the form $\langle \alpha \rangle$ is called a **principal ideal**.

Definition *If $\psi: R \to S$ is a* **homomorphism**, *then*

$$\ker(\psi) = \{\alpha \in R \mid \psi\alpha = 0\}$$

is an ideal of R. □

If R is a ring and $\mathcal{I}$ is an ideal in R then for each $\alpha \in R$, we can form the **coset**

$$\alpha + \mathcal{I} = \{\alpha + \iota \mid \iota \in \mathcal{I}\}$$

It is easy to see that $\alpha + \mathcal{I} = \beta + \mathcal{I}$ if and only if $\alpha - \beta \in \mathcal{I}$, and that any two cosets $\alpha+\mathcal{I}$ and $\beta+\mathcal{I}$ are either disjoint or identical. The collection of all (distinct) cosets is a ring itself, with addition and multiplication defined by

$$(a+\mathcal{I})+(b+\mathcal{I})=(a+b)+\mathcal{I}$$

and

$$(a+\mathcal{I})(b+\mathcal{I})=ab+\mathcal{I}$$

The ring of cosets of $\mathcal{I}$ is called a **factor ring** and is denoted by $R/\mathcal{I}$.

Isomorphism theorems similar to those for groups also hold for rings. Here is the first isomorphism theorem.

Theorem 0.4.1 (The First Isomorphism Theorem) *Let R be a ring. Let $\psi\colon R\to R'$ be a ring homomorphism with kernel I. Then I is an ideal of R and the map $\overline{\psi}\colon R/I\to \operatorname{im}(\psi)$ defined by $\overline{\psi}(\alpha+I)=\psi\alpha$ is an isomorphism. Hence $R/I\approx$ im ψ. In particular, ψ is injective if and only if* $\ker(\psi)=\{0\}$.□

Definition *An ideal $\mathcal{I}$ of a ring R is* **maximal** *if $\mathcal{I}\neq R$ and if whenever $\mathcal{I}\subseteq\mathcal{J}\subseteq R$ for any ideal $\mathcal{J}$, then $\mathcal{J}=\mathcal{I}$ or $\mathcal{J}=R$. An ideal $\mathcal{I}$ is* **prime** *if $\mathcal{I}\neq R$ and if $\alpha\beta\in\mathcal{I}$ implies $\alpha\in\mathcal{I}$ or $\beta\in\mathcal{I}$.* □

It is not hard to see that a maximal ideal in a commutative ring with identity is prime. This also follows from the next theorem.

Theorem 0.4.2 *Let R be a commutative ring with identity and let $\mathcal{I}$ be an ideal of R.*
1) *$R/\mathcal{I}$ is a field if and only if $\mathcal{I}$ is maximal.*
2) *$R/\mathcal{I}$ is an integral domain if and only if $\mathcal{I}$ is prime.* □

Theorem 0.4.3 *Any commutative ring R with identity contains a maximal ideal.*
Proof. Since R is not the zero ring, the ideal $\{0\}$ is a proper ideal of R. Hence, the set $\mathcal{S}$ of all proper ideals of R is nonempty. If

$$\mathcal{C}=\{\mathcal{I}_i\mid i\in I\}$$

is a chain of proper ideals in R then the union $\mathcal{J}=\bigcup_{i\in I}\mathcal{I}_i$ is also an ideal. Furthermore, if $\mathcal{J}=R$ is not proper, then $1\in\mathcal{J}$ and so $1\in\mathcal{I}_i$, for some $i\in I$, which implies that $\mathcal{I}_i=R$ is not proper. Hence, $\mathcal{J}\in\mathcal{S}$. Thus, any chain in $\mathcal{S}$ has an upper bound in $\mathcal{S}$ and so Zorn's lemma implies that $\mathcal{S}$ has a maximal element. This shows that R has a maximal ideal. □

The Characteristic of a Ring

Let R be a ring and let $r \in R$. For any positive integer n, we define

$$nr = \underbrace{r + r + \cdots + r}_{n \text{ terms}}$$

and for any negative integer n, we set $nr = -((-n)r)$.

The **characteristic** $\text{char}(R)$ of a ring R is the smallest positive integer n for which $n1 = 0$ (or equivalently, $nr = 0$ for all $r \in R$), should such an integer exist. If it does not, we say that R has characteristic 0. If $\text{char}(R) = 0$ then R contains a copy of the integers $\mathbb{Z}$, in the form $\mathbb{Z} \cdot 1 = \{n1 \mid n \in \mathbb{Z}\}$. If $\text{char}(R) = r$, then R contains a copy of $\mathbb{Z}_r = \{0, 1, \ldots, n-1\}$.

Theorem 0.4.4 *The characteristic of an integral domain R is either 0 or a prime. In particular, a finite field has prime characteristic.*
Proof. If $r = \text{char}(R)$ is not 0 and if $r = st$, where s and t are positive integers, then $0 = r \cdot 1 = (s \cdot 1)(t \cdot 1)$ and so one of $s \cdot 1$ or $t \cdot 1$ is equal to 0. But since r is the smallest such positive integer, it follows that either $s = r$ or $t = r$. Hence, r is prime.□

If F is a field, the intersection of all of its subfields is the smallest subfield of F and is referred to as the **prime subfield** of F.

Theorem 0.4.5 *Let F be a field. If $\text{char}(F) = 0$, the prime subfield of F is isomorphic to the rational numbers $\mathbb{Q}$. If $\text{char}(F) = p$ is prime, the prime field of F is isomorphic to $\mathbb{Z}_p$.*
Proof. If $\text{char}(F) = 0$, consider the map $\phi\colon \mathbb{Q} \to F$ defined by

$$\phi\left(\frac{p}{q}\right) = \frac{p \cdot 1}{q \cdot 1}$$

This is easily seen to be a ring homomorphism. For example

$$\begin{aligned}
\phi\left(\frac{p}{q} + \frac{r}{s}\right) &= \phi\left(\frac{ps + rq}{qs}\right) \\
&= \frac{(ps + rq) \cdot 1}{qs \cdot 1} \\
&= \frac{(p \cdot 1)(s \cdot 1) + (r \cdot 1)(q \cdot 1)}{(q \cdot 1)(s \cdot 1)} \\
&= \frac{(p \cdot 1)}{(q \cdot 1)} + \frac{(r \cdot 1)}{(s \cdot 1)} \\
&= \phi\left(\frac{p}{q}\right) + \phi\left(\frac{r}{s}\right)
\end{aligned}$$

Now, $\phi(p/q) = 0$ if and only if $p \cdot 1 = 0$ in F, and since $\text{char}(F) = 0$, we see that $p=0$ and so ϕ is a monomorphism. Thus, the subfield $\phi(\mathbb{Q})$ is isomorphic to $\mathbb{Q}$. Clearly, any subfield of F must contain the elements $p \cdot 1$, where $p \in \mathbb{Z}$ and therefore also the elements $\phi(p/q)$ and so $\phi(\mathbb{Q})$ is the prime subfield of F.

Now suppose that $\text{char}(F)=p$ is a prime. The map $\phi: \mathbb{Z}_p \to F$ defined by $\phi(z)=z\cdot 1$ is a ring homomorphism and is also injective since $z<p$. Hence, $\phi(\mathbb{Z}_p)$ is a subfield of F isomorphic to $\mathbb{Z}_p$. Since any subfield of F must contain $\phi(\mathbb{Z}_p)$, this is the prime subfield of F.□

The following result is of considerable importance for the study of fields of nonzero characteristic.

Theorem 0.4.6 Let R be a commutative ring with identity of *prime* characteristic p. Then

$$(\alpha+\beta)^{p^n}=\alpha^{p^n}+\beta^{p^n},(\alpha-\beta)^{p^n}=\alpha^{p^n}-\beta^{p^n}$$

Proof. Since the binomial formula holds in any commutative ring with identity, we have

$$(\alpha+\beta)^p=\sum_{k=0}^{p}\binom{p}{k}\alpha^k\beta^{p-k}$$

where

$$\binom{p}{k}=\frac{p(p-1)\cdots(p-k+1)}{k!}$$

But $p \mid \binom{p}{k}$ for $0<k<p$, and so $\binom{p}{k}=0$ in R. The binomial formula therefore reduces to

$$(\alpha+\beta)^p=\alpha^p+\beta^p$$

Repeated use of this formula gives $(\alpha+\beta)^q=\alpha^q+\beta^q$. The second formula is proved similarly.□

These formulas are very significant. They say that the **Frobenius map** $\sigma_p: R \to R^p$ is a surjective ring homomorphism. When $R=F$ is a field of characteristic p, then σ_p is an isomorphism and $F \approx F^p$.

0.5 Integral Domains

Theorem 0.5.1 *Let R be an integral domain. Let $\alpha, \beta \in R$.*

1) *We say that α* **divides** *β and write $\alpha \mid \beta$ if $\beta=\rho\alpha$ for some $\rho \in R$. If ρ and α are nonunits and $\beta=\rho\alpha$ then α* **properly divides** *β.*
   a) *A unit divides every element of R.*
   b) *$\alpha \mid \beta$ if and only if $\langle\beta\rangle \subseteq \langle\alpha\rangle$.*
   c) *$\alpha \mid \beta$ properly if and only if $\langle\beta\rangle \subset \langle\alpha\rangle \subset R$.*

2) *If* $\alpha = u\beta$ *for some unit* u *then* α *and* β *are* **associates** *and we write* $\alpha \sim \beta$.
   a) $\alpha \sim \beta$ *if and only if* $\alpha \mid \beta$ *and* $\beta \mid \alpha$.
   b) $\alpha \sim \beta$ *if and only if* $\langle \alpha \rangle = \langle \beta \rangle$.
3) *A nonzero element* $\rho \in R$ *is* **irreducible** *if* ρ *is not a unit and if* ρ *has no proper divisors. Thus, a nonunit* ρ *is irreducible if and only if* $\rho = \alpha\beta$ *implies that either* α *or* β *is a unit.*
4) *A nonzero element* $\pi \in R$ *is* **prime** *if* π *is not a unit and whenever* $\pi \mid \alpha\beta$ *then* $\pi \mid \alpha$ *or* $\pi \mid \beta$.
   a) *Every prime element is irreducible.*
   b) $\pi \in R$ *is prime if and only if* $\langle \pi \rangle$ *is a nonzero prime ideal.*
5) *Let* $\alpha, \beta \in R$. *An element* $d \in R$ *is called a* **greatest common divisor (gcd)** *of* α *and* β, *written* (α, β) *or* $\gcd(\alpha, \beta)$, *if* $d \mid \alpha$ *and* $d \mid \beta$ *and if whenever* $e \mid \alpha$, $e \mid \beta$ *then* $e \mid d$. *If* $\gcd(\alpha, \beta)$ *is a unit, we say that* α *and* β *are* **relatively prime**. The greatest common divisor of two elements, if it exists, is unique up to associate. □

Theorem 0.5.2 *An integral domain* R *is a field if and only if it has no ideals other than the zero ideal and* R *itself. Any nonzero homomorphism* $\sigma: F \to E$ *of fields is a monomorphism.* □

Theorem 0.5.3 *Every finite integral domain is a field.* □

Field of Quotients

If R is an integral domain, we may form the set

$$R' = \{\alpha/\beta \mid \alpha, \beta \in R, \beta \neq 0\}$$

where $\alpha/\beta = a/b$ if and only if $\alpha b = a\beta$. We define addition and multiplication on R' in the "obvious way"

$$\frac{\alpha}{\beta} + \frac{a}{b} = \frac{\alpha b + \beta a}{\beta b}, \quad \frac{\alpha}{\beta} \cdot \frac{a}{b} = \frac{\alpha a}{\beta b}$$

It is easy to see that these operations are well-defined and that R' is actually a field, called the **field of quotients** of the integral domain R. It is the *smallest* field containing R (actually, an isomorphic copy of R), in the sense that if F is a field and $R \subseteq F$ then $R \subseteq R' \subseteq F$. The following fact will prove useful.

Theorem 0.5.4 *Let* R *be an integral domain with field of quotients* R'. *Then any monomorphism* $\sigma: R \to F$ *from* R *into a field* F *has a unique extension to a monomorphism* $\overline{\sigma}: R' \to F$.
Proof. Define $\overline{\sigma}(\alpha/\beta) = \sigma\alpha/\sigma\beta$, which makes sense since $\beta \neq 0$ implies $\sigma\beta \neq 0$. One can easily show that $\overline{\sigma}$ is well-defined. Since $\sigma\alpha/\sigma\beta = 0$ if and only if $\sigma\alpha = 0$, which in turn holds if and only if $\alpha/\beta = 0$, we see that $\overline{\sigma}$ is injective. Uniqueness is clear since $\sigma|_R$ (σ restricted to R) uniquely determines σ on R'. □

0.6 Unique Factorization Domains

Definition *An integral domain R is a* **unique factorization domain (ufd)** *if*

1) *Any nonunit $r \in R$ can be written as a product $r = p_1 \cdots p_n$ where p_i is irreducible for all i. We refer to this as the* **factorization property** *for R.*
2) *This factorization is* **essentially unique** *in the sense that if $r = p_1 \cdots p_n = q_1 \cdots q_m$ are two factorizations into irreducible elements then $m = n$ and there is some permutation π for which $p_i \sim q_{\pi(i)}$ for all i.* □

If $r \in R$ is not irreducible, then $r = st$ where s and t are nonunits. Evidently, we may continue to factor as long as at least one factor is not irreducible. An integral domain R has the factorization property precisely when this factoring process always stops after a finite number of steps.

Actually, the uniqueness part of the definition of a ufd is equivalent to some very important properties.

Theorem 0.6.1 *Let R be an integral domain for which the factorization property holds. The following conditions are equivalent and therefore imply that R is a unique factorization domain.*

1) *Factorization in R is essentially unique.*
2) *Every irreducible element of R is prime.*
3) *Any two elements of R, not both zero, have a greatest common divisor.* □

Corollary 0.6.2 *In a unique factorization domain, the concepts of prime and irreducible are equivalent.* □

0.7 Principal Ideal Domains

Definition *An integral domain R is called a* **principal ideal domain (pid)** *if every ideal of R is principal.* □

Theorem 0.7.1 *Every principal ideal domain is a unique factorization domain.* □

We remark that the ring $\mathbb{Z}[x]$ is a ufd (as we prove in Chapter 1) but not a pid (the ideal $\langle 2, x\rangle$ is not principal) and so the converse of the previous theorem is not true.

Theorem 0.7.2 *Let R be a principal ideal domain and let $\mathcal{I}$ be an ideal of R.*

1) *$\mathcal{I}$ is maximal if and only if $\mathcal{I} = \langle \rho \rangle$ where ρ is irreducible.*
2) *$\mathcal{I}$ is prime if and only if $\mathcal{I} = \{0\}$ or $\mathcal{I}$ is maximal.*
3) *The following are equivalent:*
   a) *$R/\langle \rho \rangle$ is a field*
   b) *$R/\langle \rho \rangle$ is an integral domain*
   c) *ρ is irreducible*

d) ρ *is prime.* □

0.8 Euclidean Domains

Roughly speaking, a Euclidean domain is an integral domain in which we can perform "division with remainder."

Definition *An integral domain* R *is a* **Euclidean domain** *if there is a function* $\sigma\colon (R-\{0\}) \to \mathbb{N}$ *with the property that given any* α, $\beta \in R$, $\beta \neq 0$, *there exist* $q, r \in R$ *satisfying*

$$\alpha = q\beta + r$$

where $r = 0$ *or* $\sigma r < \sigma\beta$. □

Theorem 0.8.1 *A Euclidean domain is a principal ideal domain* (*and hence also a unique factorization domain*).

Proof. Let $\mathcal{I}$ be an ideal in the Euclidean domain R and let $\alpha \in \mathcal{I}$ be minimal with respect to the value of σ. Thus, $\sigma\alpha \leq \sigma\beta$ for all $\beta \in \mathcal{I}$. If $s \in \mathcal{I}$ then

$$s = r\alpha + q$$

where $q = 0$ or $\sigma q < \sigma r$. But $q = s - r\alpha \in \mathcal{I}$ and so the latter is not possible, leaving $q = 0$ and $s \in \langle\alpha\rangle$. Hence, $\mathcal{I} = \langle\alpha\rangle$. □

Theorem 0.8.2 *If* F *is a field, then* $F[x]$ *is a Euclidean domain with* $\sigma(p(x)) = \deg(p(x))$. *Hence* $F[x]$ *is also a principal ideal domain and a unique factorization domain.*

Proof. This follows from ordinary division of polynomials; to wit, if $f(x)$ $g(x) \in F[x], g(x) \neq 0$, then there exist $q(x), r(x) \in F[x]$ such that

$$f(x) = q(x)g(x) + r(x)$$

where $\deg(r(x)) < \deg(g(x))$. □

0.9 Tensor Products

Tensor products are used only in the optional Section 5.6, on linear disjointness.

Definition *Let* U, V *and* W *be vector spaces over a field* F. *A function* $f\colon U \times V \to W$ *is* **bilinear** *if it is linear in both variables separately, that is, if*

$$f(ru + su', v) = rf(u, v) + sf(u', v)$$

and

$$f(u, rv + sv') = rf(u, v) + sf(u, v')$$

The set of all bilinear functions from $U \times V$ *to* W *is denoted by* $\mathcal{B}(U, V; W)$. *A bilinear function* $f\colon U \times V \to F$, *with values in the base field* F, *is called a* **bilinear form** *on* $U \times V$. □

Example 0.9.1

1) *A real inner product* $\langle,\rangle: V \times V \to \mathbb{R}$ *is a bilinear form on* $V \times V$.
2) *If* A *is an algebra, the product map* $\mu: A \times A \to A$ *defined by* $\mu(a,b) = ab$ *is bilinear.* □

We will denote the set of all linear transformations from $U \times V$ to W by $\mathcal{L}(U \times V, W)$. There are many definitions of the tensor product. We choose a universal definition.

Theorem 0.9.1 *Let* U *and* V *be vector spaces over the same field* F. *There exists a unique vector space* $U \otimes V$ *and bilinear map* $t: U \times V \to U \otimes V$ *with the following property. If* $f: U \times V \to W$ *is any bilinear function from* $U \times V$ *to a vector space* W *over* F, *then there is a unique linear transformation* $\tau: U \otimes V \to W$ *for which*

$$\tau \circ t = f$$

□

This theorem says that to each *bilinear* function $f: U \times V \to W$, there corresponds a unique *linear* function $\tau: U \otimes V \to W$, through which f can be factored (that is, $f = \tau \circ t$). The vector space $U \otimes V$, whose existence is guaranteed by the previous theorem, is called the **tensor product** of U and V over F. We denote the image of (u, v) under the map t by $t(u, v) = uu \otimes v$.

If $X = \operatorname{im}(t) = \{u \otimes v \mid u \in U,\ v \in V\}$ is the image of the tensor map t then the uniqueness statement in the theorem implies that X spans $U \otimes V$. Hence, every element of $\alpha \in U \otimes V$ is a finite sum of elements of the form $u \otimes v$

$$\alpha = \sum_{\text{finite}} a_i (u_i \otimes v_i)$$

We establish a few basic properties of the tensor product.

Theorem 0.9.2 If $\{u_1, \dots, u_n\} \subseteq U$ is linearly independent and $\{v_1, \dots, v_n\} \subseteq V$ then

$$\sum u_i \otimes v_i = 0 \Rightarrow v_i = 0 \text{ for all } i$$

Proof. Consider the dual vectors $\delta_i \in U^*$ to the vectors u_i, where $\delta_i u_j = \delta_{i,j}$. For linear functionals $\epsilon_i: V \to F$, we define a bilinear form $f: U \times V \to F$ by

$$f(u, v) = \sum_{j=1}^{n} \delta_j(x) \epsilon_j(y)$$

Since there exists a unique linear functional $\tau: U \otimes V \to F$ for which $\tau \circ t = f$, we have

$$0 = \tau\Big(\sum_i u_i \otimes v_i\Big) = \sum_i \tau \circ t(u_i, v_i)$$
$$= \sum_i f(u_i, v_i) = \sum_i \sum_j \delta_j(u_i)\epsilon_j(v_i) = \sum_i \epsilon_i(v_i)$$

Since the ϵ_i's are arbitrary, we deduce that $v_i = 0$ for all i. □

Corollary 0.9.3 *If $u \neq 0$ and $v \neq 0$, then $u \otimes v \neq 0$.* □

Theorem 0.9.4 Let $\mathcal{B} = \{e_i \mid i \in I\}$ be a basis for U and let $\mathcal{C} = \{f_j \mid j \in J\}$ be a basis for V. Then $\mathcal{D} = \{e_i \otimes f_j \mid i \in I,\ j \in J\}$ is a basis for $U \otimes V$.
Proof. To see that the $\mathcal{D}$ is linearly independent, suppose that

$$\sum_{i,j} r_{i,j}(e_i \otimes f_j) = 0$$

This can be written

$$\sum_i e_i \otimes \Big(\sum_j r_{i,j} f_j\Big) = 0$$

Theorem 0.9.2 implies that

$$\sum_j r_{i,j} f_j = 0$$

for all i, and hence $r_{i,j} = 0$ for all i and j. To see that $\mathcal{D}$ spans $U \otimes V$, let $u \otimes v \in U \otimes V$. Since $u = \sum r_i e_i$, and $v = \sum s_j f_j$, we have

$$u \otimes v = \sum_i r_i e_i \otimes \sum_j s_j f_j = \sum_j s_j \Big(\sum_i r_i e_i \otimes f_j\Big)$$
$$= \sum_j s_j \Big(\sum_i r_i (e_i \otimes f_j)\Big) = \sum_{i,j} r_i s_j (e_i \otimes f_j)$$

Since any vector in $U \otimes V$ is a finite sum of vectors $u \otimes v$, we deduce that $\mathcal{D}$ spans $U \otimes V$. □

Corollary 0.9.5 *For finite dimensional vector spaces,*

$$\dim(U \otimes V) = \dim(U) \cdot \dim(V)$$ □

Exercises

1. The relation of being associates in an integral domain is an equivalence relation.
2. Prove that the characteristic of an integral domain is either 0 or a prime, and that a finite field has prime characteristic.

3. If $\mathrm{char}(F) = 0$, the prime subfield of F is isomorphic to the rational numbers $\mathbb{Q}$. If $\mathrm{char}(F) = p$ is prime, the prime field of F is isomorphic to $\mathbb{Z}_p$.
4. If $F < E$ show that E and F must have the same characteristic.
5. Let F be a field of characteristic p. The Frobenius map $\sigma\colon F \to F$ defined by $\sigma\alpha = \alpha^p$ is a homomorphism. Show that $F \approx F^p = \{a^p \mid a \in F\}$. What if F is a finite field?
6. Consider the polynomial ring $F[x_1, x_2, \dots]$ where $x_i^2 = x_{i-1}$. Show that the factorization process need not stop in this ring.
7. Let $R = \mathbb{Z}[\sqrt{-5}] = \{a + b\sqrt{-5} \mid a, b \in \mathbb{Z}\}$. Show that this integral domain is not a unique factorization domain by showing that $6 \in R$ has essentially two different factorizations in R. Show also that the irreducible element 2 is not prime.
8. Let R be a pid. Then an ideal $\mathcal{I}$ of R is maximal if and only if $\mathcal{I} = \langle p \rangle$ where p is irreducible. Also, $R/\langle p \rangle$ is a field if and only if p is irreducible.
9. Prove that $\langle x \rangle$ and $\langle 2, x \rangle$ are both prime ideals in $\mathbb{Z}[x]$ and that $\langle x \rangle$ is properly contained in $\langle 2, x \rangle$.
10. Describe the divisor chain condition in terms of principal ideals.

Part I—Field Extensions

Chapter 1
Polynomials

In this chapter, we discuss properties of polynomials that will be needed in the sequel. Since we assume that the reader is familiar with the basic properties of polynomials, some of the present material may constitute a review.

1.1 Polynomials over a Ring

We will be concerned in this book mainly with polynomials over a field F, but it is useful to make a few remarks about polynomials over a ring R as well, especially since many polynomials encountered in practice are defined over the integers. Let $R[x]$ denote the ring of polynomials in the single variable x over R. If

$$f(x) = a_0 + a_1 x + \cdots + a_n x^n$$

where $a_i \in R$ and $a_n \neq 0$ then n is called the **degree** of $f(x)$, written $\deg(f(x))$ or $\deg(f)$ and a_n is called the **leading coefficient** of $f(x)$. A polynomial is **monic** if its leading coefficient is 1. The degree of the zero polynomial is defined to be $-\infty$. If R is a ring, the units of $R[x]$ are the units of R, since no polynomial of positive degree can have an inverse in $R[x]$. Note that the units in $R[x]$ are the units in R.

In general, if $f(x) = a_0 + a_1 x + \cdots + a_n x^n \in R[x]$ is a polynomial over a ring R and if $\sigma : R \to S$ is a ring homomorphism, then we denote the polynomial $\sigma(a_0) + \sigma(a_1)x + \cdots + \sigma(a_n)x^n \in S[x]$ by $\sigma f(x)$ or by $f^\sigma(x)$ and the function that sends $f(x)$ to $f^\sigma(x)$ by σ^*, that is,

$$\sigma^*(f(x)) = f^\sigma(x)$$

We may refer to σ^* as the **extension** of σ to $R[x]$. It is easy to see that σ^* is also a ring homomorphism.

One of the most useful examples of ring homomorphisms in this context is the projection maps $\pi_p : R \to R/\langle p \rangle$, where p is a prime in R, defined by $\pi_p(\alpha) = \alpha + \langle p \rangle$. It is not hard to see that π_p is a surjective ring homomorphism, and that $R/\langle p\rangle$ is an integral domain. The maps π_p are also referred to as **localization maps.**

Note that the units of $R[x]$ are the units of R.

Definition *Let R be a ring. A nonzero polynomial $f(x)\in R[x]$ is* **irreducible** *over R if $f(x)$ is not a unit and whenever $f(x)=p(x)q(x)$ for $p(x),q(x)\in R[x]$, then one of $p(x)$ and $q(x)$ is a unit in $R[x]$. A polynomial that is not irreducible is said to be* **reducible.** □

We can simplify this definition for polynomials over a field. A polynomial over a field is irreducible if and only if it has positive degree and cannot be factored into the product of two polynomials of positive degree.

Many important properties that a ring R may possess carry over to the ring of polynomials $R[x]$.

Theorem 1.1.1 *Let R be a ring.*
1) *If R is an integral domain, then so is $R[x]$*
2) *If R is a unique factorization domain, then so is $R[x]$.*
3) *If R is a principal ideal domain, $R[x]$ need not be a principal ideal domain.*
4) *If F is a field, then $F[x]$ is a principal ideal domain.*

Proof. For part 3), the ring $\mathbb{Z}$ of integers is a principal ideal domain, but $\mathbb{Z}[x]$ is not, since the ideal $\langle 2,x\rangle$ is not principal.□

1.2 Primitive Polynomials and Irreducibility

We now consider polynomials over a unique factorization domain.

Content and Primitivity

If $p(x)$ is a polynomial over the integers, it is often useful to factor out the positive greatest common divisor of the coefficients, so that the remaining coefficients are relatively prime. For polynomials over an arbitrary unique factorization domain, the greatest common divisor is not unique and there is no way to single one out in general.

Definition *Let $f(x)\in R[x]$ where R is a unique factorization domain. Any greatest common divisor of the coefficients of $f(x)$ is called a* **content** *of $f(x)$. A polynomial with content 1 is said to be* **primitive**. *Let $c(f)$ denote the set of all contents of $f(x)$. Thus, $c(f)$ is the set of all associates of any one of its elements. For this reason, one often speaks of "the" content of a polynomial.*□

A content of $f(x)=a_0+a_1x+\cdots+a_nx^n$ can be obtained by factoring each coefficient a_k of f into a product of powers of distinct primes and then taking the product of each prime π that appears in any of these factorizations, raised to the *smallest* power to which π appears in all of the factorizations.

There is no reason why we cannot apply this same procedure to a polynomial over R', the field of quotients of R. If $f(x) \in R'[x]$, then each coefficient of $f(x)$ can be written as a product of integral powers of distinct primes.

Definition *Let $f(x) = a_0 + a_1x + \cdots + a_dx^d \in R'[x]$, where R' is the field of quotients of a unique factorization domain R. Let $p_1, \ldots, p_n$ be a complete list of the distinct primes dividing any coefficient a_k of $f(x)$. Then each coefficient a_k can be written in the form*

$$a_k = p_1^{e_1(a_k)} \cdots p_n^{e_n(a_k)}$$

where $e_i(a_k) \in \mathbb{Z}$. Let $m_i = \min(e_i(a_k) \mid k = 0, \ldots, d\}$ be the smallest exponent of p_i among the factorizations of the coefficients of $f(x)$. The element

$$\Pi = p_1^{m_1} \cdots p_m^{m_n}$$

is a **content** *of $f(x)$, and so is any element $u\Pi$, where u is a unit in R. The set of all contents of f is denoted by $c(f)$. A polynomial $f(x)$ is* **primitive** *if $1 \in c(f)$.*□

Note the following simple facts about content.

Lemma 1.2.1 *For any $f(x) \in R'[x]$ and $\alpha \in R'$*

$$c(\alpha f(x)) = \alpha c(f(x))$$

It follows that
1) *α is a content of $f(x)$ if and only if $f(x) = \alpha p(x)$, where $p(x)$ is primitive in $R[x]$.*
2) *If $f(x)$ is primitive, then $f(x) \in R[x]$.*
3) *$f(x) \in R[x]$ if and only if $c(f) \in R$.*□

We now come to a key result concerning primitive polynomials.

Theorem 1.2.2 *Let R be a unique factorization domain, with field of quotients R'.*
1) **(Gauss's lemma)** *The product of primitive polynomials is primitive.*
2) *If $f(x), g(x) \in R'[x]$ then $c(fg) = c(f)c(g)$.*
3) *If a polynomial $f(x) \in R[x]$ can be factored*

$$f(x) = p(x)h(x)$$

where $p(x)$ is primitive and $h(x) \in R'(x)$ then, in fact, $h(x) \in R[x]$.

Proof. To prove Gauss's lemma, let $f(x), g(x) \in R[x]$ and suppose that fg is not primitive. Then there exists a prime $p \in R$ for which $p \mid fg$. Consider the localization map π_p^*. The condition $p \mid fg$ is equivalent to $\pi_p(fg) = 0$, that is,

$\pi_p(f)\pi_p(g) = 0$ and since $R/\langle p \rangle$ is an integral domain, one of the factors must be 0, that is, one of f or g must be divisible by p, and hence not primitive.

To prove part 2), observe that if c_f is a content of $f(x)$ and c_g is a content of $g(x)$ then $f = c_f f'$ and $g = c_g g'$, where f' and g' are primitive over R. Hence, by Gauss's lemma, if U is the set of units of R, then

$$c(fg) = c(c_f c_g f'g') = c_f c_g c(f'g') = c_f c_g U = c(f)c(g)$$

As to part 3), we have

$$c(f) = c(ph) = c(p)c(h) = c(h)$$

and since $c(f) \in R$, so is $c(h)$, whence $h(x) \in R[x]$. □

Irreducibility over R and R′

If $f(x) \in R[x]$, then it can also be thought of as a polynomial over $R'[x]$. We would like to relate the irreducibility of f over R to its irreducibility over R'. Let us say that a factorization $f(x) = p(x)q(x)$ is **over** a set S if $p(x)$ and $q(x)$ have coefficients in S.

The relationship between irreducibility over R and over R' would be quite simple were it not for the presence of irreducible constants in R, which are *not* irreducible over R'.

To formulate a clear description of the situation, let us make the following nonstandard (not found in other books) definition. We say that a factorization of the form $f(x) = p(x)q(x)$, where $\deg(p) > 0$ and $\deg(q) > 0$, is a **degreewise factorization** of $f(x)$ and that $f(x)$ is **degreewise reducible**.

Now, if

$$f(x) = p(x)q(x)$$

is a degreewise factorization over R, then it is also a degreewise factorization over R'. Conversely, if this is a degreewise factorization over R', then we can move the content of $q(x)$ to the other factor and write

$$f(x) = [c_q p(x)]r(x)$$

where $r(x)$ is primitive. Theorem 1.1.1 implies that $c_q p(x)$ is also in $R[x]$ and so this is a degreewise factorization of $f(x)$ over R. Thus, $f(x)$ has a degreewise factorization over R if and only if it has a degreewise factorization over R'. Note also that the corresponding factors in the two factorizations have the same degree.

It follows that $f(x)$ is irreducible over R' if and only if it is degreewise irreducible over R. But degreewise irreducibility over a field R' is the only kind of irreducibility.

Theorem 1.2.3 *Let R be a unique factorization domain, with field of quotients R'. Let $f(x) \in R[x]$.*

1) *$f(x)$ is degreewise irreducible over R if and only if it is irreducible over R'.*
2) *If $f(x)$ is primitive, then it is irreducible over R if and only if it is irreducible over R'.* □

1.3 The Division Algorithm and its Consequences

The familiar division algorithm for polynomials over a field F can be easily extended to polynomials over a commutative ring with identity, provided that we divide only by polynomials with leading coefficient a unit. We leave proof of the following to the reader.

Theorem 1.3.1 (Division algorithm) *Let R be a commutative ring with identity. Let $g(x) \in R[x]$ have an invertible leading coefficient (which happens if $f(x)$ is monic, for example). Then for any $f(x) \in R[x]$, there exist unique $q(x), r(x) \in R[x]$ such that*

$$f(x) = q(x)g(x) + r(x)$$

where $\deg(r) < \deg(g)$. □

This theorem has some very important immediate consequences. Dividing $f(x)$ by $x - \alpha$, where $\alpha \in R$ gives

$$f(x) = q(x)(x - \alpha) + r$$

where $r \in R$. Hence, α is a root of $f(x)$ if and only if $x - \alpha$ is a factor of $f(x)$ over R.

Corollary 1.3.2 *Let R be a commutative ring with identity and let $f(x) \in R[x]$. Then α is a root of $f(x)$ if and only if $x - \alpha$ is a factor of $f(x)$ over R.* □

Also, since the usual degree formula

$$\deg(fg) = \deg(f) + \deg(g)$$

holds when R is an integral domain, we get an immediate upper bound on the number of roots of a polynomial.

Corollary 1.3.3 *If R is an integral domain, then a nonzero polynomial $f(x) \in R[x]$ can have at most* $\deg(f)$ *distinct roots in R.* □

Note that if R is not an integral domain then the preceding result fails. For example, in $\mathbb{Z}_8$, the four elements $1, 3, 5$ and 7 are roots of the polynomial $x^2 - 1$.

From this, we get the following fundamental fact concerning finite multiplicative subgroups of a field.

Corollary 1.3.4 *Let F^* be the multiplicative group of all nonzero elements of a field F. If G is a finite subgroup of F^*, then G is cyclic. In particular, if F is a finite field then F^* is cyclic.*
Proof. If $|G| = m$, then every element of G satisfies the polynomial $x^m - 1$. But G cannot have an exponent $e < m$, for then every one of the m elements of G would be a root of the polynomial $x^e - 1$, of degree less than m. Hence, the smallest exponent of G is the order of G and Theorem 0.2.2 implies that G is cyclic.□

Polynomials as Functions

In the customary way, a polynomial $p(x) \in R[x]$ can be thought of as a function on R. Of course, the zero polynomial is also the zero function. However, the converse is not true! For example, the nonzero polynomial $p(x) = x^2 - x$ in $\mathbb{Z}_2[x]$ is the zero function on $\mathbb{Z}_2$.

This raises the question of how to decide, based on the zero set of a polynomial, when that polynomial must be the zero polynomial.

If R is an integral domain, then Corollary 1.3.3 ensures that if $p(x)$ has degree at most d but has more than d zeros, then it must be the zero polynomial. The previous example shows that we cannot improve on this statement. It follows that if the zero set of $p(x)$ is *infinite*, then $p(x)$ must be the zero polynomial. We can make no such blanket statements in the context of finite rings, as the previous example illustrates.

Now let us consider polynomials in more than one variable. We can no longer claim that if a polynomial $p(x_1, \ldots, x_n)$ has an infinite zero set, then it must be the zero polynomial. For example, the nonzero polynomial $p(x, y) = x - y$ has the infinite zero set $\{(x, y) \mid x = y\}$.

It is not hard to prove by induction that if R is infinite and $p(x_1, \ldots, x_n)$ is the zero function, that is, $p(x_1, \ldots, x_n)$ has zero set R, then $p(x_1, \ldots, x_n)$ is the zero polynomial. We leave the details to the reader. Again, we cannot strengthen this to finite rings, as the polynomial $p(x, y) = (x^2 - x)y$ in $\mathbb{Z}_2[x, y]$ shows.

However, we can improve upon this. There is a middle ground between "an infinite set of zeros" and "zero set equal to all of R" that is sufficient to guarantee that $p(x_1,\ldots,x_n)$ is the zero polynomial. This middle ground is "an infinite *subfield* worth of zeros."

Theorem 1.3.5 *Let $p(x_1,\ldots,x_n)$ be a polynomial over L and let $F < L$, where F is infinite. If $p(a_1,\ldots,a_n) = 0$ for all $a_i \in F$, then $p(x_1,\ldots,x_n)$ is the zero polynomial.*
Proof. Write

$$p(x_1,\ldots,x_n) = \sum_{i_1,\ldots,i_n} \lambda_{i_1,\ldots,i_n} x_1^{i_1}\cdots x_n^{i_n}$$

where $\lambda_{i_1,\ldots,i_n} \in L$. Let $\{\beta_i\}$ be a basis for L as a vector space over F. Then

$$\lambda_{i_1,\ldots,i_n} = \sum_k a_{i_1,\ldots,i_n,k}\beta_k$$

for $a_{i_1,\ldots,i_n,k} \in F$ and so

$$\begin{aligned} p(x_1,\ldots,x_n) &= \sum_{i_1,\ldots,i_n} \left[\sum_k a_{i_1,\ldots,i_n,k}\beta_k \right] x_1^{i_1}\cdots x_n^{i_n} \\ &= \sum_k \left[\sum_{i_1,\ldots,i_n} a_{i_1,\ldots,i_n,k} x_1^{i_1}\cdots x_n^{i_n} \right] \beta_k \end{aligned}$$

Hence, the independence of the β_i's implies that the polynomial

$$\sum_{i_1,\ldots,i_n} a_{i_1,\ldots,i_n,k} x_1^{i_1}\cdots x_n^{i_n}$$

in $F[x_1,\ldots,x_n]$ is the zero function on F. As we have remarked, this implies that $a_{i_1,\ldots,i_n,k} = 0$ for all $i_1,\ldots,i_n$ and k. Hence, $\lambda_{i_1,\ldots,i_n} = 0$ for all $i_1,\ldots,i_n$ and $p(x_1,\ldots,x_n)$ is the zero polynomial.□

Common Divisors and Greatest Common Divisors

In defining the greatest common divisor of two polynomials, it is customary (in order to obtain uniqueness) to require that it be monic.

Definition *Let $f(x)$ and $g(x)$ be polynomials over F. The* **greatest common divisor** *of $f(x)$ and $g(x)$, denoted by $(f(x), g(x))$ or $\gcd(f(x), g(x))$, is the unique monic polynomial $p(x)$ over F for which*
1) $p(x) \mid f(x)$ and $p(x) \mid g(x)$.
2) If $r(x) \in F[x]$ and $r(x) \mid f(x)$ and $r(x) \mid g(x)$ then $r(x) \mid p(x)$. □

The existence of greatest common divisors is easily proved using the fact that $F[x]$ is a principal ideal domain. Since the ideal

$$I = \langle f(x), g(x) \rangle = \{a(x)f(x) + b(x)g(x) \mid a(x), b(x) \in F[x]\}$$

is principal, we have $I = \langle p(x) \rangle$, for some monic $p(x) \in F[x]$. Since $f(x), g(x) \in \langle p(x) \rangle$, it follows that $p(x) \mid f(x)$ and $p(x) \mid g(x)$. Moreover, since $p(x) \in \langle f(x), g(x) \rangle$, there exist $a(x), b(x) \in K[x]$ such that

$$p(x) = a(x)f(x) + b(x)g(x)$$

Hence, if $q(x) \mid f(x)$ and $q(x) \mid g(x)$, then $q(x) \mid p(x)$ and so $p(x) = \gcd(f(x), g(x))$.

As to uniqueness, if $p(x)$ and $q(x)$ are both greatest common divisors of $f(x)$ and $g(x)$ then each divides the other and since they are both monic, we conclude that $p(x) = q(x)$.

Greatest Common Divisor Is Field Independent

The definition of greatest common divisor seems at first to depend on the field F, since all divisions are over F. However, this is not the case.

To see this, note that for any field K containing the coefficients of $f(x)$ and $g(x)$, the ideal

$$I_K = \langle f(x), g(x) \rangle = \{a(x)f(x) + b(x)g(x) \mid a(x), b(x) \in K[x]\}$$

is principal and so $I_K = \langle r_K(x) \rangle$, where $r_K(x)$ is the gcd with respect to the field K. But if $K < L$, then $I_K \subseteq I_L$ and so $r_K(x) \in I_L$. This implies two things. First, $r_L(x) \mid r_K(x)$ because $r_L(x)$ generates I_L and second, $r_K(x) \mid r_L(x)$ because $r_L(x)$ is the *greatest* common divisor of $f(x)$ and $g(x)$ in $L[x]$. Hence, $r_L(x) = r_K(x)$.

Thus, if K is the smallest field containing the coefficients of $f(x)$ and $g(x)$, then $r_K(x)$ is the same polynomial as $r_L(x)$, for any field L containing the coefficients of $f(x)$ and $g(x)$. In other words, the gcd can be computed using any field containing the coefficients of $f(x)$ and $g(x)$. This also shows that the gcd of $f(x)$ and $g(x)$ has coefficients in the field K.

Theorem 1.3.6 *Let $f(x), g(x) \in F[x]$. Let K be the smallest field containing the coefficients of $f(x)$ and $g(x)$.*

1) *The greatest common divisor $d(x)$ of $f(x)$ and $g(x)$ does not depend on the the base field F.*
2) *Hence, $d(x)$ has coefficients in K.*
3) *There exist polynomials $a(x), b(x) \in K[x]$ such that*

$$d(x) = a(x)f(x) + b(x)g(x) \qquad \square$$

This result has a somewhat surprising corollary: If $f(x), g(x) \in F[x]$ have a nonconstant common factor in *any extension* E of F, then $d(x) = \gcd(f, g)$ is nonconstant and so $f(x)$ and $g(x)$ have a nonconstant factor over *every* field containing the coefficients of $f(x)$ and $g(x)$.

Corollary 1.3.7 *Let $f(x), g(x) \in F[x]$ and let $F < E$. Then $f(x)$ and $g(x)$ have a nonconstant common factor over F if and only if they have a nonconstant common factor over E.* □

Now we can make sense of the notion that two polynomials are relatively prime without mentioning a specific field.

Definition *The polynomials $f(x)$ and $g(x)$ are* **relatively prime** *if they have no nonconstant common factors, that is, if* $\gcd(f(x), g(x)) = 1$. *In particular, $f(x)$ and $g(x)$ are relatively prime if and only if there exist polynomials $a(x)$ and $b(x)$ over the smallest field containing the coefficients of $f(x)$ and $g(x)$ for which*

$$a(x)f(x) + b(x)g(x) = 1$$

□

Roots and Common Roots

It is a fundamental fact that every nonconstant polynomial $f(x) \in F[x]$ has a root in *some* field.

Theorem 1.3.8 *Let F be a field, and let $f(x) \in F[x]$ be a nonconstant polynomial. Then there exists an extension E of F and an $\alpha \in E$ such that $f(\alpha) = 0$.*

Proof. We may assume that $f(x)$ is irreducible. Consider the field

$$E = \frac{F[x]}{\langle f(x) \rangle}$$

The field F is isomorphic to a subfield of E, by identifying $\alpha \in F$ with $\alpha + \langle f(x) \rangle \in E$. Under this identification, $x + \langle f(x) \rangle$ is a root of $f(x)$ in E.

Thus, we have shown that F can be *embedded* in a field E in which $f(x)$ (with its coefficients embedded as well) has a root. While this is not quite the statement of the theorem, it is possible to show that there is a "true" extension of F that has a root of $f(x)$, using simple techniques from the next chapter.□

Repeated application of Theorem 1.3.8 gives the following corollary.

Corollary 1.3.9 *Let $f(x) \in F[x]$. There exists an extension of F over which $f(x)$* **splits**, *that is, factors into linear factors.*□

Corollary 1.3.10 *Two polynomials $f(x), g(x) \in F[x]$ have a nonconstant common factor over some extension of F if and only if they have a common root over some extension of F. Put another way, $f(x)$ and $g(x)$ are relatively prime if and only if they have no common roots in any extension F.*□

Since distinct irreducible polynomials are relatively prime, we get the following corollary.

Corollary 1.3.11 *If $f(x)$ and $g(x)$ are distinct irreducible polynomials over F then they have no common roots in any extension E of F.* □

1.4 Splitting Fields

If a polynomial $f(x) \in F[x]$ factors into linear factors

$$f(x) = a(x - \alpha_1)(x - \alpha_2)\cdots(x - \alpha_n)$$

in an extension field E, that is, if $\alpha_1, \ldots, \alpha_n \in E$, we say that $f(x)$ **splits** in E.

Definition *Let $\mathcal{F} = \{f_i(x) \mid i \in I\}$ be family of polynomials over a field F. A* **splitting field** *for $\mathcal{F}$ is an extension field E of F with the following properties:*
1) *Each $f_i(x) \in \mathcal{F}$ splits over E, and thus has a full set of* $\deg(f_i)$ *roots in E*
2) *E is the smallest field satisfying $F < K < E$ that contains the roots of each $f_i(x) \in \mathcal{F}$ mentioned in part 1).*□

Theorem 1.4.1 *Every finite family of polynomials over a field F has a splitting field.*

Proof. According to Corollary 1.3.9, there is an extension $F < E$ in which a given polynomial $p(x)$ has a full set of roots $\alpha_1, \ldots, \alpha_n$. The smallest subfield of E containing F and these roots is a splitting field for $p(x)$. If $\mathcal{F}$ is a finite family of polynomials, then a splitting field for $\mathcal{F}$ is a splitting field for the product of the polynomials in $\mathcal{F}$. □

We will see in the next chapter that any family of polynomials has a splitting field. We will also see that any two splitting fields S_1 and S_2 for a family of polynomials over F are isomorphic by an isomorphism that fixes each element of the base field F.

1.5 The Minimal Polynomial

Let $F < E$. An element $\alpha \in E$ is said to be **algebraic** over F if α is a root of some polynomial over F. An element that is not algebraic over F is said to be **transcendental** over F.

If α is algebraic over F, the set of all polynomials satisfied by α

$$\mathcal{I}_\alpha = \{g(x) \in F[x] \mid g(\alpha) = 0\}$$

is a nonzero ideal in $F[x]$ and is therefore generated by a unique *monic* polynomial $p(x)$, called the **minimal polynomial** of α over F and denoted by $p_\alpha(x)$, $p_{\alpha,F}(x)$ or $\min(\alpha, F)$. The following theorem characterizes minimal polynomials in a variety of useful ways. Proof is left to the reader.

Theorem 1.5.1 *Let $F < E$ and let $\alpha \in E$ be algebraic over F. Then among all polynomials in $F[x]$, the polynomial* $\min(\alpha, F)$ *is*

1) the unique monic irreducible polynomial $p(x)$ for which $p(\alpha) = 0$

2) the unique monic polynomial $p(x)$ of smallest degree for which $p(\alpha) = 0$

3) the unique monic polynomial $p(x)$ with the property that $f(\alpha) = 0$ if and only if $p(x) \mid f(x)$.

In other words, $\min(\alpha, F)$ *is the unique monic generator of the ideal $\mathcal{I}_\alpha$.* □

Definition *Let $F < E$. Then $\alpha, \beta \in E$ are said to be* **conjugates** *over F if they have the same minimal polynomial over F.* □

1.6 Multiple Roots

Let us now explore the issue of multiple roots of a polynomial.

Definition *Let α be a root of $f(x) \in F[x]$. The* **multiplicity** *of α is the largest positive integer n for which $(x - \alpha)^n$ divides $f(x)$. If $n = 1$, then α is a* **simple root** *and if $n > 1$, then α is a* **multiple root** *of $f(x)$.* □

Definition *An irreducible polynomial $f(x) \in F[x]$ is* **separable** *if it has no multiple roots in any extension of F. An irreducible polynomial that is not separable is* **inseparable**. □

We should make a comment about this definition. It is not standard. For example, Lang defines a polynomial to be separable if it has no multiple roots, saying nothing about irreducibility. Hence, $p(x) = x^2$ is not separable under this definition. Jacobson defines a polynomial to be separable if its irreducible factors have no multiple roots. Hence, $p(x)$ is separable under this definition. However, van der Waerden, who first proposed the term "separable", gave the definition we have adopted, which does require irreducibility. Hence, for us, the question of whether $p(x) = x^2$ is separable is not applicable, since $p(x)$ is not irreducible. The only inconvenience with this definition is that we cannot say that if $f(x)$ is separable over F, then it is also separable over an extension E of F. Instead we must say that the irreducible factors of $f(x)$ are separable over E.

Although, as we will see, all irreducible polynomials over a field of characteristic zero or a finite field are separable, the concept of separability (that is, inseparability) plays a key role in the theory of more "unusual" fields.

Theorem 1.6.1 *A polynomial $f(x) \in F[x]$ has no multiple roots if and only if $f(x)$ and its derivative $f'(x)$ are relatively prime.*

Proof. Over a splitting field E for $f(x)$, we have

$$f(x) = (x - \alpha_1)^{e_1} \cdots (x - \alpha_n)^{e_n}$$

where the α_i's are distinct. It is easy to see that $f(x)$ and $f'(x)$ have no nontrivial common factors over E if and only if $e_i = 1$ for all $i = 1, \ldots, n$.□

Corollary 1.6.2 *An* irreducible *polynomial $f(x)$ is separable if and only if $f'(x) \neq 0$.*
Proof. Since $\deg(f'(x)) < \deg(f(x))$ and $f(x)$ is irreducible, it follows that $f(x)$ and $f'(x)$ are relatively prime if and only if $f'(x) \neq 0$. □

If $\operatorname{char}(F) = 0$ then $f'(x) \neq 0$ for any nonconstant $f(x)$. Thus, we get the following corollary.

Corollary 1.6.3 *All irreducible polynomials over a field of characteristic 0 are separable.*□

What Do Inseparable Polynomials Look Like?

When $\operatorname{char}(F) = p \neq 0$, inseparable polynomials are precisely the polynomials of the form $g(x^{p^d})$ for some $d \geq 1$. After all, if $f(x)$ is inseparable (and therefore irreducible by definition), then $f'(x) = 0$, and this can happen only if the exponents of each term in $f(x)$ are multiples of the characteristic p. Hence, $f(x)$ must have the form $g(x^p)$. But we can say more.

Corollary 1.6.4 *Let* $\operatorname{char}(F) = p \neq 0$. *An irreducible polynomial $f(x)$ over F is inseparable if and only if $f(x)$ has the form*

$$f(x) = g(x^{p^d})$$

where $d > 0$ and $g(x)$ is a nonconstant polynomial. In this case, the integer d can be chosen so that $g(x)$ is separable, in which case every root of $f(x)$ has multiplicity p^d. In this case, the number d is called the **radical exponent** *of $p(x)$.*
Proof. As we mentioned, if $f(x) = \sum a_i x^i$ is inseparable then $f'(x) = 0$, which implies that $i a_i = 0$ for all i, which in turn implies that $p \mid i$ for all i such that $a_i \neq 0$. Hence, $f(x) = q(x^p)$.

If $q(x)$ has no multiple roots, we are done. If not, then we may repeat the argument with the irreducible polynomial $q(x)$, eventually obtaining the equation $f(x) = g(x^{p^d})$, where $g(x)$ is separable.

For the converse, suppose that $f(x) = g(x^{p^d})$ for some $d > 0$. Let K be a field in which both $f(x)$ and $g(x)$ split. Thus,

$$g(x) = (x - \alpha_1) \cdots (x - \alpha_k)$$

for $\alpha_i \in K$ and so

$$f(x) = (x^{p^d} - \alpha_1)\cdots(x^{p^d} - \alpha_k)$$

Since $f(x)$ splits in K, there exist roots $\beta_i \in K$ for each of the factors $x^{p^d} - \alpha_i$, and so $\alpha_i = \beta_i^{p^d}$. Hence,

$$f(x) = (x^{p^d} - \beta_1^{p^d})\cdots(x^{p^d} - \beta_k^{p^d}) = (x - \beta_1)^{p^d}\cdots(x - \beta_k)^{p^d}$$

This shows that $f(x)$ is inseparable. Finally, if $f(x) = g(x^{p^d})$, where $g(x)$ is inseparable, then the α_i's above are distinct and so are the β_i's. Hence, each root of $f(x)$ has multiplicity p^d.□

We can now prove that all irreducible polynomials over a finite field are separable.

Corollary 1.6.5 *All irreducible polynomials over a finite field are separable.*
Proof. First, we show that a finite field F of characteristic p has p^n elements, for some $n > 0$. To see this, note that F is an extension of its prime subfield $\mathbb{Z}_p$ and if the dimension of F as a vector space over $\mathbb{Z}_p$ is n, then F has $q = p^n$ elements.

It follows that the multiplicative group F^* of nonzero elements of F has order $q - 1$ and so $a^q = a$ for all $a \in F$. In particular, any element of F is a pth power of some other element of F. Thus, if $f(x)$ is not separable, then $f(x) = g(x^p)$. Hence

$$\begin{aligned} g(x^p) &= a_0 + a_1x^p + \cdots + a_nx^{pn} \\ &= b_0^p + b_1^px^p + \cdots + b_n^px^{np} \\ &= (b_0 + b_1x + \cdots + b_nx^n)^p \end{aligned}$$

is not irreducible.□

The next example shows that inseparable polynomials do exist.

Example 1.6.1 Let F be a field of characteristic 2 and consider the field $F(t)$ of all rational functions in the variable t. The polynomial $f(x) = x^2 - t^2$ is irreducible over $F(t^2)$, since it has no linear factors over $F(t^2)$. However, in $F(t)$ we have $f(x) = (x - t)^2$ and so t is a double root of $f(x)$. □

1.7 Testing for Irreducibility

We next discuss some methods for testing a polynomial for irreducibility. Note first that $p(x) \in F[x]$ is irreducible if and only if $p(x + a)$ is irreducible, for $a \in F$. This is often a useful device in identifying irreducibility.

Localization

Sometimes it is possible to identify irreducibility by changing the base ring. In particular, suppose that R and S are rings and $\sigma: R \to S$ is a ring homomorphism. If a polynomial $p(x) \in R[x]$ is degreewise reducible, then

$$p(x) = f(x)g(x)$$

where $\deg(f) < \deg(p)$ and $\deg(g) < \deg(p)$. Applying σ gives

$$p^\sigma(x) = f^\sigma(x)g^\sigma(x)$$

and since the degree cannot increase, if $\deg(p^\sigma) = \deg(p)$, then we can conclude that $p^\sigma(x)$ is degreewise reducible over S. Hence, if $p^\sigma(x)$ is degreewise irreducible, then so is $p(x)$. This situation is a bit too general, and we take S to be a field.

Theorem 1.7.1 *Let R be a ring and let F be a field. Let $\sigma: R \to F$ be a ring homomorphism. A polynomial $p(x) \in R[x]$ is degreewise irreducible (not the product of two polynomials of smaller degree) over R if*
1) $\deg(p^\sigma) = \deg(p)$
2) $p^\sigma(x)$ *is irreducible over* F.□

The following special case is sometimes called **localization**. Recall that if R is a ring and $p \in R$ is a prime, then the canonical projection map $\pi_p: R \to R/\langle p \rangle$ is defined by $\pi_p(\alpha) = \alpha + \langle p \rangle$. This map is a surjective ring homomorphism.

Corollary 1.7.2 (Localization) *Let R be a principal ideal domain and let*

$$f(x) = a_0 + a_1 x + \cdots + a_n x^n$$

be a polynomial over R. Let $p \in R$ be a prime that does not divide a_n. If $\pi_p f(x)$ is irreducible over $R/\langle p \rangle$, then $f(x)$ is degreewise irreducible over R.□

Example 1.7.1 Let $p(x) = x^3 + 6x^2 + 5x + 1 \in \mathbb{Z}[x]$. Since $p(x)$ has degree 3, it is reducible if and only if it has an integer root. We could simply start checking integers, but localization saves a lot of time. By localizing to $\mathbb{Z}_3$, we get $\pi_3 p(x) = x^3 + 2x + 1$, and we need only check for a root in $\mathbb{Z}_3 = \{0, 1, 2\}$. Since none of these is a root, $\pi_3 p(x)$ and therefore $p(x)$, is degreewise irreducible. But since $p(x)$ is primitive, it is just plain irreducible.□

It is interesting to point out that there are polynomials $f(x)$ for which $\pi_p f(x)$ is reducible for *all* primes p, and yet $f(x)$ is irreducible over $\mathbb{Z}$. Thus, the method of localization cannot be used to prove that a polynomial is *reducible.*

Example 1.7.2 Let $f(x) = x^4 + ax^2 + b^2$, for $a, b \in \mathbb{Z}$. We claim that $\pi_p f(x)$ is reducible for all primes p. If $p = 2$, then $f(x)$ is one of the following polynomials

$$x^4 + x^2 + 1 = (x^2 + x + 1)^2$$
$$x^4 + 1 = (x^2 + 1)^2$$
$$x^4 + x^2 = x^2(x^2 + 1)$$

or x^4, each of which is reducible modulo 2. Now assume that $p > 2$. In the field $\mathbb{Z}_p$, let c satisfy $2c = a$, in which case $f(x) = x^4 + 2cx^2 + b$, which can be written in any of the following ways

$$f(x) \equiv (x^2 + c)^2 - (c^2 - b^2)$$
$$f(x) \equiv (x^2 + b)^2 - (2b - 2c)x^2$$
$$f(x) \equiv (x^2 - b)^2 - (-2b - 2c)x^2$$

Each of these has the potential of being the difference of two squares, which is reducible. In fact, this will happen if any of $c^2 - b^2, 2b - 2c$ and $-2b - 2c$ is a square modulo p.

Since the multiplicative group $\mathbb{Z}_p^*$ of nonzero elements of $\mathbb{Z}_p$ is cyclic (a fact about finite fields that we will prove later), we can write $\mathbb{Z}_p^* = \langle \beta \rangle$. Note that the group homomorphism $\sigma_2 : \alpha \mapsto \alpha^2$ has kernel $\{\pm 1\}$ and so exactly half of the elements of $\mathbb{Z}_p^*$ are squares, and these are the even powers of β. So, if $2b - 2c$ and $-2b - 2c$ are nonsquares, that is, odd powers of β, then their product

$$(2b - 2c)(-2b - 2c) \equiv 4(c^2 - b^2)$$

is a square, and therefore so is $c^2 - b^2$ modulo p.

Now, we can choose a and b so that $f(x)$ is irreducible over $\mathbb{Z}$. For example, $f(x) = x^4 + 1$ is irreducible over $\mathbb{Z}$.□

Eisenstein's Criterion

The following is the most famous criterion for irreducibility.

Theorem 1.7.3 (Eisenstein's criterion) *Let R be an integral domain and let $p(x) = a_0 + a_1 x + \cdots + a_n x^n \in R[x]$. If there exists a prime $p \in R$ satisfying*

$$p \mid a_i \text{ for } 0 \le i < n,\ p \nmid a_n,\ p^2 \nmid a_0$$

then $p(x)$ is degreewise irreducible. In particular, if $p(x)$ is primitive, then it is irreducible.

Proof. Let $\pi : \mathbb{Z} \to \mathbb{Z}_p$ be the canonical projection map. Suppose that $p(x) = f(x)g(x)$ where $\deg(f) < \deg(p)$ and $\deg(g) < \deg(p)$. Since $\pi(a_i) = 0$ for all $i < n$, it follows that

$$\pi(a_n)x^n = \pi p(x) = \pi f(x) \pi g(x)$$

Since $\pi(a_n) \neq 0$, this implies that $\pi f(x)$ and $\pi g(x)$ are monomials of positive degree (since R is an integral domain). In particular, the constant terms πf_0 and πg_0 are 0 in $R/\langle p \rangle$, that is, $p \mid f_0$ and $p \mid g_0$ and therefore $p^2 \mid a_0$, which is a contradiction. Hence, $p(x)$ is degreewise irreducible.□

Eisenstein's criterion can be useful as a theoretical tool.

Corollary 1.7.4 *Let R be an integral domain that contains at least one prime. For every positive integer n, there is an irreducible polynomial $p_n(x)$ of degree n over R.*
Proof. According to Eisenstein's criterion, the primitive polynomial $p(x) = x^n - p$ is irreducible, where p is a prime.□

Exercises

1. Prove that if R is an integral domain, then so is $R[x_1, \ldots, x_n]$.
2. **(Chinese Remainder Theorem)** Let $p_1(x), \ldots, p_n(x)$ be pairwise relatively prime polynomials over a field F. Let $f_1(x), \ldots, f_n(x)$ be polynomials over F. Prove that the system of congruences

$$\begin{gathered} g(x) \equiv f_1(x) \bmod p_1(x) \\ \vdots \\ g(x) \equiv f_n(x) \bmod p_n(x) \end{gathered}$$

   has a unique solution modulo the polynomial $P(x) = \prod p_i(x)$.
3. Let E, F be fields with $F < E$. Prove that if $f(x) = g(x)h(x)$ is a factorization of polynomials over $E[x]$, where two of the three polynomials have coefficients in F, then the third also has coefficients in F.
4. Let R be a unique factorization domain. Prove that $c(\alpha p(x)) = \alpha c(p(x))$ for any $p(x) \in R[x]$ and $\alpha \in R$.
5. Prove that if $n > 1$ then the ring $F[x_1, \ldots, x_n]$ is not a principal ideal domain.
6. Verify the division algorithm (Theorem 1.3.1) for commutative rings with identity. *Hint*: try induction on $\deg(f)$.
7. Let $p(x), q(x) \in F[x]$. Prove that there exist polynomials $a(x)$, $b(x) \in F[x]$ with $\deg(a) < \deg(q)$ and $\deg(b) < \deg(p)$ for which

$$a(x)p(x) + b(x)q(x) = 0$$

   if and only if $p(x)$ and $q(x)$ are not relatively prime.
8. Let F^* be the multiplicative group of all nonzero elements of a field F. We have seen that if G is a finite subgroup of F^*, then G is cyclic. Prove that if F is an infinite field then no infinite subgroup G of F^* is cyclic.
9. Prove Theorem 1.5.1.
10. Show that the following are irreducible over $\mathbb{Q}$.
    a) $x^4 + 8x^3 + 28x^2 + 48x + 34$
    b) $x^3 + 3x^2 + 5x + 2$
    c) $x^3 + 3x^2 - 6x + 9$

d) $x^4 - x + 1$

11. For p prime show that $p(x) = 1 + x + x^2 + \cdots + x^{p-1}$ is irreducible over $\mathbb{Z}[x]$. *Hint*: apply Eisenstein to the polynomial $p(x+1)$.
12. Prove that for p prime, $p(x) = x^n + px + p^2$ is irreducible over $\mathbb{Z}[x]$.
13. If R is an infinite integral domain and $p(x_1, \ldots, x_n)$ is a polynomial in several variables over R, show that $p(x_1, \ldots, x_n)$ is zero as a function if and only if it is zero as a polynomial.
14. Let p be a prime. Show that the number of monic irreducible polynomials of degree 3 over $\mathbb{Z}_p$ is $(1/3)(p^3 - p)$.
15. There is a simple (but not necessarily practical) algorithm for factoring any polynomial over $\mathbb{Q}$, due to Kronecker. In view of Theorem 1.2.3, it suffices to consider polynomials with integer coefficients. A polynomial of degree n is completely determined by specifying $n+1$ of its values. This follows from the *Lagrange Interpolation Formula*

$$p(x) = \sum_{i=0}^{n} p(i) \left[\prod_{j \neq i} \frac{x - j}{i - j} \right]$$

Let $f(x)$ be a polynomial of degree $n > 1$ over $\mathbb{Z}$. If $f(x)$ has a nonconstant factor $p(x)$ of degree at most $n/2$, what can you say about the values $p(i)$ for $i = 0, \ldots, [\![n/2]\!]$? Construct an algorithm for factoring $f(x)$ into irreducible factors. Use this method to find a linear factor of the polynomial $f(x) = 1 - x - x^2 - 2x^3$ over $\mathbb{Z}$.

16. Prove that if $f(x)/g(x) = p(x)/q(x)$, where each rational expression is in lowest common terms (no common nonconstant factors in the numerator and denominator) then $f(x) \sim p(x)$ and $g(x) \sim q(x)$.
17. Let $p(x)$ be a polynomial over F with multiple roots. Show that there is a polynomial $q(x)$ over F whose distinct roots are the same as the distinct roots of $p(x)$, but that occur in $q(x)$ only as simple roots.

Reciprocal Polynomials

If $f(x)$ is a polynomial of degree d, we define the **reciprocal polynomial** by $f_R(x) = x^d f(x^{-1})$. Thus, if

$$f(x) = a_n x^n + a_{n-1} x^{n-1} + \cdots + a_1 x + a_0$$

then

$$f_R(x) = a_0 x^n + a_1 x^{n-1} + \cdots + a_{n-1} x + a_n$$

If a polynomial satisfies $f(x) = f_R(x)$, we say that $f(x)$ is **self-reciprocal**.

18. Show that $\alpha \neq 0$ is a root of $f(x)$ if and only if α^{-1} is a root of $f_R(x)$.
19. Show that the reciprocal of an irreducible polynomial $f(x)$ with nonzero constant term is also irreducible.

20. Let $p(x) = a_0 + a_1 x + \cdots + a_n x^n$. Prove that if p is a prime for which $p \mid a_i$ for $i > 0$, $p^2 \nmid a_n$, $p \nmid a_0$ then $p(x)$ is irreducible.
21. Show that if a polynomial $f(x) \neq 1 + x$ is self-reciprocal and irreducible, then $\deg(f)$ must be even. *Hint*: check the value of $f(-1)$.
22. Suppose that $f(x) = p(x)q(x) \in F[x]$, where $p(x)$ and $q(x)$ are irreducible, and $f(x)$ is self-reciprocal. Show that either
    a) $p(x) = \alpha p_R(x)$ and $q(x) = \alpha q_R(x)$ with $\alpha = \pm 1$, or
    b) $p(x) = \alpha q_R(x)$ and $q(x) = \alpha^{-1} p_R(x)$ for some $\alpha \in F$.

Chapter 2
Field Extensions

In this chapter, we will describe several types of field extensions and study their basic properties.

2.1 The Lattice of Subfields of a Field

If E is an extension field of F, then E can be viewed as a vector space over F. The dimension of E over F is denoted by $[E : F]$ and called the **degree** of E over F.

A sequence of fields $E_1, \ldots, E_n$ for which $E_i < E_{i+1}$ is referred to as a **tower** of fields, and we write

$$E_1 < E_2 < \cdots < E_n$$

The fact that dimension is multiplicative over towers is fundamental.

Theorem 2.1.1 *Let $F < K < E$. Then*

$$[E : F] = [E : K][K : F]$$

Moreover, if $A = \{\alpha_i \mid i \in I\}$ is a basis for E over K and $B = \{\beta_j \mid j \in J\}$ is a basis for K over F, then the set of products $C = \{\alpha_i\beta_j \mid i \in I, j \in J\}$ is a basis for E over F.

Proof. For the independence of C, suppose that $\sum_{i,j} a_{i,j}\alpha_i\beta_j = 0$. Then

$$0 = \sum_{i,j} a_{i,j}\alpha_i\beta_j = \sum_i \left[\sum_j a_{i,j}\beta_j\right]\alpha_i$$

and the independence of A over K implies that $\sum_j a_{i,j}\beta_j = 0$ for all i, and the independence of B over F implies that $a_{i,j} = 0$ for all i and j. Hence, C is linearly independent. Next, if $\gamma \in E$ then there exist $a_i \in K$ such that $\gamma = \sum a_i\alpha_i$. Since each a_i is a linear combination of the β_j's, it follows that γ is a linear combination of the products $\alpha_i\beta_j$. Hence C spans E over F. □

The Composite of Fields

If F and E are subfields of a field K, then the intersection $F \cap E$ is clearly a field. The **composite** FE of F and E is defined to be the smallest subfield of K containing both F and E. The composite FE is also equal to the intersection of all subfields of K containing E and F.

More generally, the composite $\bigvee E_i$ of a family $\mathcal{E} = \{E_i \mid i \in I\}$ of fields, all of which are contained in a single field K, is the smallest subfield of K containing all members of the family.

> *Note that the composite of fields is defined only when the fields are all contained in one larger field. Whenever we form a composite, it is with the tacit understanding that the relevant fields are so contained.*

A **monomial** over a family $\mathcal{E} = \{E_i \mid i \in I\}$ of fields with $E_i < E$ is simply a product of a finite number of elements from the union $\bigcup E_i$.

The set of all finite sums of monomials over $\mathcal{E}$ is the smallest subring R of E containing each field E_i and the set of all quotients of elements of R (the quotient field of R) is the composite $\bigvee E_i$. Thus, each element of $\bigvee E_i$ involves only a finite number of elements from the union $\bigcup E_i$ and is therefore contained in a composite of a finite number of fields from the family $\mathcal{E}$.

The collection of all subfields of a field K forms a complete lattice $\mathcal{L}$ (under set inclusion), with meet being intersection and join being composite. The bottom element in $\mathcal{L}$ is the prime subfield of K (see Chapter 0) and the top element is K itself.

2.2 Types of Field Extensions

Field extensions $F < E$ can be classified into several types, as shown in Figure 2.2.1. The goal of this chapter is to explore the properties of these various types of extensions.

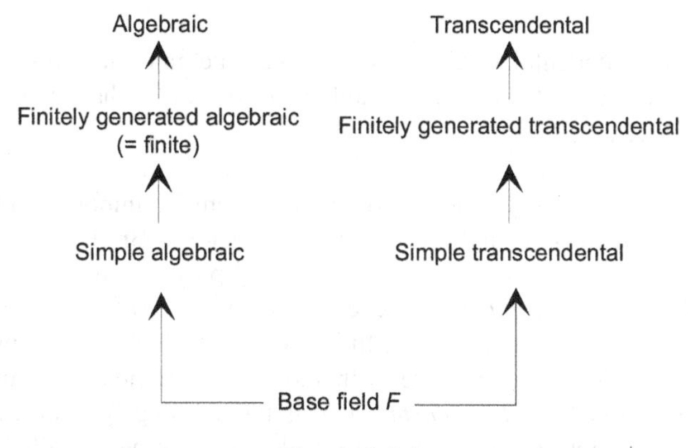

Figure 2.2.1

It is worth noting that some types of extensions are defined in terms of the individual elements in the extension, whereas others are more "global" in nature. For instance, an extension $F < E$ is *algebraic* if each element $\alpha \in E$ is algebraic over F. Other characterizations involve properties of the field E as a whole. For instance, $F < E$ is *normal* if E is the splitting field of a family of polynomials over F.

Let us begin with the basic definitions (which will be repeated as we discuss each type of extension in detail). Recall that if $F < E$, then an element $\alpha \in E$ is said to be **algebraic** over F if α is a root of some nonzero polynomial over F. An element that is not algebraic over F is said to be **transcendental** over F.

If $F < E$ and if S is a subset of E, the smallest subfield of E containing both F and S is denoted by $F(S)$. When $S = \{\alpha_1, \ldots, \alpha_n\}$ is a finite set, it is customary to write $F(\alpha_1, \ldots, \alpha_n)$ for $F(S)$.

Definition *Let $F < E$. Then*

1) *E is* **algebraic** *over F if every element $\alpha \in E$ is algebraic over F. Otherwise, E is* **transcendental** *over F.*
2) *E is* **finitely generated** *over F if $E = F(S)$, where $S \subseteq E$ is a finite set.*
3) *E is a* **simple** *extension of F if $E = F(\alpha)$, for some $\alpha \in E$. In this case, α is called a* **primitive element** *of E.*
4) *E is a* **finite extension** *of F if $[E : F]$ is finite.*□

To save words, it is customary to say that the *extension* $F < E$ is algebraic, transcendental, finitely generated, finite or simple, as the case may be, if E has this property as an extension of F.

The reader may have encountered a different meaning of the term *primitive* in connection with elements of a finite field. We will discuss this alternative meaning when we discuss finite fields later in the book.

Note that a transcendental extension may have algebraic elements not in the base field. For example, the transcendental extension $\mathbb{Q} < \mathbb{R}$ has many algebraic elements, such as $\sqrt{2}$.

In later chapters, we will study two other extremely important classes of extensions: the *separable* and the *normal* extensions. Briefly, an algebraic element $\alpha \in E$ is *separable* over F if its minimal polynomial is separable and an extension $F < E$ is *separable* if every element of E is separable over F. When $\text{char}(F) = 0$ or when F is a finite field, all algebraic extensions are separable, but such is not the case with more unusual fields. As mentioned earlier, an extension E of F is *normal* if it is the splitting field of a family of polynomials. An extension that is both separable and normal is called a *Galois extension.*

Distinguished Extensions

We will have much to say about towers of fields of the form $F < K < E$. Let us refer to such a tower as a **2-tower**, where K is the **intermediate field**, $F < K$ is the **lower step**, $K < E$ is the **upper step** and $F < E$ is the **full extension**.

Following Lang, we will say that a class $\mathcal{C}$ of field extensions is **distinguished** provided that it has the following properties

1) **The Tower Property**
For any 2-tower $F < K < E$, the full extension is in $\mathcal{C}$ if and only if the upper and lower steps are in $\mathcal{C}$. In symbols,

$$(F < E) \in \mathcal{C} \Leftrightarrow (F < K) \in \mathcal{C} \text{ and } (K < E) \in \mathcal{C}$$

2) **The Lifting Property**
The class $\mathcal{C}$ is closed under lifting by an arbitrary field, that is,

$$(F < E) \in \mathcal{C} \text{ and } F < K \Rightarrow (K < EK) \in \mathcal{C}$$

provided, of course, that EK is defined. The tower $K < EK$ is the **lifting** of $F < E$ by K.

Note that if $\mathcal{C}$ is distinguished, then it also has the following property:

3) **Closure under finite composites**
If EK is defined, then

$$(F < E) \in \mathcal{C} \text{ and } (F < K) \in \mathcal{C} \Rightarrow (F < EK) \in \mathcal{C}$$

This follows from the fact that $F < EK$ can be decomposed into

$$F < E < EK$$

and the first step is in $\mathcal{C}$, the second step is in $\mathcal{C}$ since it is the lifting of $F < E$ by K, and so the full extension is in $\mathcal{C}$.

Figure 2.2.2 illustrates these properties.

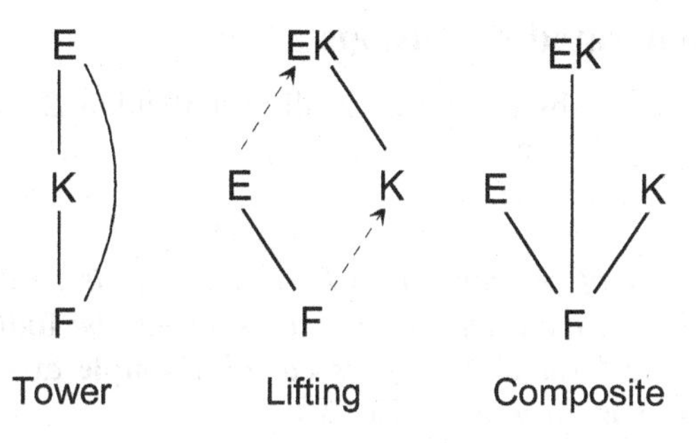

Figure 2.2.2

Consider a tower $\mathcal{T}$ of field extensions

$$F_1 < F_2 < \cdots$$

We say that the tower $\mathcal{T}$ is in $\mathcal{C}$, or has property $\mathcal{C}$, if all extensions of the form $F_i < F_j$, where $i < j$, are in $\mathcal{C}$. To illustrate the terminology, an *algebraic tower* is a tower in which each extension $F_i < F_j$, where $i < j$, is algebraic.

If a class $\mathcal{C}$ has the tower property, then the following are equivalent for a *finite* tower $\mathcal{T} = (F_1 < F_2 < \cdots < F_n)$:

1) $\mathcal{T}$ is in $\mathcal{C}$
2) The full extension $F_1 < F_n$ is in $\mathcal{C}$
3) Each step $F_i < F_{i+1}$ is in $\mathcal{C}$.

If a class $\mathcal{C}$ of extensions has the property that

$$(F < E_i) \in \mathcal{C} \Rightarrow (F < \bigvee E_i) \in \mathcal{C}$$

for any family $\{E_i\}$ of fields (provided, as always, that the composite is defined), we say that $\mathcal{C}$ is **closed under arbitrary composites**. This property does not follow from closure under *finite* composites.

Here is a list of the common types of extensions and their distinguishedness. We will verify these statements in due course.

Distinguished

Algebraic extensions
Finite extensions
Finitely generated extensions
Separable extensions

Not Distinguished
Simple extensions (lifting property holds, upper and lower steps simple)
Transcendental extensions
Normal extensions (lifting property holds, upper step normal)

2.3 Finitely Generated Extensions

If $F < E$ and if S is a subset of E, the smallest subfield of E containing F and S is denoted by $F(S)$. When $S = \{\alpha_1, \ldots, \alpha_n\}$ is a finite set, it is customary to write $F(\alpha_1, \ldots, \alpha_n)$ for $F(S)$.

Definition *Any field of the form* $E = F(\alpha_1, \ldots, \alpha_n)$ *is said to be* **finitely generated** *over* F *and the extension* $F < E$ *is said to be* **finitely generated**. *Any extension of the form* $F < F(\alpha)$ *is called a* **simple** *extension and* α *is called a* **primitive element** *in* $F(\alpha)$. □

The reader may have encountered a different meaning of the term *primitive* in connection with elements of a finite field. We will discuss this alternative meaning when we discuss finite fields later in the book.

Note that for $1 \leq k \leq n-1$,

$$F(\alpha_1, \ldots, \alpha_n) = [F(\alpha_1, \ldots, \alpha_k)](\alpha_{k+1}, \ldots, \alpha_n)$$

and so a finitely generated extension $F < F(\alpha_1, \ldots, \alpha_n)$ can be decomposed into a tower of simple extensions

$$F < F(\alpha_1) < F(\alpha_1, \alpha_2) < \cdots < F(\alpha_1, \alpha_2, \ldots, \alpha_n)$$

It is evident that $F(\alpha_1, \ldots, \alpha_n)$ consists of all quotients of polynomials in the α_i's:

$$F(\alpha_1, \ldots, \alpha_n) = \left\{ \frac{f(\alpha_1, \ldots, \alpha_n)}{g(\alpha_1, \ldots, \alpha_n)} \,\middle|\, f, g \in F[x_1, \ldots, x_n], g(\alpha_1, \ldots, \alpha_n) \neq 0 \right\}$$

The class of finite extensions is our first example of a distinguished class.

Theorem 2.3.1 *The class of all finitely generated extensions is distinguished.*
Proof. For the tower property, if $F < F(S) < F(S)(T)$ is a 2-tower in which each step is finitely generated, that is, if S and T are finite sets, then since $F(S)(T) = F(S \cup T)$, the full extension is finitely generated by $S \cup T$ over F.

Also, if $F < K < F(S)$, where S is finite, then since $F(S) = K(S)$, the upper step is $K < K(S)$, which is finitely generated by S. However, the proof that the lower step $F < K$ is finitely generated is a bit testy and we must postpone it until we have discussed transcendental extensions in the next chapter.

For the lifting property, if $E = F(S)$, where S is finite and if $F < K$, with EK defined, then

$$EK = K(F(S)) = K(S)$$

and so the composite EK is finitely generated over K by S. □

2.4 Simple Extensions

Let us take a closer look at simple extensions $F < F(\alpha)$.

Simple Extensions Are Not Distinguished

The class of simple extensions has all of the properties required of distinguished extensions *except* that the lower and upper steps being simple does not imply that the full extension is simple. That is, if each step in a 2-tower is simple

$$F < F(\alpha) < F(\alpha)(\beta) = F(\alpha, \beta)$$

this does not imply that the full extension is simple.

Example 2.4.1 Let s and t be independent variables and let p be a prime. In the tower

$$\mathbb{Z}_p(s^p, t^p) < \mathbb{Z}_p(s, t^p) < \mathbb{Z}_p(s, t)$$

each step is simple but the full extension is not. We leave proof of this as a (nontrivial) exercise.□

On the other hand, if the full extension is simple $F < K < F(\alpha)$, then the upper step is $K < K(\alpha)$, which is simple. Also, the lower step is simple, but the nontrivial proof requires us to consider the algebraic and transcendental cases separately, which we will do at the appropriate time.

As to lifting, if $F < F(\alpha)$ is simple and $F < K$, then the lifting is $K < K(\alpha)$, which is simple. Thus, the lifting property holds.

Simple Algebraic Extensions

Suppose that $F < F(\alpha)$ is a simple extension, where α is algebraic over F. We have seen that the minimal polynomial $p_\alpha(x) = \min(\alpha, F)$ of α over F is the unique monic polynomial of smallest degree satisfied by α. Also, $p_\alpha(x)$ is irreducible.

Now, $F(\alpha)$ is the field of all rational expressions in α

$$F(\alpha) = \left\{ \frac{f(\alpha)}{g(\alpha)} \;\middle|\; f, g \in F[x], g(\alpha) \neq 0 \right\}$$

but we can improve upon this characterization considerably. Since $g(\alpha) \neq 0$, it follows that $p_\alpha(x) \nmid g(x)$ and the irreducibility of $p_\alpha(x)$ implies that

$g(x) \not| \, p_\alpha(x)$. Hence, $g(x)$ and $p_\alpha(x)$ are relatively prime and there exist polynomials $a(x)$ and $b(x)$ for which

$$a(x)g(x) + b(x)p_\alpha(x) = 1$$

Evaluating at α gives

$$1 = a(\alpha)g(\alpha) + b(\alpha)p_\alpha(\alpha) = a(\alpha)g(\alpha)$$

and so the inverse of $g(\alpha)$ is the *polynomial* $a(\alpha)$. It follows that

$$F(\alpha) = \left\{ f(\alpha) \,\middle|\, f(x) \in F[x] \right\}$$

Moreover, if $\deg(f) > \deg(p_\alpha)$, then

$$f(x) = q(x)p_\alpha(x) + r(x)$$

where $r(x) = 0$ or $\deg(r) < \deg(p_\alpha)$. Hence,

$$f(\alpha) = q(\alpha)p_\alpha(\alpha) + r(\alpha) = r(\alpha)$$

Thus,

$$F(\alpha) = \left\{ f(\alpha) \,\middle|\, f(x) \in F[x], \deg(f) < \deg(p_\alpha) \right\}$$

In words, $F(\alpha)$ is the set of all polynomials in α over F of degree less than the degree of the minimal polynomial of α, where multiplication is performed modulo $p_\alpha(x)$.

The map $\phi: F[x] \to F(\alpha)$ defined by

$$\phi(f(x)) = [f(x) \bmod p_\alpha(x)]|_{x=\alpha}$$

is easily seen to be a surjective ring homomorphism. In fact, it is the composition of two surjective ring homomorphisms: the first is projection modulo $p_\alpha(x)$ and the second is evaluation at α.

The kernel of ϕ is the ideal $\langle p_\alpha(x) \rangle$ generated by $p_\alpha(x)$, since

$$\begin{aligned} \ker(\phi) &= \{ f(x) \in F[x] \mid [f(\alpha) \bmod p_\alpha(x)] = 0 \} \\ &= \{ f(x) \in F[x] \mid p_\alpha(x) \mid f(\alpha) \} \\ &= \langle p_\alpha(x) \rangle \end{aligned}$$

It follows that

$$F(\alpha) \approx \frac{F[x]}{\langle p_\alpha(x) \rangle}$$

This has a couple of important consequences. First, if we restrict attention to polynomials of degree less than $\deg(p_\alpha(x))$, then α can be treated as an "independent" variable. Also, if $\alpha, \beta \in E$ are conjugate (have the same minimal polynomial) over F, then the substitution map $\sigma: F(\alpha) \to F(\beta)$ defined by

$$\sigma(f(\alpha)) = f(\beta)$$

is an isomorphism from $F(\alpha)$ to $F(\beta)$.

Let us summarize.

Theorem 2.4.1 *Let $F < E$ and let $\alpha \in E$ be algebraic over F.*

1) *Then*

$$F(\alpha) = \left\{ f(\alpha) \,\middle|\, f(x) \in F[x], \deg(f) < \deg(p_\alpha) \right\}$$

where multiplication is performed modulo $p_\alpha(x)$.

2) *Moreover,*

$$F(\alpha) \approx \frac{F[x]}{\langle p_\alpha(x) \rangle}$$

3) *The extension $F < F(\alpha)$ is finite and*

$$d = [F(\alpha) : F] = \deg(p_\alpha(x))$$

In fact, the set $\mathcal{B} = \{1, \alpha, \ldots, \alpha^{d-1}\}$ is a vector space basis for E over F.

4) *If the elements $\alpha, \beta \in E$ are conjugate over F then $F(\alpha) \approx F(\beta)$.*$\square$

We have seen that a simple extension $F < F(\alpha)$, where α is algebraic, is finite. Conversely, if $F < F(\alpha)$ is finite and simple, then for any $\beta \in F(\alpha)$, the sequence $1, \beta, \beta^2, \ldots$ is linearly dependent and so β is algebraic. Hence, all elements of $F(\alpha)$ are algebraic and so $F < F(\alpha)$ is an algebraic extension.

Theorem 2.4.2 *The following are equivalent for a simple extension $F < F(\alpha)$*

1) *α is algebraic*
2) *$F < F(\alpha)$ is algebraic*
3) *$F < F(\alpha)$ is finite.*

In this case, $[F(\alpha) : F] = \deg(\min(\alpha, F))$.$\square$

Characterizing Simple Algebraic Extensions

Simple algebraic extensions can be characterized in terms of the number of intermediate fields.

Theorem 2.4.3 *Let $E = F(\alpha_1, \ldots, \alpha_n)$ be finitely generated over F by algebraic elements over F.*

1) *Then $E = F(\alpha)$ for some algebraic element $\alpha \in E$ if and only if there is only a finite number of intermediate fields $F < K < E$ between E and F.*

2) *In this case, if* E *is an infinite field, then* $E = F(\alpha)$ *where* α *has the form*

$$\alpha = a_1\alpha_1 + \cdots + a_n\alpha_n$$

for $a_i \in F$.

Proof. Suppose first that $F < F(\alpha)$ for some algebraic element $\alpha \in E$. For each intermediate field $F < K < F(\alpha)$, the minimal polynomial $\min(\alpha, F)$ is also a polynomial over K and is satisfied by α. Hence, $\min(\alpha, K) \mid \min(\alpha, F)$. But $\min(\alpha, F)$ has only a finite number of monic factors. Therefore, this part of the proof will be complete if we show that there is only one intermediate field with minimal polynomial $\min(\alpha, K)$.

Suppose that K and L have the property that

$$p(x) = \min(\alpha, K) = \min(\alpha, L)$$

Then the coefficients of $p(x)$ lie in $K \cap L$. Since $p(x)$ is irreducible over K, it is also irreducible over $K \cap L$ and so

$$p(x) = \min(\alpha, K) = \min(\alpha, L) = \min(\alpha, K \cap L)$$

But $K \cap L < K$ and so

$$[F(\alpha) : K \cap L] = \deg(p(x)) = [F(\alpha) : K]$$

which implies that $[K : K \cap L] = 1$ and so $K = K \cap L$. Similarly, $L = K \cap L$ and so $K = L$. This shows that K is uniquely determined by the polynomial $\min(\alpha, K)$ and so there are only finitely many intermediate fields $F < K < F(\alpha)$.

For the converse, if F is a finite field, then so is E, since it is finite-dimensional over F and so the multiplicative group E^* of nonzero elements of E is cyclic. If α generates this group, then $E = F(\alpha)$ is simple. Now suppose that F is an infinite field and there are only finitely many intermediate fields between E and F. Consider the intermediate fields $F(\alpha_1 + a\alpha_2)$, for all $a \in F$. By hypothesis, $F(\alpha_1 + a\alpha_2) = F(\alpha_1 + b\alpha_2)$ for some $a \neq b \in F$. Hence, $\alpha_1 + b\alpha_2 \in F(\alpha_1 + a\alpha_2)$, implying that

$$\alpha_2 = \frac{1}{a-b}[(\alpha_1 + a\alpha_2) - (\alpha_1 + b\alpha_2)] \in F(\alpha_1 + a\alpha_2)$$

and

$$\alpha_1 = (\alpha_1 + a\alpha_2) - a\alpha_2 \in F(\alpha_1 + a\alpha_2)$$

Hence, $F(\alpha_1, \alpha_2) \subseteq F(\alpha_1 + a\alpha_2)$. The reverse inclusion is evident and so $F(\alpha_1, \alpha_2) = F(\alpha_1 + a\alpha_2)$. Hence,

$$F(\alpha_1, \alpha_2, \ldots, \alpha_n) = F(\alpha_1 + a\alpha_2, \alpha_3, \ldots, \alpha_n)$$

We can repeat this process to eventually arrive at a primitive element of the desired form.□

In view of the previous theorem, it is clear that if $F < K < F(\alpha)$, where α is *algebraic*, then the lower step $F < K$ is also simple. (Note that $F < K$ is a finite extension and therefore finitely generated by the elements of a basis for K over F, whose elements are algebraic over F.)

Simple Transcendental Extensions

If t is transcendental over F, then $F(t)$ is the field of all rational expressions in α:

$$F(t) = \left\{ \frac{f(t)}{g(t)} \mid f, g \in F[x], g \neq 0 \right\}$$

The fact that t is transcendental implies that there are no algebraic dependencies in these rational expressions and $F(t)$ is, in fact, isomorphic to the field of rational functions in a single variable.

Theorem 2.4.4 *Let $F < E$ and let $t \in E$ be transcendental over F. Then $F(t)$ is isomorphic to the field of all rational functions $F(x)$ in a single variable x.*
Proof. The evaluation homomorphism ϕ: $F(x) \to E$ defined by

$$\phi\left(\frac{f(x)}{g(x)}\right) = \frac{f(t)}{g(t)}$$

is easily seen to be an isomorphism. To see that ϕ is injective, note that $f(t)/g(t) = 0$ implies $f(t) = 0$, which implies that $f(x) = 0$, since otherwise t would be algebraic.□

Simple transcendental extensions fail rather misreably to be distinguished. For example, the lifting of the transcendental extension $F < F(t)$ by $F(t^2)$ is $F(t^2) < F(t)$, which is algebraic. Also, in the tower $F < F(t^2) < F(t)$, the upper step is algebraic.

Let $F(s,t)$ be the field of rational functions in two independent variables. Then each step in the 2-tower

$$F < F(s) < F(s)(t) = F(s,t)$$

is simple, but the extension $F < F(s,t)$ is not simple. The proof is left as an exercise. (Intuitively speaking, we cannot expect a *single* rational function in x and y to be able to express both x and y individually.)

On the other hand, the lower step $F < K < F(t)$ of a transcendental extension is simple and transcendental (provided that $K \neq F$). This result is known as *Luroth's theorem* and will be proved in the next chapter.

Thus, simple transcendental extensions fail to be distinguished on every count except that the lower step in a simple transcendental extension is simple and transcendental.

More on Simple Transcendental Extensions

The fact that the upper step in the tower $F < F(t^2) < F(t)$ is algebraic is not an isolated case. Suppose that $F < F(t)$ is transcendental. Then any $s \in F(t) \setminus F$ is a nonconstant rational function in t

$$s = \frac{f(t)}{g(t)}$$

where we can assume that f and g are relatively prime. It turns out that s carries with it the full "transcendental nature" of the extension $F < F(t)$. To be more precise, consider the polynomial

$$p(x) = g(x)s - f(x) \in F(s)[x]$$

Then t is a root of $p(x)$ and so t is algebraic over $F(s)$. In other words, the upper step in the tower

$$F < F(s) < F(t)$$

is algebraic and finitely generated (by t) and therefore finite, by Theorem 2.4.2. As to the lower step, if it were also algebraic, it would be finite and so by the multiplicativity of degree, $F < F(t)$ would be finite and therefore algebraic. Since this is not the case, we deduce that $F < F(s)$ is transcendental, which means that s does not satisfy any nonzero polynomial over F.

We can now show that $p(x)$ is irreducible over $F(s)$. Since s is transcendental over F, we have $F(s) \approx F(y)$, where y is an independent variable. It follows that $F(s)[x] \approx F(y)[x]$ and so it is sufficient to show that the polynomial

$$h(y, x) = g(x)y - f(x) \in F(y)[x]$$

is irreducible over $F(y)$. However, this follows from the fact that $p(x)$ is irreducible as a polynomial over the ring $F[y]$, that is, as a polynomial in $F[y][x] = F[y, x] = F[x][y]$. To see this, note that any factorization in $F[x][y]$ has the form

$$p(y, x) = a(x)[b(x)y + c(x)]$$

where $a(x), b(x)$ and $c(x)$ are over F. But $f(x)$ and $g(x)$ are relatively prime and so $a(x)$ must be a unit in $F[x]$, which implies that $p(y, x)$ is irreducible over $F[y, x]$.

Hence, $p(x)$ is irreducible over $F(s)$ and

$$[F(t) : F(s)] = \deg(p) = \max(\deg(f), \deg(g))$$

Theorem 2.4.5

1) Consider the extension $F < F(t)$, where t is transcendental over F. Let

$$s = \frac{f(t)}{g(t)} \in F(t)$$

be any element of $F(t) \setminus F$, where $f(t)$ and $g(t)$ are relatively prime. Then in the tower

$$F < F(s) < F(t)$$

the lower step is transcendental (and so s is transcendental over F) and the upper step is algebraic, with

$$[F(t) : F(s)] = \max(\deg(f), \deg(g))$$

2) If t is transcendental over F, then $F(t)$ is algebraic over any intermediate field K other than F itself.

Proof. Part 1) has already been proved. As to part 2), if $F < K < F(t)$ where $K \neq F$, then let $s \in K \setminus F$. In the tower $F < F(s) < K < F(t)$, we know that $F(s) < F(t)$ is algebraic and simple and thus finite. It follows that $K < F(t)$ is also finite, hence algebraic.□

We should note that this theorem does not hold for nonsimple extensions. Specifically, just because an extension $F < E$ is generated by transcendental elements does not mean that all of the elements of $E \setminus F$ are transcendental. For example, the extension $\mathbb{Q} < \mathbb{Q}(t, \sqrt{2}t)$, where t is transcendental over $\mathbb{Q}$, is generated by transcendental elements t and $\sqrt{2}t$, but some elements of $\mathbb{Q}(t, \sqrt{2}t) \setminus \mathbb{Q}$ are algebraic over $\mathbb{Q}$. We will have more to say about this in Chapter 3.

2.5 Finite Extensions

If $F < E$ and $[E : F]$ is finite, we say that E is a **finite extension** of F or that $F < E$ is **finite**.

Theorem 2.5.1 *An extension is finite if and only if it is finitely generated by algebraic elements.*

Proof. If $F < E$ is finite and if $\{\alpha_1, \ldots, \alpha_n\}$ is a basis for E over F, then $E = F(\alpha_1, \ldots, \alpha_n)$ is finitely generated over F. Moreover, for each α_k, the sequence $1, \alpha_k, \alpha_k^2, \ldots$ over powers is linearly dependent over F, and so α_k is algebraic over F. Thus, $F < E$ is algebraic.

For the converse, assume that $E = F(\alpha_1, \ldots, \alpha_n)$, where each α_i is algebraic over F. Each step in the tower

$$F < F(\alpha_1) < F(\alpha_1, \alpha_2) < \cdots < F(\alpha_1, \ldots, \alpha_n) = E$$

is simple and algebraic, hence finite by Theorem 2.4.2. It follows that E is finite over F. □

Suppose that $E = F(\alpha_1, \ldots, \alpha_n)$ is finitely generated by algebraic elements α_i over F and consider the tower

$$F < F(\alpha_1) < F(\alpha_1, \alpha_2) < \cdots < F(\alpha_1, \ldots, \alpha_n) = E$$

Our results on simple algebraic extensions show that any element of $F(\alpha_1)$ is a polynomial in α_1 over F. Further, any element of $F(\alpha_1, \alpha_2)$ is a polynomial in α_2 over $F(\alpha_1)$, and hence a polynomial in the two variables α_1 and α_2. Continuing in this way, we conclude that E is the set of all polynomials over F in $\alpha_1, \ldots, \alpha_n$.

Theorem 2.5.2 *The class of finite extensions is distinguished. Moreover, if $\mathcal{B}$ is a finite basis for E over F and if $F < K$, then $\mathcal{B}$ spans EK over K, in particular,*

$$[EK : K] \leq [E : F]$$

Proof. The multiplicativity of degree shows that the tower property holds. As to lifting, let $F < E$ be finite, with basis $\{\alpha_1, \ldots, \alpha_n\}$ and let $F < K$. Then $E = F(\alpha_1, \ldots, \alpha_n)$, where each α_i is algebraic over F and so also over K. Since $EK = K(\alpha_1, \ldots, \alpha_n)$ is finitely generated by elements algebraic over K, it is a finite extension of K.

For the statement concerning degree, let $\mathcal{B} = \{\beta_1, \ldots, \beta_n\}$ be a basis for E over F. If $F < K$, then the lifting is $K < EK = K(\beta_1, \ldots, \beta_n)$ and each β_i is algebraic over K. It follows that EK is the set of polynomials over K in $\beta_1, \ldots, \beta_n$. However, any monomial in the β_i's is a linear combination (over F) of $\beta_1, \ldots, \beta_n$ and so EK is the set of linear combinations of $\beta_1, \ldots, \beta_n$ over K. In other words, $\mathcal{B}$ spans EK over K.□

We will see much later in the book that if $F < E$ is finite, and also normal and separable, then $[EK : K]$ actually divides $[E : F]$.

Note that if E is a splitting field for $p(x) \in F[x]$, then E is generated by the set of distinct roots $\alpha_1, \ldots, \alpha_n$ of $p(x)$. Thus $E = F(\alpha_1, \ldots, \alpha_n)$ is finitely generated by algebraic elements and so is a finite extension of F, of degree at most $d!$, where $d = \deg(p)$.

2.6 Algebraic Extensions

We now come to algebraic extensions.

Definition *An extension E of F is* **algebraic** *over F if every element $\alpha \in E$ is algebraic over F. Otherwise, E is* **transcendental** *over F.* □

Theorem 2.6.1 *A finite extension is algebraic.*
Proof. As we have said before, if $F < E$ is finite and $\alpha \in E$, then the sequence of powers $1, \alpha, \alpha^2, \ldots$ is linearly dependent over F and therefore some nontrivial polynomial in α must equal 0, implying that α is algebraic over F. □

Corollary 2.6.2 *The following are equivalent for an extension $F < E$*
1) $F < E$ is finite
2) $F < E$ is finitely generated by algebraic elements
3) $F < E$ is algebraic and finitely generated. □

Theorem 2.6.3 *Let $F < E$. The set K of all elements of E that are algebraic over F is a field, called the* **algebraic closure of** F **in** E.
Proof. Let $\alpha, \beta \in K$. The field $F(\alpha, \beta)$ is finitely generated over F by algebraic elements and so is algebraic over F, that is, $F(\alpha, \beta) \subseteq K$. This implies that α^{-1}, $\alpha \pm \beta$ and $\alpha\beta$ all lie in K, and so K is a subfield of E. □

Theorem 2.6.4 *The class of algebraic extensions is distinguished. It is also closed under the taking of arbitrary composites.*
Proof. For the tower property, let $F < K < E$. If the full extension $F < E$ is algebraic then so is the lower step $F < K$. Also, since any polynomial over F is a polynomial over K, the upper step $K < E$ is also algebraic. Conversely, suppose that $F < K$ and $K < E$ are algebraic and let $\alpha \in E$ have minimal polynomial $p(x) = \sum a_i x^i$ over K. Consider the tower of fields

$$F < F(a_1, \ldots, a_n) < F(a_1, \ldots, a_n, \alpha)$$

Since α is algebraic over $F(a_1, \ldots, a_n)$ and each a_i, being in K, is algebraic over F, we deduce that each step in the tower is finite and so $F < F(a_1, \ldots, a_n, \alpha)$ is finite. Hence, α is algebraic over F.

For the lifting property, let $F < E$ be algebraic and let $F < K$. Let $K < A < EK$, where A is the algebraic closure of K in EK. Then since each $\alpha \in E$ is algebraic over F it is a fortiori algebraic over K and so $E < A$. Clearly, $K < A$ and so $EK < A$. It follows that $EK = A$ is algebraic over K.

Finally, if $\{E_i\}$ is a family of fields, each algebraic over F, then so is $\bigvee E_i$, since an element of $\bigvee E_i$ is also an element of a composite of only a finite number of members of the family. □

The algebraic closure of the rational numbers $\mathbb{Q}$ in the complex numbers $\mathbb{C}$ is called the field $\mathcal{A}$ of **algebraic numbers**. We saw in the previous chapter that there is an irreducible polynomial $p_n(x) \in \mathbb{Z}[x]$ of every positive degree n.

Hence, $\mathcal{A}$ is an infinite algebraic extension of $\mathbb{Q}$, showing that the converse of Theorem 2.6.1 does not hold: algebraic extensions need not be finite.

Note that if $\alpha \in \mathbb{R}$ is an algebraic number, it also satisfies a polynomial over the integers. Thus, the algebraic numbers can be defined as the set of complex roots of polynomials over the integers. The subfield of all complex roots of *monic* polynomials over the integers is called the field of **algebraic integers**.

We note finally that if $F < E$ is algebraic and if $E = F(S)$ for some $S \subseteq E$ then each element of E is a polynomial in finitely many elements from S. This follows from the fact that each $\alpha \in F(S)$ is a rational function in finitely many elements of S and so there exists a finite subset $S_0 \subseteq S$ such that $\alpha \in F(S_0)$. Hence, our discussion related to finitely generated algebraic extensions applies here.

2.7 Algebraic Closures

Definition *A field E is said to be* **algebraically closed** *if any nonconstant polynomial with coefficients in E splits in E.* $\square$

Note that an algebraically closed field E cannot have a nontrivial algebraic extension $E < K$, since any $\alpha \in K$ is algebraic over E and its minimal polynomial over E must split over E, whence $\alpha \in E$.

Theorem 2.7.1 *Let F be a field. Then there is an extension E of F that is algebraically closed.*

Proof. The following proof is due to Emil Artin. The first step is to construct an extension field F_1 of F, with the property that all nonconstant polynomials in $F[x]$ have a root in F_1. To this end, for each nonconstant polynomial $p(x) \in F[x]$, let X_p be an independent variable and consider the ring $\mathcal{R}$ of all polynomials in the variables X_p over the field F. Let $\mathcal{I}$ be the ideal generated by the polynomials $p(X_p)$. We contend that $\mathcal{I}$ is not the entire ring $\mathcal{R}$. For if it were, then there would exist polynomials $q_1, \ldots, q_n \in \mathcal{R}$ and $p_1, \ldots, p_n \in \mathcal{I}$ such that

$$q_1 p_1(X_{p_1}) + \cdots + q_n p_n(X_{p_n}) = 1$$

This is an algebraic expression over F in a finite number of independent variables. But there is an extension field E of F in which each of the polynomials $p_1(x), \ldots, p_n(x)$ has a root, say $\alpha_1, \ldots, \alpha_n$. Setting $X_{p_i} = \alpha_i$ and setting any other variables appearing in the equation above equal to 0 gives $0 = 1$. This contradiction implies that $\mathcal{I} \neq \mathcal{R}$.

Since $\mathcal{I} \neq \mathcal{R}$, there exists a maximal ideal $\mathcal{J}$ such that $\mathcal{I} \subseteq \mathcal{J} \subset \mathcal{R}$. Then $F_1 = \mathcal{R}/\mathcal{J}$ is a field in which each polynomial $p(x) \in F[x]$ has a root, namely $X_p + \mathcal{J}$. (We may think of F_1 as an extension of F by identifying $\alpha \in F$ with $\alpha + \mathcal{J}$.)

Using the same technique, we may define a tower of field extensions

$$F < F_1 < F_2 < \cdots$$

such that each nonconstant polynomial $p(x) \in F_i[x]$ has a root in F_{i+1}. The union $E = \bigcup F_i$ is an extension field of F. Moreover, any polynomial $p(x) \in E[x]$ has all of its coefficients in F_i for some i and so has a root in F_{i+1}, hence in E. It follows that every polynomial $p(x) \in E[x]$ splits over E. Hence E is algebraically closed. □

Definition *Let $F < E$. Then E is an* **algebraic closure** *of F if $F < E$ is algebraic and E is algebraically closed. We will denote an algebraic closure of a field F by $\overline{F}$.* □

We can now easily establish the existence of algebraic closures.

Theorem 2.7.2 *Let $F < E$ where E is algebraically closed. Let $F < A < E$ where A is the algebraic closure of F in E. Then A is the only algebraic closure of F that is contained in E. Thus, any field has an algebraic closure.*
Proof. We have already seen that A is an algebraic extension of F. By hypothesis, any $p(x) \in A[x]$ splits in E and so all of its roots lie in E. Since these roots are algebraic over A, they are also algebraic over F and thus lie in A. Hence $p(x)$ splits in A and so A is algebraically closed.

As to uniqueness, if $F < B < E$ with B an algebraic closure of F, then since $F < B$ is algebraic, we have $B < A$. But if the inclusion is proper, then there is an $\alpha \in A \setminus B$. It follows that $\min(\alpha, F)$ does not split over B, a contradiction to the fact that B is algebraically closed. Hence, $B = A$. The final statement of the theorem follows from Theorem 2.7.1.□

We will show a bit later in the chapter that all algebraic closures of a field F are isomorphic, which is one reason why the notation $\overline{F}$ is (at least partially) justified.

Here is a characterization of algebraic closures.

Theorem 2.7.3 *Let $F < E$. The following are equivalent.*

1) *E is an algebraic closure of F.*
2) *E is a maximal algebraic extension of F, that is, $F < E$ is algebraic and if $E < K$ is algebraic then $K = E$.*
3) *E is a minimal algebraically closed extension of F, that is, if $F < K < E$ where K is algebraically closed, then $K = E$.*
4) *$F < E$ is algebraic and every nonconstant polynomial over F splits over E.*

Proof. To see that 1) implies 2), suppose that $\overline{F}$ is an algebraic closure of F and $F < \overline{F} < K$ is algebraic. Hence, any $\alpha \in K$ is algebraic over F. But $\min(\alpha, F)$ splits over $\overline{F}$ and so $\overline{F}$ contains a full set of roots of $\min(\alpha, F)$. Hence, $\alpha \in \overline{F}$, which shows that $K = \overline{F}$. Thus, $\overline{F}$ is a maximal algebraic extension of F.

Conversely, let E be a maximal algebraic extension of F and let $p(x) \in E[x]$. Let K be the splitting field for $p(x)$ over E. Thus, $F < E < K$ is an algebraic tower, since K is generated over E by the finite set of roots of $p(x)$. Hence, the maximality of E implies that $K = E$, and so $p(x)$ splits in E, which says that E is algebraically closed and therefore an algebraic closure of F.

To see that 1) implies 3), suppose that $F < K < \overline{F}$ where K is algebraically closed. Since $K < \overline{F}$ is algebraic, it follows that $K = \overline{F}$. Conversely, suppose that E is a minimal algebraically closed extension of F. Let A be the algebraic closure of F in E. Thus, $F < A < E$, with $F < A$ algebraic. If A is not algebraically closed, then there is a polynomial $p(x)$ over A that does not split over A. But $p(x)$ is also a polynomial over E and therefore splits over E. Hence, each of its roots in E is algebraic over A and therefore also over F, and so lies in A, which is a contradiction. Hence, A is algebraically closed and so the minimality of E implies that $A = E$, whence E is an algebraic closure of F.

Finally, it is clear that 1) implies 4). If 4) holds, then $F < E$ is algebraic and if $F < E < K$ is algebraic, then let $\alpha \in K \setminus E$ have minimal polynomial $p(x)$ over F. This polynomial splits over E and so $\alpha \in E$, which implies that $K = E$, whence E is a maximal algebraic extension of F and so 2) holds.□

2.8 Embeddings and Their Extensions

Homomorphisms between fields play a key role in the theory. Since a field F has no ideals other than $\{0\}$ and F, it follows that any nonzero ring homomorphism $\sigma: F \to R$ from F into a ring R must be a monomorphism, that is, an *embedding* of F into R.

A bit of notation: Let $f: A \to B$ be a function.

1) The restriction of f to $C \subseteq A$ is denoted by $f|_C$.
2) The image of A under f is denoted by fA or by A^f.
3) The symbol $\hookrightarrow$ denotes an embedding. Thus, $\sigma: F \hookrightarrow L$ signifies that σ is an embedding of F into L.
4) If $p(x) = \sum a_i x^i \in F[x]$ and if $\sigma: F \hookrightarrow E$ is an embedding, the polynomial $\sum \sigma(a_i) x^i \in E[x]$ is denoted by $(\sigma p)(x)$ or $p^\sigma(x)$.

Definition *Let $\sigma: F \hookrightarrow L$ be an embedding of F into L and let $F < E$. Referring to Figure 2.8.1, an embedding $\overline{\sigma}: E \hookrightarrow L$ for which $\overline{\sigma}|_F = \sigma$ is called an* **extension** *of σ to E. An embedding of E that extends the identity map $\iota: F \to F$ is called an embedding* **over** *F, or an F-***embedding**.

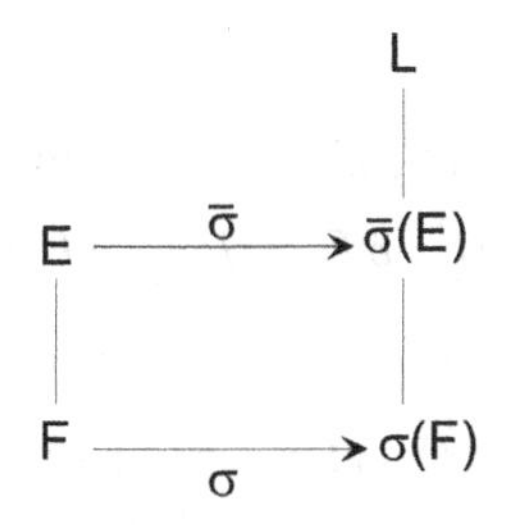

Figure 2.8.1

The set of all embeddings of F into L is denoted by $\hom(F,L)$. *The set of all embeddings of E into L that extend σ is denoted by* $\hom_\sigma(E,L)$ *and the set of all embeddings over F is denoted by* $\hom_F(E,L)$. $\square$

Embeddings play a central role in Galois theory, and it is important to know when a given embedding $\sigma: F \hookrightarrow L$ can be extended to a larger field E, and how many such embeddings are possible. We will discuss the former issue here, and the latter issue in the next chapter.

The Properties of Embeddings

Embeddings preserve many properties. For example, an embedding maps roots to roots and preserves composites.

Lemma 2.8.1

1) **(Embeddings preserve factorizations and roots)** If $\sigma: F \hookrightarrow L$ and $f(x) \in F[x]$, then $f(x) = p(x)q(x)$ if and only if $f^\sigma(x) = p^\sigma(x)q^\sigma(x)$. Also, $\alpha \in F$ is a root of $p(x)$ if and only if $\sigma\alpha$ is a root of $p^\sigma(x)$.
2) **(Embeddings preserve the lattice structure)** If $\sigma: K \hookrightarrow L$ and if $\{E_i \mid i \in I\}$ is a family of subfields of K then
$$\sigma(\bigcap E_i) = \bigcap \sigma E_i \quad \text{and} \quad \sigma(\bigvee E_i) = \bigvee \sigma E_i$$
3) **(Embeddings preserve adjoining)** If $\sigma: K \hookrightarrow L$ and if $F < K$ and $S \subseteq K$ then
$$\sigma(F(S)) = F^\sigma(\sigma S)$$
4) **(Embeddings preserve being algebraic)** Let $\sigma: F \hookrightarrow L$ and let $F < E$ be algebraic. If $\bar{\sigma}: E \hookrightarrow L$ is an extension of σ, then $\sigma F < \bar{\sigma} E$ is algebraic.
5) **(Embeddings preserve algebraic closures)** Let $\sigma: F \hookrightarrow L$ and let E be an algebraic closure of F. If $\bar{\sigma}: E \hookrightarrow L$ is an extension of σ, then $\bar{\sigma}E$ is an algebraic closure of σF.

Proof. We leave the proof of parts 1), 4) and 5) to the reader. For part 2), since σ is injective, it preserves intersections. But
$$\bigvee E_i = \bigcap\{H \mid E_i < H < K \text{ for all } i \in I\}$$

and so

$$\begin{aligned}\sigma(\bigvee E_i) &= \bigcap\{\sigma H \mid E_i < H < K \text{ for all } i \in I\} \\ &= \bigcap\{H' \mid \sigma E_i < H' < \sigma K \text{ for all } i \in I\} \\ &= \bigvee \sigma E_i\end{aligned}$$

For part 3), $\sigma(FE)$ contains σF and σE and so it contains the smallest field containing these two sets, that is $\sigma(FE) \supseteq \sigma F \sigma E$. On the other hand, if K is a field for which $\sigma F < K$ and $\sigma E < K$, then $F < \sigma^{-1}K$ and $E < \sigma^{-1}K$, whence $FE < \sigma^{-1}K$, and so $\sigma(FE) < K$. In other words, $\sigma(FE)$ is contained in any field containing σF and σE, including the composite $\sigma F \sigma E$.□

Even though the next result has a simple proof, the result is of major importance. If $F < E$ is algebraic and $\sigma: E \hookrightarrow E$ over F, then since σ permutes the roots of any polynomial over F and since every element of E is a root of a polynomial over F, it follows that every element of E is the image of some element of E, that is, the embedding σ must be surjective, and hence an automorphism.

Theorem 2.8.2 *If $F < E$ is algebraic and $\sigma: E \hookrightarrow E$ over F, then σ is an automorphism of E. In symbols,*

$$\hom_F(E, E) = \mathrm{Aut}_F(E)$$

Proof. Let $\alpha \in E$ and let S be the set of roots of the minimal polynomial $\min(\alpha, F)$ that lie in E. Then $\sigma|_S$ is a permutation on S and so there is a $\beta \in S$ for which $\sigma\beta = \alpha$. Hence, σ is surjective and thus an automorphism of E.□

Extensions in the Simple Case

Consider the case of a simple algebraic extension. Suppose that $\sigma: F \hookrightarrow L$, where L is algebraically closed. Let α be algebraic over F. We can easily extend σ to $F(\alpha)$, using the minimal polynomial $p(x)$ of α over F.

The key point is that any extension $\overline{\sigma}$ of σ is completely determined by its value on α and this value must be a root β of $p^\sigma(x)$, where $p(x) = \min(\alpha, F)$. In fact, we must have

$$\overline{\sigma}(f(\alpha)) = f^\sigma(\beta)$$

for any $f(x) \in F[x]$. Moreover, it is easy to see that this condition defines an isomorphism $\sigma_\beta: F(\alpha) \hookrightarrow F^\sigma(\beta)$ over σ.

Theorem 2.8.3 *Let $F < E$ and let $\alpha \in E$ be algebraic over F, with minimal polynomial $p(x)$. Let $\sigma: F \hookrightarrow L$, where L is algebraically closed.*

1) *If β is a root of $p^\sigma(x)$ in L, then σ can be extended to an embedding $\sigma_\beta: F(\alpha) \hookrightarrow L$ over σ for which $\sigma_\beta\alpha = \beta$.*

2) *Any extension of* σ *to* $F(\alpha)$ *must have the form* σ_β*, as described in part 1).*
3) *The number of extensions of* σ *to* $F(\alpha)$ *is equal to the number of distinct roots of* $\min(\alpha, F)$ *in* $\overline{F}$.□

The previous theorem shows that the *cardinality* of $\hom_\sigma(F(\alpha), L)$ depends only on α through its minimal polynomial, and furthermore, that it does not depend on either σ or L! We will explore this issue further in the next chapter.

Extensions in the Algebraic Case

The simple case, together with Zorn's lemma, is just what we need to prove that if $\sigma: F \hookrightarrow L$, with L algebraically closed and if $F < E$ is algebraic, then there is at least one extension of σ to E.

Theorem 2.8.4 *Let* $F < E$ *be algebraic.*
1) *Any embedding* $\sigma: F \hookrightarrow L$*, where* L *is algebraically closed, can be extended to an embedding* $\overline{\sigma}: E \hookrightarrow L$.
2) *Moreover, if* $\alpha \in E$ *and* $p(x) = \min(\alpha, F)$ *and* $\beta \in L$ *is a root of* $p^\sigma(x)$*, then we can choose* $\overline{\sigma}$ *so that* $\overline{\sigma}\alpha = \beta$*. (See Figure 2.8.2.)*

Proof. Let $\mathcal{E}$ be the set of all embeddings $\tau \in \hom_\sigma(K, L)$ for which $\alpha \in K$ and $\tau\alpha = \beta$ where $F < K < E$. Since σ can be extended to an embedding of $F(\alpha)$ into L in such a way that $\sigma(\alpha) = \beta$, it follows that $\mathcal{E}$ is not empty.

The set $\mathcal{E}$ is a partially ordered set under the order defined by saying that $(\tau: K \hookrightarrow L) \leq (\tau': K' \hookrightarrow L)$ if $K < K'$ and τ' is an extension of τ. If $\mathcal{C} = \{\tau_i: K_i \to L\}$ is a chain in $\mathcal{E}$, the map $\tau: \bigcup K_i \to L$ defined by the condition $\tau|_{K_i} = \tau_i$, is an upper bound for $\mathcal{C}$ in $\mathcal{E}$. Zorn's lemma implies the existence of a maximal extension $\tau: K \to L$. We contend that $K = E$, for if not, there is an element $\gamma \in E \setminus K$. But γ is algebraic over K and so we may extend τ to $K(\gamma)$, contradicting the maximality of τ.□

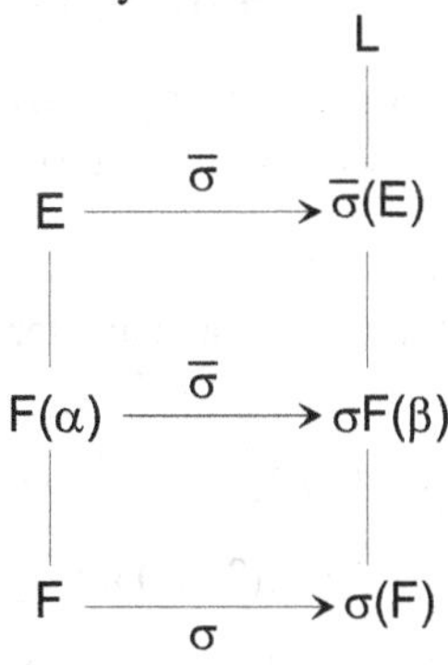

Figure 2.8.2

As a corollary, we can establish the essential uniqueness of algebraic closures.

Corollary 2.8.5 *Any two algebraic closures of a field* F *are isomorphic.*

Proof. Let K and L be algebraic closures of F. The identity map $\iota: F \to F$ can be extended to an embedding $\tau: K \to L$. Since K is algebraically closed so is τK. But L is an algebraic extension of τK and so $L = \tau K$. Hence, τ is an isomorphism. □

Independence of Embeddings

Next, we come to a very useful result on independence of embeddings. We choose a somewhat more general setting, however. A **monoid** is a nonempty set M with an associative binary operation and an identity element. If M and M' are monoids, a **homomorphism** of M into M' is a map $\psi: M \to M'$ such that $\psi(\alpha\beta) = \psi(\alpha)\psi(\beta)$ and $\psi(1) = 1$.

Definition *Let M be a monoid and let K be a field. A homomorphism $\chi: M \to K^*$, where K^* is the multiplicative group of all nonzero elements of K is called a* **character** *of M in K.* □

Note that an embedding $\sigma: E \hookrightarrow L$ of fields is a character, when restricted to E^*.

Theorem 2.8.6 (*E. Artin*) *Any set $\mathcal{T}$ of distinct characters of M in K is linearly independent over K.*

Proof. Suppose to the contrary that

$$\alpha_1\chi_1 + \cdots + \alpha_n\chi_n = 0$$

for $\chi_i \in \mathcal{T}$ and $\alpha_i \in K$, not all 0. Look among all such nontrivial linear combinations of the χ_i's for one with the fewest number of nonzero coefficients and, by relabeling if necessary, assume that these coefficients are $\alpha_1, \ldots, \alpha_r$. Thus,

$$\alpha_1\chi_1(g) + \cdots + \alpha_r\chi_r(g) = 0 \tag{2.8.1}$$

for all $g \in M$ and this is the "shortest" such nontrivial equation (hence $\alpha_i \neq 0$ for all i). Note that since $\chi_i(g) \in K^*$, we have $\chi_i(g) \neq 0$ for all $g \in M$. Hence, $r > 1$.

Let us find a shorter relation. Since χ_1 is a character, $\chi_1(f) \neq 0$ for all $f \in M$. Multiplying by $\chi_1(f)$ gives

$$\alpha_1\chi_1(f)\chi_1(g) + \alpha_2\chi_1(f)\chi_2(g) + \cdots + \alpha_r\chi_1(f)\chi_r(g) = 0$$

On the other hand, replacing g by fg in (2.8.1) gives,

$$\alpha_1\chi_1(f)\chi_1(g) + \alpha_2\chi_2(f)\chi_2(g) + \cdots + \alpha_r\chi_r(f)\chi_r(g) = 0$$

Subtracting the two equations cancels the first term, and we get

$$\alpha_2[\chi_1(f) - \chi_2(f)]\chi_2(g) + \cdots + \alpha_r[\chi_1(f) - \chi_r(f)]\chi_r(g) = 0$$

Now, since $\chi_1 \neq \chi_r$, there is an $f \in M$ for which $\chi_1(f) - \chi_r(f) \neq 0$ and we have a shorter nontrivial relation of the form (2.8.1). This contradiction proves the theorem.□

Corollary 2.8.7 (Dedekind independence theorem) *Let E and L be fields. Any set of distinct embeddings of E into L is linearly independent over L.* □

2.9 Splitting Fields and Normal Extensions

Let us repeat a definition from Chapter 1.

Definition *Let $\mathcal{F} = \{f_i(x) \mid i \in I\}$ be a family of polynomials in $F[x]$. A* **splitting field** *for $\mathcal{F}$ over F is an extension field E of F with the property that each $f_i(x)$ splits in E and that E is generated by the set of all roots of the polynomials in $\mathcal{F}$.* □

The next theorem says that splitting fields not only exist, but are essentially unique.

Theorem 2.9.1 (Existence and uniqueness of splitting fields) *Let $\mathcal{F}$ be a family of polynomials over F.*

1) *In any algebraic closure $\overline{F}$ of F, there is a unique splitting field for $\mathcal{F}$.*
2) *If $F < S_1 < K_1$ and $F < S_2 < K_2$ are algebraic, where S_1 is the splitting field for $\mathcal{F}$ in K_1 and S_2 is the splitting field for $\mathcal{F}$ in K_2 then any embedding $\sigma: K_1 \hookrightarrow K_2$ over F maps S_1 onto S_2.*
3) *Any two splitting fields for $\mathcal{F}$ are isomorphic over F.*

Proof. For part 1), if $\mathcal{F}$ is a family of polynomials over F, then every member of $\mathcal{F}$ splits in $\overline{F}$ and so $\overline{F}$ contains the field S generated over F by the roots in $\overline{F}$ of the polynomials in $\mathcal{F}$, that is, $\overline{F}$ contains a splitting field for $\mathcal{F}$. It is clear that this splitting field is unique in $\overline{F}$, because any splitting field in $\overline{F}$ must be generated, in $\overline{F}$, by the roots of all polynomials in $\mathcal{F}$.

For part 2), if R is the family of roots of $\mathcal{F}$ contained in S_1 then $S_1 = F(R)$ and so

$$\sigma(S_1) = \sigma(F(R)) = F(\sigma R)$$

But σR is precisely the set of roots of $\mathcal{F}$ in S_2 and so $F(\sigma R)$ is the splitting field for $\mathcal{F}$ in K_2, that is, $\sigma(S_1) = F(\sigma R) = S_2$. Part 3) follows immediately from part 2).□

σ-Invariance and Normal Extensions

Speaking very generally, if $f: A \to A$ is any function on a set A and if $S \subseteq A$ has the property that $f(S) \subseteq S$, then S is said to be **invariant under *f***, or ***f*-invariant**. This notion occurs in many contexts, including the present one, although the term "invariant" is seldom used in the present context.

Suppose that $F < K < \overline{F}$ is algebraic and that $\sigma: K \hookrightarrow \overline{F}$ is an embedding over F of K into the algebraic closure $\overline{F}$. Then K is σ-invariant if $\sigma K \subseteq K$. However, since $F < K$ is algebraic, any embedding of K into itself is an automorphism of K and so K is σ-invariant if and only if $\sigma K = K$, that is, if and only if σ is an *automorphism* of K.

Suppose that K is σ-invariant for *all* embeddings $\sigma: K \hookrightarrow \overline{F}$ over F. Then it is not hard to see that any irreducible polynomial $p(x)$ over F that has one root α in K must split over K. For if β is also a root of $p(x)$ in $\overline{F}$, then there is an embedding $\sigma \in \hom_F(K, \overline{F})$ for which $\sigma\alpha = \beta$. Hence, the σ-invariance of K implies that $\beta \in K$. Put another way, we can say that K is the splitting field for the family

$$\operatorname{MinPoly}(K, F) = \{\min(\alpha, F) \mid \alpha \in K\}$$

Thus, for $F < K$ algebraic, we have shown that 1) $\Rightarrow$ 2) $\Rightarrow$ 3), where

1) K is σ-invariant for all embeddings $\sigma: K \hookrightarrow \overline{F}$ over F
2) If an irreducible polynomial over F has one root in K, then it splits over K.
3) K is a splitting field, specifically for the family $\operatorname{MinPoly}(K, F)$.

On the other hand, suppose that K is a splitting field of a family $\mathcal{F}$ of polynomials over F. Thus, $K = F(R)$, where R is the set of roots of the polynomials in $\mathcal{F}$. But any embedding $\sigma: K \hookrightarrow \overline{F}$ over F sends roots to roots and so sends R to itself. Hence,

$$\sigma(K) = \sigma(F(R)) = F^{\sigma}(\sigma R) = F(R) = K$$

Since σ is an embedding of K into itself over F and $F < K$ is algebraic, it follows that σ is an automorphism of K. Thus, 1)-3) are equivalent.

Theorem 2.9.2 *Let $F < K < \overline{F}$, where $\overline{F}$ is an algebraic closure of F. The following are equivalent.*

1) *K is a splitting field for a family $\mathcal{F}$ of polynomials over F.*
2) *K is invariant under every embedding $\sigma: K \hookrightarrow \overline{F}$ over F. (It follows that every embedding of K into $\overline{F}$ over F is an automorphism of K.)*
3) *Every irreducible polynomial over F that has one root in K splits in K.* □

Definition *An algebraic extension $F < E$ that satisfies any (and hence all) of the conditions in the previous theorem is said to be a* **normal extension** *and we write $F \lhd E$. We also say that E is* **normal over** *F.* □

Corollary 2.9.3 *If $F \lhd E$ is a finite normal extension, then E is the splitting field of a finite family of irreducible polynomials.*

Proof. Let $E = F(\alpha_1, \ldots, \alpha_n)$. Since $E \lhd F$, each minimal polynomial $\min(\alpha_i, F)$ splits in E. Clearly, E is generated by the roots of the finite family $\mathcal{F} = \{\min(\alpha_i, F)\}$ and so E is the splitting field of $\mathcal{F}$. □

Note that the extension $F < \overline{F}$ is normal, since *any* nonconstant $p(x) \in F[x]$ splits in E.

Normal Extensions Are Not Distinguished

As it happens, the class of normal extensions is not distinguished, but it does enjoy some of the associated properties.

Example 2.9.1 It is not hard to see that any extension of degree 2 is normal. The extension $\mathbb{Q} < \mathbb{Q}(\sqrt[4]{2})$ is not normal since $\mathbb{Q}(\sqrt[4]{2})$ contains exactly two of the four roots of the irreducible polynomial $x^4 - 2$. On the other hand,

$$\mathbb{Q} < \mathbb{Q}(\sqrt{2}) < \mathbb{Q}(\sqrt[4]{2})$$

has each step of degree 2 and therefore each step is normal.□

Here is what we can say on the positive side.

Theorem 2.9.4

1) **(Full extension normal implies upper step normal)** *Let $F < K < E$. If $F \lhd E$ is normal then $K \lhd E$.*
2) **(Lifting of a normal extension is normal)** *If $F \lhd E$ and $F < K$ then $K \lhd EK$.*
3) **(Arbitrary composites and intersections of normal are normal)** *If $\{E_i\}$ is a family of fields, and $F \lhd E_i$ then $F \lhd \bigvee E_i$ and $F \lhd \bigcap E_i$.*

Proof. Part 1) follows from the fact that a splitting field for a family of polynomials over F is also a splitting field for the same family of polynomials over K.

For part 2), let E be a splitting field for a family $\mathcal{F}$ of polynomials over F and let R be the set of roots in E of all polynomials in $\mathcal{F}$. Then $E = F(R)$. Hence, $EK = K(R)$, which shows that EK is a splitting field for the family $\mathcal{F}$, thought of as a family of polynomials over K. Hence, $K \lhd EK$.

For part 3), let $\sigma\colon \bigvee E_i \hookrightarrow \overline{F}$ over F. Then σ is an embedding when restricted to each E_i and so $\sigma E_i = E_i$, whence

$$\sigma\left(\bigvee E_i\right) = \bigvee \sigma E_i = \bigvee E_i$$

and so σ is an automorphism of $\bigvee E_i$. Similarly, if $\sigma\colon \bigcap E_i \hookrightarrow \overline{F}$ over F then

$$\sigma\left(\bigcap E_i\right) = \bigcap \sigma E_i = \bigcap E_i$$

□

Normal Closures

If $F < E$ is not normal, then there is a *smallest* extension N of E (in a given algebraic closure $\overline{F}$) for which N is normal over F. Perhaps the simplest way to see this is to observe that $F < \overline{F}$ is normal and the intersection of normal extensions is normal, so

$$N = \bigcap\{K \mid E < K < \overline{F} \text{ and } F \lhd K\}$$

Definition *Let $F < E < \overline{F}$. The* **normal closure** *of E over F in $\overline{F}$ is the smallest intermediate field $E < K < \overline{F}$ for which $F \lhd K$. The normal closure is denoted by* $\text{nc}(E/F)$.□

Theorem 2.9.5 *Let $F < E < \overline{F}$ be algebraic, with normal closure* $\text{nc}(E/F)$.

1) The normal closure $\text{nc}(E/F)$ *exists and is equal to*

$$N = \bigcap\{K \mid E < K < \overline{F} \textit{ and } F \lhd K\}$$

2)

$$\text{nc}(E/F) = \bigvee_{\sigma \in \text{hom}_F(E,\overline{F})} \sigma E$$

3) $\text{nc}(E/F)$ *is the splitting field in $\overline{F}$ of the family*

$$\text{MinPoly}(E, F) = \{\min(\alpha, F) \mid \alpha \in E\}$$

4) If $E = F(S)$, where $S \subseteq \overline{F}$, then $\text{nc}(E/F)$ *is the splitting field in $\overline{F}$ of the family*

$$\text{MinPoly}(S, F) = \{\min(\alpha, F) \mid \alpha \in S\}$$

5) If $F < E$ is finite, then $F <$ $\text{nc}(E/F)$ *is also finite.*

Proof. We prove only part 2), leaving the rest for the reader. Let $E < L < \overline{F}$ with $F \lhd L$. Since $E < L$ is algebraic, any embedding $\sigma \in \text{hom}_F(E, \overline{F})$ can be extended to an embedding $\tau: L \to \overline{F}$ over F. Since $F \lhd L$, τ is an automorphism of L. It follows that $\sigma E \subseteq L$ and so $\bigvee \sigma E < L$. On the other hand, if $J = \bigvee \sigma E$, then $F \lhd J$, since if $\tau \in \text{hom}_F(J, \overline{F})$ then $\tau\sigma$ runs over all elements of $\text{hom}_F(J, \overline{F})$ as σ does and so

$$\tau J = \tau\left(\bigvee \sigma E\right) = \bigvee(\tau\sigma E) < \bigvee \sigma E = J$$

Hence, $F \lhd J$ and J is the smallest normal extension of F in E, that is, $J = \text{nc}(E/F)$. □

Exercises

1. Prove that $\sqrt{5}, \sqrt{7} \in \mathbb{Q}(\sqrt{5} + \sqrt{7})$.

2. Prove that if E is an algebraic extension of the real field $\mathbb{R}$ and $E \neq \mathbb{R}$, then E is isomorphic to the complex numbers $\mathbb{C}$.
3. Prove that every finite field F of characteristic p is a simple extension of its prime subfield $\mathbb{Z}_p$.
4. Let $F < E$. Suppose that $F < F(S)$ is finite, where S is a subset of E. Is it true that $F(S) = F(S_0)$ for some finite subset S_0 of S?
5. If $F < E$ and E is algebraically closed, is E necessarily an algebraic closure of F?
6. Suppose that $\operatorname{char}(F) = 0$ and that $F < E < \overline{F}$. Let $\alpha \in E$. Prove that if $\min(\alpha, F)$ has only one distinct root in $\overline{F}$, then $\alpha \in F$ and the multiplicity of α is 1. What can be said if $\operatorname{char}(F) \neq 0$?
7. Let $F < E$ be a **quadratic extension**, that is, an extension of degree 2. Show that E has a basis over F of the form $\{1, a\}$ where $a^2 \in F$.
8. a) Find all automorphisms of $\mathbb{Q}$.
   b) Is there an isomorphism $\sigma: \mathbb{Q}(\sqrt{2}) \to \mathbb{Q}(\sqrt{3})$ over $\mathbb{Q}$ for which $\sigma(\sqrt{2}) = \sqrt{3}$?
   c) Is there an isomorphism $\sigma: \mathbb{Q}(\sqrt{2}) \to \mathbb{Q}(\sqrt{2})$ over $\mathbb{Q}$ other than the identity?
9. Show that the automorphism $\sigma: \mathbb{Q}(\sqrt{2}) \to \mathbb{Q}(\sqrt{2})$ over $\mathbb{Q}$ that sends $\sqrt{2}$ to $-\sqrt{2}$ is not continuous.
10. Prove that if $F < E$ is algebraic and has only a finite number of intermediate fields, then $F < E$ is a finite extension.
11. Let R be an integral domain containing a field F. Then R is a vector space over F. Show that if $[R : F] < \infty$ then R must be a field. Find a counterexample when R is a commutative ring with identity but not an integral domain.
12. If $F < E$ is algebraic and R is a ring such that $F \subseteq R \subseteq E$, show that R is a field. Is this true if $F < E$ is not algebraic?
13. Let $F < E$ and $F < K$ be finite extensions and assume that EK is defined. Show that $[EK : F] \leq [E : F][K : F]$, with equality if $[E : F]$ and $[K : F]$ are relatively prime.
14. Let $\sigma: K \hookrightarrow E$ and let $F < K \cap E$. Show that σ is F-linear if and only if $\sigma(a) = a$ for all $a \in F$.
15. Find an extension $F < E$ that is algebraic but not finite.
16. The algebraic closure of $\mathbb{Q}$ in $\mathbb{C}$, that is, the set of all complex roots of polynomials with integer coefficients, is called the field A of **algebraic numbers**. Prove that $\mathbb{Q} < A$ is algebraic and infinite by showing that if $p_1, \ldots, p_m$ are distinct primes, then

    $$[\mathbb{Q}(\sqrt{p_1}, \ldots, \sqrt{p_m} : \mathbb{Q}] = 2^m$$

    Hint: use induction on m.
17. Prove that any extension of degree 2 is normal.
18. Let $F < E$ be a finite Galois extension and let $\alpha, \beta \in E$ have degrees m and n over F, respectively. Suppose that $[F(\alpha, \beta) : F] = mn$.

a) Show that if α_i is a conjugate of α and β_j is a conjugate of β, then there is a $\sigma \in G_F(E)$ such that $\sigma\alpha = \alpha_i$ and $\sigma\beta = \beta_j$. Hence, the conjugates of $\alpha + \beta$ are $\alpha_i + \beta_j$.
b) Show that if the difference of two conjugates of α is never equal to the difference of two conjugates of β then $F(\alpha, \beta) = F(\alpha + \beta)$.

19. Let F be an infinite field and let $F < E$ be an algebraic extension. Show that $|E| = |F|$.
20. Let $F < \overline{F}$ where $\overline{F}$ is an algebraic closure of F and let $G = \mathrm{Aut}_F(\overline{F})$ be the group of all automorphisms of $\overline{F}$ fixing F pointwise. Assume that all irreducible polynomials over F are separable. Let

$$\mathrm{fix}(G) = \{a \in \overline{F} \mid \sigma a = a \text{ for all } \sigma \in G\}$$

be the fixed field of F under G. Evidently $F < \mathrm{fix}(G) < \overline{F}$. Prove that $\mathrm{fix}(G) = F$.
21. (For readers familiar with complex roots of unity) Let p be a prime and let $\alpha \neq 1$ be a complex pth root of unity. Show that $\min(\alpha, \mathbb{Q}) = 1 + x + x^2 + \cdots + x^{p-1}$. What is the splitting field for $x^p - 1$ over $\mathbb{Q}$?
22. Let F be a field of characteristic $p \neq 0$ and let $\alpha \in F$. Show that the following are equivalent: a) $\alpha^{p^k} \in F$, b) $F(\alpha^{p^k}) = F$, c) $[F(\alpha)]^{p^k} \subseteq F$ where $[F(\alpha)]^{p^k} = \{s^{p^k} \mid s \in F(\alpha)\}$.
23. Let $F < E$ be a finite normal extension and let $p(x) \in F[x]$ be irreducible. Suppose that the polynomials $f(x)$ and $g(x)$ are monic irreducible factors of $p(x)$ over E. Show that there exists a $\sigma \in \mathrm{Aut}_F(E)$ for which $f^\sigma(x) = g(x)$.
24. Show that an extension $F < E$ is algebraic if and only if any subalgebra S of E over F is actually a subfield of E.
25. Let $F \triangleleft E$. Can *all* automorphisms of F be extended to an automorphism of E?
26. Suppose that F and E are fields and $\sigma\colon F \to E$ is an embedding. Construct an extension of F that is isomorphic to E.
27. Let $F < E$ be algebraic.
    a) Finish the proof of Theorem 2.9.5.
    b) Show that any two normal closures $F < E < K_1 < L_1$ and $F < E < K_2 < L_2$, where L_i is an algebraic closure of F are isomorphic.
28. With reference to Example 2.4.1, let s and t be independent variables and let p be a prime. Show that, in the tower

$$\mathbb{Z}_p(s^p, t^p) < \mathbb{Z}_p(s, t^p) < \mathbb{Z}_p(s, t)$$

each step is simple but the full extension is not.
29. Consider the field $F(x, y)$ of rational functions in two (independent) variables. Show that the extension $F < F(x, y)$ is not simple.

Constructions

The goal of the following series of exercises is to prove that certain constructions are not possible using straight edge and compass alone. In particular, not all angles can be trisected, a circle cannot be "squared" and a cube cannot be "doubled." The first step is to define the term *constructible.*

Definition *We assume the existence of two distinct points P_1 and P_2 in the plane and take the distance between these points to be one unit. A point, line or circle in the plane is said to be* **constructible** *if it can be obtained by a finite number of applications of the following rules:*
1) P_1 and P_2 are constructible.
2) The line through any two constructible points is constructible.
3) The circle with center at one constructible point and passing through another constructible point is constructible.
4) The points of intersection of any two constructible lines or circles are constructible. □

30. Show that if a line L and point P are constructible, then the line through P perpendicular to L is also constructible.
31. Show that if a line L and point P are constructible, then the line through P parallel to L is also constructible.
32. Taking the constructible line through P_1 and P_2 as the x-axis and the point P_1 as the origin, the y-axis is also constructible. Show that any point (a, b) with integer coordinates is constructible.
33. Show that the perpendicular bisector of any line segment connecting two constructible points is constructible. Show that the circle through two constructible points P and Q with center equal to the midpoint of P and Q is constructible.
34. If P, Q and R are constructible points and L is a constructible line through R then a point S can be constructed on L such that the distance from S to R is the same as the distance from P to Q. (Thus, given distances can be marked off on constructible lines.)

Constructible Numbers

Definition *A real number r is* **constructible** *if its absolute value is the distance between two constructible points.* □

35. Show that any integer is constructible.
36. Prove that a point (a, b) is constructible if and only if its coordinates a and b are constructible real numbers.
37. Prove that the set of numbers that are constructible forms a subfield of the real numbers containing $\mathbb{Q}$. Hint: to show that the product of two constructible numbers is constructible or that the inverse of a nonzero constructible number is constructible, use similar triangles.

38. Prove that if $\alpha > 0$ is constructible, then so is $\sqrt{\alpha}$. Hint: first show that a circle of diameter $1+\alpha$, with center on the x-axis and going through the origin P_1 is constructible. Mark off α units along the x-axis and draw the perpendicular.

The two previous exercises prove the following theorem.

Theorem C1 *If the elements of a field $F < \mathbb{R}$ are constructible, and if $\alpha \in F$, then $F(\sqrt{\alpha}) = \{a + b\sqrt{\alpha} \mid a, b \in F\}$ is constructible.* □

Theorem C2 *Let F be a subfield of $\mathbb{R}$ and let $E > F$ be a quadratic extension. Then $E = F(\sqrt{\alpha})$ for some $\alpha \in F$.*
Proof. Exercise. □

It follows from the two previous theorems that if F is constructible and if $F < E$ is a **quadratic extension,** that is, $[E : F] = 2$, then E is constructible. Any tower $F_1 < F_2 < \cdots < F_n$, where each extension has degree 2 is a **quadratic tower**. Thus, if $\mathbb{Q} < E_1 < E_2 < \cdots < E_n$ is a quadratic tower, then every element of E_n is constructible.

The converse of this statement also happens to be true.

Theorem C3 (Constructible numbers) *The set of constructible real numbers is the set of all numbers that lie in some quadratic tower*

$$\mathbb{Q} < E_1 < E_2 < \cdots < E_n$$

with base $\mathbb{Q}$. In particular, the degree of a constructible number over $\mathbb{Q}$ must be a power of 2.
Proof. Exercise. □

Constructible Angles

Now consider what it means to say that an angle of $\theta°$ is constructible. Informally, we will take this to mean that we may construct a line L through the origin that makes an angle of θ with the x-axis. Formally, the angle (real number) θ is **constructible** if the real number $\cos(\theta)$ is constructible.

39. Show that such a line L making angle θ with the x-axis is constructible if and only if the real number $\cos(\theta)$ is constructible. (This is an informal demonstration, since we have not formally defined angles.)
40. Show that a 60° angle is constructible.
41. Show that a 20° angle is not constructible. *Hint*: Verify the formula

$$\cos(3\theta) = 4\cos^3(\theta) - 3\cos(\theta)$$

Let $\alpha = \cos 20°$ and show that α is a root of

$$p(x) = 8x^3 - 6x - 1$$

Show that $p(x)$ is irreducible over $\mathbb{Q}$ and so $[\mathbb{Q}(\alpha) : \mathbb{Q}] = 3$.

42. Prove that every constructible real number is algebraic over $\mathbb{Q}$. Assuming that π is transcendental over $\mathbb{Q}$, show that any circle with a constructible radius cannot be "squared," that is, a square cannot be constructed whose area is that of a unit circle.
43. Verify that it is impossible to "double" any cube whose side length r is constructible, that is, it is impossible to construct an edge of a cube whose volume is twice that of a cube with side length r.

Chapter 3

Embeddings and Separability

3.1 Recap and a Useful Lemma

Let us recall a few facts about separable polynomials from Chapter 1.

Definition *An irreducible polynomial* $p(x) \in F[x]$ *is* **separable** *if it has no multiple roots in any extension of* F. *An irreducible polynomial that is not separable is* **inseparable**.□

Definition *If* $F < E$, *then an algebraic element* $\alpha \in E$ *is* **separable** *if its minimal polynomial* $\min(\alpha, F)$ *is separable. Otherwise, it is* **inseparable**. *Also, the* **radical exponent** *of* α *over* F *is the radical exponent of* $\min(\alpha, F)$. □

Theorem 3.1.1

1) *An irreducible polynomial* $p(x)$ *is separable if and only if* $p'(x) \neq 0$.
2) *If* F *is a field of characteristic* 0, *or a finite field, then all irreducible polynomials over* F *are separable.*
3) *Let* $\operatorname{char}(F) = p \neq 0$. *An irreducible polynomial* $f(x)$ *over* F *is inseparable if and only if* $f(x)$ *has the form*

$$f(x) = g(x^{p^d})$$

   where $d > 0$ *and* $g(x)$ *is a nonconstant polynomial. In this case, the integer* d *can be chosen so that* $g(x)$ *is separable and then every root of* $f(x)$ *has multiplicity* p^d, *where* d *is called the* **radical exponent** *of* $f(x)$. The radical exponent of $p(x)$ can be characterized as the largest integer d for which $p(x) = q(x^{p^d})$.
4) *Let* $\operatorname{char}(F) = p \neq 0$. *If* α *has radical exponent* d *then* α^{p^d} *is separable over* F, *and is the smallest power of* p *for which* α^{p^d} *is separable over* F.□

In Chapter 2, we considered the problem of extending the domain of an embedding of F to a larger field E that is algebraic over F. Here is a brief summary of what we discussed.

Theorem 3.1.2

1) **(Simple extensions)** *Let $F < E$ and let $\alpha \in E$ be algebraic over F, with minimal polynomial $p(x)$. Let $\sigma: F \hookrightarrow L$, where L is algebraically closed.*
   a) If β is a root of $p^\sigma(x)$ in L, then σ can be extended to an embedding $\sigma_\beta: F(\alpha) \hookrightarrow L$ over σ for which $\sigma_\beta \alpha = \beta$.
   b) Any extension of σ to $F(\alpha)$ must have the form σ_β.
   c) The number of extensions of σ to $F(\alpha)$ is equal to the number of distinct roots of $\min(\alpha, F)$ in $\overline{F}$.

2) **(Algebraic extensions)** *Let $F < E$ be algebraic. Any embedding $\sigma: F \hookrightarrow L$, where L is algebraically closed, can be extended to an embedding $\overline{\sigma}: E \hookrightarrow L$. Moreover, if $\alpha \in E$, $p(x) = \min(\alpha, F)$ and $\beta \in L$ is a root of $p^\sigma(x)$, then we can choose $\overline{\sigma}$ so that $\overline{\sigma}\alpha = \beta$.*□

A Useful Lemma

Before proceeding, we record a useful lemma. If F is a field and $S \subseteq F$ then S^n denotes the set $\{s^n \mid s \in S\}$.

Lemma 3.1.3 *Let $F < E$ be algebraic with $\mathrm{char}(F) = p \neq 0$ and let $S \subseteq E$.*

1) $F(S) = F(S^{p^k})$ holds for some $k \geq 1$ if and only if it holds for all $k \geq 1$.

2) $F = F^{p^k}$ holds for some $k \geq 1$ if and only if it holds for all $k \geq 1$.

Proof. For part 1), suppose that $F(S) = F(S^{p^k})$ holds for some $k \geq 1$. Since

$$F(S) = F(S^{p^k}) < F(S^p) < F(S)$$

it follows that $E = F(S^p)$. Now, since $[F(S)]^p = F^p(S^p)$, we have for any $k \geq 1$

$$[F(S)]^{p^k} = F^{p^k}(S^{p^k})$$

and so

$$F(S^{p^k}) = F[F^{p^k}(S^{p^k})] = F([F(S)]^{p^k}) = F([F(S^p)]^{p^k}) = F(S^{p^{k+1}})$$

Hence, $E = F(S^p) = F(S^{p^k})$, for all $k \geq 1$.

For part 2), we observe that

$$F^{p^k} < F^p < F$$

and so $F = F^{p^k}$ holds for some $k \geq 1$ if and only if $F = F^p$, which holds if and only if $F = F^{p^k}$ for all $k \geq 1$. □

3.2 The Number of Extensions: Separable Degree

According to Theorem 2.8.3, the number of extensions of an embedding $\sigma: F \hookrightarrow L$ to $F(\alpha)$, where L is algebraically closed, is equal to the number of distinct roots of $\min(\alpha, F)$. Hence, as we remarked earlier, the size of $\hom_\sigma(F(\alpha), L)$ does not depend on either σ or L. The same is true for extensions of σ to any algebraic extension.

Theorem 3.2.1 *If $F < E$ is algebraic and $\sigma: F \hookrightarrow L$, where L is algebraically closed, then the cardinality of* $\hom_\sigma(E, L)$ *depends only on the extension $F < E$ and not on σ or L. In other words, if $\tau: F \hookrightarrow L'$, with L' algebraically closed, then*

$$|\hom_\sigma(E, L)| = |\hom_\tau(E, L')|$$

as cardinal numbers.

Proof. We refer the reader to Figure 3.2.1. Since for any $\overline{\sigma} \in \hom_\sigma(E, L)$, the image $\overline{\sigma}(E)$ is contained in an algebraic closure of σF, we may assume that L is an algebraic closure of σF, and similarly, that L' is an algebraic closure of τF.

Since $\tau\sigma^{-1}: \sigma F \hookrightarrow \tau F$ is an isomorphism and $F < L$ is algebraic, the map $\tau\sigma^{-1}$ can be extended to an embedding of L into L'. Since $\overline{\sigma} F < L$ is algebraic, so is its image under λ, which is $\tau_0 F < \lambda L$, and since λL is algebraically closed, we have $\lambda L = L'$, implying that $\lambda: L \to L'$ is an isomorphism.

Now, if $\overline{\sigma} \in \hom_\sigma(E, L)$, then the map $\lambda\overline{\sigma}: E \to L'$ is an embedding of E into L' extending τ on F. This defines a function from $\hom_\sigma(E, L)$ to $\hom_\tau(E, L')$ given by $\overline{\sigma} \mapsto \lambda\overline{\sigma}$. Moreover, if $\tau, \mu \in \hom_\sigma(E, L)$ are distinct, then there is a $\beta \in E$ for which $\tau(\beta) \neq \mu(\beta)$ and since λ is injective, $\lambda\tau(\beta) \neq \lambda\mu(\beta)$, which implies that the map $\overline{\sigma} \mapsto \lambda\overline{\sigma}$ is injective. Hence,

$$|\hom_\sigma(E, L)| \leq |\hom_\tau(E, L')|$$

By a symmetric argument, we have the reverse inequality and so equality holds.□

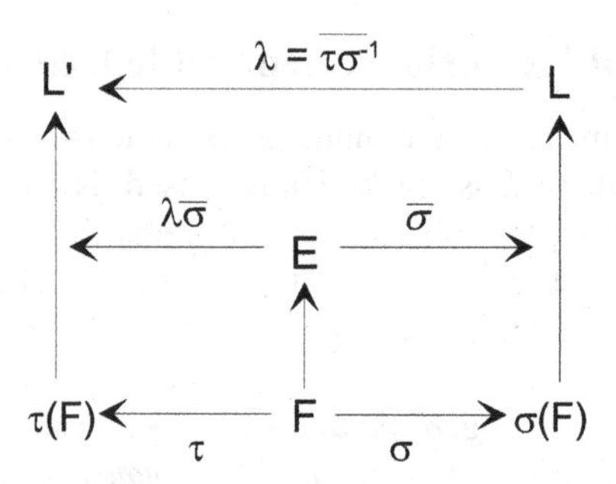

Figure 3.2.1

In view of Theorem 3.2.1, we may make the following definition.

Definition *Let $F < E$ be algebraic and let $\sigma: F \hookrightarrow L$, where L is algebraically closed. The cardinality of the set $\hom_\sigma(E, L)$ is called the* **separable degree** *of E over F and is denoted by $[E : F]_s$.*□

This new terminology allows us to rephrase the situation for simple extensions.

Theorem 3.2.2 (Simple extensions) *Let $F < E$ and let $\alpha \in E$ be algebraic over F, with minimal polynomial $p(x)$. Let $\sigma: F \hookrightarrow L$, where L is algebraically closed. Then*

1) *If α is separable then*

$$[F(\alpha) : F]_s = [F(\alpha) : F]$$

2) *If α is inseparable with radical exponent d, then*

$$[F(\alpha) : F]_s = \frac{1}{p^d}[F(\alpha) : F]$$

In either case, $|\hom_\sigma(F(\alpha), L)|$ divides $[F(\alpha) : F]$.□

Properties of Separable Degree

Like the ordinary degree, the separable degree is multiplicative.

Theorem 3.2.3 *If $F < K < E$ is algebraic then*

$$[E : F]_s = [E : K]_s[K : F]_s$$

as cardinal numbers.

Proof. The set $\hom_j(K, \overline{E})$ of extensions of the inclusion map $j: F \to \overline{E}$ to an embedding $\overline{j}: K \to \overline{E}$ has cardinality $[K : F]_s$. Each extension $\sigma \in \hom_j(K, \overline{E})$ is an embedding of K into $\overline{E}$ and can be further extended to an embedding of E into $\overline{E}$. Since the resulting extensions, of which there are $[E : K]_s[K : F]_s$, are distinct extensions of j to E, we have

$$[E:F]_s \geq [E:K]_s[K:F]_s$$

On the other hand, if $\sigma \in \hom_j(E,\overline{E})$ then $\sigma|_K$ is the extension of j to K, hence an element of $\hom_j(K,\overline{E})$. Since σ is an extension of $\sigma|_K$ to E, we see that σ is obtained by a double extension of $j: F \to \overline{E}$ and so equality holds in the inequality above. □

3.3 Separable Extensions

We have discussed separable elements and separable polynomials. it is now time to discuss separable extensions.

Definition *An algebraic extension $F < E$ is* **separable** *if every element $\alpha \in E$ is separable over F. Otherwise, it is* **inseparable.** □

The goal of this chapter is to explore the properties of algebraic extensions with respect to separability. It will be convenient for our present discussion to adopt the following nonstandard (not found in other books) terminology.

Definition *An algebraic extension $F < E$ is* **degreewise separable** *if $[E:F]_s = [E:F]$. An algebraic extension $F < E$ is* **separably generated** *if $E = F(S)$ where each $\alpha \in S$ is separable over F.* □

Simple Extensions

According to Theorem 3.2.2, if $F < E$ and $\alpha \in E$, then α is separable if and only if

$$[F(\alpha):F]_s = [F(\alpha):F]$$

Hence, α is separable if and only if $F < F(\alpha)$ is degreewise separable. Moroever, if $F < F(\alpha)$ is degreewise separable, and if $\beta \in F(\alpha)$, consider the tower

$$F < F(\beta) < F(\alpha)$$

The separable degree and the ordinary (vector space) degree are multiplicative and, at least for simple extensions, the separable degree does not exceed the ordinary degree. Hence, $[F(\alpha):F]_s = [F(\alpha):F]$ implies that the same is true for each step in the tower, and so

$$[F(\beta):F]_s = [F(\beta):F]$$

which shows that β is separable over F. Thus, $F < F(\alpha)$ is a separable extension. Of course, if $F < F(\alpha)$ is separable, then α is separable.

Thus, the following are equivalent:

1) α is separable over F;

2) $F < F(\alpha)$ is degreewise separable;
3) $F < F(\alpha)$ is a separable extension.

It is an extremely useful general fact that if

$$F < K < F(\alpha)$$

then the minimal polynomial of α over K divides the minimal polynomial of α over F, that is,

$$\min(\alpha, K) \mid \min(\alpha, F)$$

This tells us that if α is separable over F, then it is also separable over any intermediate field K.

In particular, if α is separable over F, then it is separable over an intermediate field of the form

$$F < F(\alpha^{p^k}) < F(\alpha)$$

where $k \geq 1$. But α satisfies the polynomial

$$p(x) = x^{p^k} - \alpha^{p^k} = (x - \alpha)^{p^k}$$

over $F(\alpha^{p^k})$ and so $\min(\alpha, F(\alpha^{p^k}))$ divides $(x - \alpha)^{p^k}$, which implies that

$$\min(\alpha, F(\alpha^{p^k})) = x - \alpha$$

Hence, $\alpha \in F(\alpha^{p^k})$, or equivalently, $F(\alpha) = F(\alpha^{p^k})$.

For the converse, suppose that $F(\alpha) = F(\alpha^{p^k})$ for some $k \geq 1$. Then Lemma 3.1.3 implies that this holds for all $k \geq 1$, in particular, $F(\alpha) = F(\alpha^{p^d})$, where d is the radical exponent of α. But α^{p^d} is separable over F and therefore so is the element α.

We can now summarize our findings on simple extensions and separability.

Theorem 3.3.1 (Simple extensions and separability) *Let α be algebraic over F, with* $\operatorname{char}(F) = p \neq 0$. *The following are equivalent.*
1) α is separable over F.
2) $F < F(\alpha)$ is degreewise separable; that is,

$$[F(\alpha) : F]_s = [F(\alpha) : F]$$

3) $F < F(\alpha)$ is a separable extension.

4) *There is a* $k \geq 1$ *for which*

$$F(\alpha) = F(\alpha^{p^k})$$

in which case $F(\alpha) = F(\alpha^{p^k})$ *for all* $k \geq 1$.
If α *is inseparable with radical exponent* d, *then*

$$[F(\alpha) : F]_s = \frac{1}{p^d}[F(\alpha) : F]$$

□

Finite Extensions

Now let us turn to finite extensions $F < E$. It should come as no surprise that the analogue of Theorem 3.3.1 holds for finite extensions.

If $F < E$ is separable, then it is clearly separably generated. If $F < F(S) = E$ is separably generated by S, then $F < F(S_0) = E$, where $S_0 = \{\alpha_1, \ldots, \alpha_n\}$ is a finite subset of S. Thus,

$$F < F(\alpha_1) < F(\alpha_1, \alpha_2) < \cdots < F(\alpha_1, \ldots, \alpha_n) = E$$

where α_i is separable over F. But α_i is also separable over $E_{i-1} = F(\alpha_1, \ldots, \alpha_{i-1})$, since the minimal polynomial of α_i over E_{i-1} divides the minimal polynomial of α over F. Hence, each simple step above is separable and therefore degreewise separable, which implies that $F < E$ is degreewise separable.

Finally, if $F < E$ is degreewise separable and $\beta \in E$, then in the tower

$$F < F(\beta) < E$$

the lower step is simple and degreewise separable, hence separable. It follows that β is separable over F and so $F < E$ is separable. Thus, as in the simple case, separable, separably generated and degreewise separable are equivalent concepts.

As to the analogue of part 4) of Theorem 3.3.1, let $E = F(S)$, where S is a finite set. If $F < F(S)$ is separable, then any $\alpha \in E$ is separable over F and so

$$F(\alpha) = F(\alpha^{p^k}) < F(S^{p^k})$$

for any $k \geq 1$. Thus, $F(S) = F(S^{p^k})$, for any $k \geq 1$. Conversely, if $F(S) = F(S^{p^k})$ for some $k \geq 1$, then Lemma 3.1.3 implies that $F(S) = F(S^{p^k})$ for all $k \geq 1$. Since S is a finite set, we can take p^k to be the maximum of the numbers p^d, where d varies over all radical exponents of the elements of S, in which case each $\alpha_i^{p^k}$ is separable, and so $F(S^{p^k})$ is separably generated, and therefore separable.

Theorem 3.3.2 (Finite extensions and separability) *Let* $\operatorname{char}(F) = p \neq 0$. *Let* $F < E$ *be finite. The following are equivalent.*

1) $F < E$ *is separable.*

2) $F < E$ *is degreewise separable; that is,*

$$[E : F]_s = [E : F]$$

3) E *is separably generated.*

4) If $E = F(S)$ *for a finite set* $S \subseteq E$, *then* $E = F(S^{p^k})$ *for some* $k \geq 1$, *in which case* $E = F(S^{p^k})$ *for all* $k \geq 1$.

If $F < E$ *is not separable, then*

$$[E : F]_s = \frac{1}{p^e}[E : F]$$

for some integer $e \geq 1$.□

Algebraic Extensions

For arbitrary algebraic extensions $F < E$, we have the following.

Theorem 3.3.3 (Algebraic extensions and separability) *Let* $\operatorname{char}(F) = p \neq 0$ *and let* $F < E$ *be algebraic.*

1) $F < E$ *is separable if and only if it is separably generated.*

2) If $F < E$ *is separable and* $E = F(S)$, *then* $E = F(S^{p^k})$ *for all* $k \geq 1$.

Proof. For part 1), if $F < E$ is separable then E is separably generated (by itself) over F. For the converse, assume that $E = F(S)$ where each $\alpha \in S$ is separable over F and let $\beta \in E$. Then $\beta \in F(S_0)$ for some finite subset $S_0 \subseteq S$. Since $F < F(S_0)$ is finitely generated and algebraic, it is finite. Thus, Theorem 3.3.2 implies that $F < F(S_0)$ is separable. Hence β is separable over F and so $F < E$ is separable. As to part 2), we have for any $\alpha \in S$ and $k \geq 1$

$$F(\alpha) = F(\alpha^{p^k}) < F(S^{p^k})$$

which implies that $F(S) < F(S^{p^k})$ and so $F(S) = F(S^{p^k})$.□

Existence of Primitive Elements

We wish now to describe conditions under which a finite extension is simple. The most famous result along these lines is the *theorem of the primitive element*, which states that a finite separable extension is simple. We want to state some slightly more general results, and to improve the statements of these results, we need to make some further observations about separable extensions. (These remarks will be repeated and elaborated upon later in the chapter.)

Suppose that $F < E$ is a finite extension. Let S be the set of all elements of E that are separable over F. By analogy to algebraic closures, we refer to S as the **separable closure** of F in E. Note that if $\alpha, \beta \in S$, then the extension

$F < F(\alpha, \beta)$ is separably generated and therefore separable. Hence, every element of $F(\alpha, \beta)$ is separable over F and so S is a field.

We claim that the extension $S < E$ has no separable elements. For if $\alpha \in E \setminus S$ is separable over S, then for the tower

$$F < S < S(\alpha)$$

we have

$$[S(\alpha) : F]_s = [S(\alpha) : S]_s [S : F]_s = [S(\alpha) : S][S : F] = [S(\alpha) : F]$$

and so α is separable over F, which is false. On the other hand, for any $\alpha \in E \setminus S$, there is a positive integer d for which α^{p^d} is separable. It follows that $\alpha^{p^d} \in S$ and so $\min(\alpha, S)$ divides $x^{p^d} - \alpha^{p^d} = (x - \alpha)^{p^d}$. Thus, $\min(\alpha, S)$ has only one distinct root. This implies that $\hom_S(E, \overline{E}) = \{\iota\}$, since any $\sigma \in \hom_S(E, \overline{E})$ must map α to itself, for all $\alpha \in E$.

Hence, $[E : S]_s = 1$ and so

$$[E : F]_s = [S : F]_s$$

We have shown that any finite extension $F < E$ can be decomposed into a tower

$$F < S < E$$

in which the first step is separable and has the same separable degree as the entire extension. Now we can state our theorem concerning simple extensions.

Theorem 3.3.4

1) *Any extension of the form*

$$F < F(\alpha_1, \ldots, \alpha_n, \beta)$$

where α_i is separable over F and β is algebraic over F is a simple extension. Moreover, if F is infinite, this extension has infinitely many primitive elements, of the form

$$a_1\alpha_1 + \cdots + a_n\alpha_n + b\beta$$

where $a_i, b \in F$.

2) *For any finite extension $F < E$, there exists a $\beta \in E$ such that*

$$[F(\beta) : F] = [E : F]_s$$

If F is infinite, there exist infinitely many such elements β.

3) **(Theorem of the primitive element)** *If $F < E$ is finite and separable, say*

$$F < F(\alpha_1, \ldots, \alpha_n)$$

where α_i is separable over F then $F < E$ is simple. If F is infinite, there *exist infinitely many primitive elements for E over F of the form*

$$a_1\alpha_1 + \cdots + a_n\alpha_n$$

where $a_i \in F$.

4) *If* F *has characteristic* 0 *or if* F *is a finite field then any finite extension of* F *is simple.*

Proof. If F is a finite field, then so is E, since $[E : F]$ is finite. Hence $E^* = \langle \beta \rangle$ is cyclic and so $F < E = F(\beta)$ is simple. Let us now assume that F is an infinite field.

For part 1), we show that if $E = F(\alpha, \beta)$, with α separable over F and β algebraic over F, then $E = F(\gamma)$, where γ is algebraic over F. The argument can be repeated to obtain a primitive element in the more general case.

Let $f(x) = \min(\alpha, F)$ and $g(x) = \min(\beta, F)$ and suppose that the roots of $f(x)$ are $\alpha = \alpha_1, \ldots, \alpha_s$ and the roots of $g(x)$ are $\beta = \beta_1, \ldots, \beta_t$. Since α is separable, the roots of $f(x)$ are distinct. However, the roots of $g(x)$ need not be distinct. We wish to show that for infinitely many values of $a \in F$, the elements $a\alpha + \beta$ are primitive. To do this, we need only show that $\alpha \in F(a\alpha + \beta)$, for then $F(\alpha, \beta) = F(a\alpha + \beta)$.

The polynomial $h(x) = g(a\alpha + \beta - ax)$ has coefficients in $F(a\alpha + \beta)$ and has α as a root, and similarily for $f(x)$. Thus, f and h have the common factor $x - \alpha$ in some extension of F. Moreover, since f is separable, α is a simple root and so no higher power of $x - \alpha$ is a factor of f. Therefore, if we can choose $a \in F$ so that f and h have no other common roots in any extension of F, it follows that $\gcd(f, h) = x - \alpha$, which must therefore be a polynomial over $F(a\alpha + \beta)$. In particular, $\alpha \in F(a\alpha + \beta)$, as desired.

The roots of h are the values of x for which $a\alpha + \beta - ax = \beta_k$ and we need only choose a so that none of the roots $\alpha_2, \ldots, \alpha_s$ satisfy this equation, that is, we need only choose a so that

$$a \neq \frac{\beta_k - \beta}{\alpha - \alpha_j}$$

for $j = 2, \ldots, s$ and $k = 1, \ldots, t$.

Part 2) follows from part 1) by considering the separable closure S of F in E. Since $F < S$ is separable, with $[S : F]_s = [E : F]_s$, we can apply part 1) to the separable extension $F < S$. Part 3) is a direct consequence of part 1), as is part 4).□

Example 3.4.1 Consider the extension $\mathbb{Q} < \mathbb{Q}(i, 2^{1/2})$. Here we have

$$f(x) = \min(i, \mathbb{Q}) = x^2 + 1$$

and

$$g(x) = \min(2^{1/2}, \mathbb{Q}) = x^2 - 2$$

and so $\alpha = i, \alpha_2 = -i$ and $\beta = \beta_1 = 2^{1/2}$ and $\beta_2 = -2^{1/2} = -\beta$. According to the previous theorem, $a\alpha + \beta = ai + 2^{1/2}$ is primitive provided that

$$a \notin \left\{ \frac{-\beta_k - \beta}{\alpha - \alpha_2} \,\middle|\, k = 1, 2 \right\} = \{0, 2^{1/2} i\}$$

In particular, we can choose any nonzero $a \in \mathbb{Q}$.□

Separable Extensions Are Distinguished

We may now establish that the class of separable extensions is distinguished.

Theorem 3.3.7

1) *The class of separable extensions is distinguished.*
2) *It is also closed under the taking of arbitrary composites.*
3) *If $F < E$ is separable, then so is $F < \mathrm{nc}(E/F)$, where $\mathrm{nc}(E/F)$ is a normal closure of E over F.*

Proof. For the tower property, if the full extension in $F < K < E$ is separable, then so is $F < K$. As to $K < E$, for any $\alpha \in K \setminus E$, we have

$$\min(\alpha, K) \mid \min(\alpha, F)$$

and so α separable over F implies α separable over K. Hence, $K < E$ is separable. Suppose now that $F < K$ and $K < E$ are separable and let $\alpha \in E$. Let $C \subseteq K$ be the set of coefficients of $p(x) = \min(\alpha, K)$. Then $p(x) = \min(\alpha, F(C))$ and so α is separable over $F(C)$. It follows that each step in the tower $F < F(C) < F(C, \alpha)$ is finite and separable, implying that α is separable over F. Hence, $F < E$ is separable.

For the lifting property, let $F < E$ be separable and let $F < K$. Since every element of E is separable over F it is also separable over the larger field K. Hence $EK = K(E)$ is separably generated and is therefore separable.

The fact that separable extensions are closed under the taking of arbitrary composites follows from the finitary property of arbitrary composites. That is, each element of an arbitrary composite involves elements from only a finite number of the fields in the composite and so is an element of a finite composite, which is separable.

Finally, a normal closure $\mathrm{nc}(E/F)$ is a splitting field in $\overline{F}$ of the family

$$\mathrm{MinPoly}(E, F) = \{\min(\alpha, F) \mid \alpha \in E\}$$

and so is generated over F by the roots of these minimal polynomials, each of which is separable over F. Hence, $\mathrm{nc}(E/F)$ is separably generated over F and is therefore separable over F.□

3.4 Perfect Fields

Definition *A field F is* **perfect** *if every irreducible polynomial over F is separable.* □

It is clear from the definitions that if F is perfect then any algebraic extension of F is separable. Conversely, suppose that every algebraic extension of F is separable. If $p(x) \in F[x]$ is irreducible and α is a root of $p(x)$ in some extension of F then $F < F(\alpha)$ is algebraic and so α is separable over F, that is, $p(x)$ is separable. Thus, F is perfect.

Theorem 3.4.1 *A field F is perfect if and only if every algebraic extension of F is separable over F.* □

Theorem 3.4.2 *Every field of characteristic* 0 *and every finite field is perfect.* □

Theorem 3.4.3 *Let F be a field with* $\mathrm{char}(F) = p \neq 0$. *The following are equivalent.*
1) *F is perfect.*
2) *$F = F^{p^k}$ for some $k \geq 1$.*
3) *The Frobenius map σ_{p^k} is an automorphism of F, for some $k \geq 1$.*

If this holds, then 2) and 3) hold for all $k \geq 1$.

Proof. Suppose F is perfect. Let $\alpha \in F$ and consider the polynomial $p(x) = x^p - \alpha \in F[x]$. If β is a root of $p(x)$ in a splitting field then $\beta^p = \alpha$ and so

$$p(x) = x^p - \beta^p = (x - \beta)^p$$

Now, if $q(x) = (x - \beta)^e$ is an irreducible factor of $p(x)$ over F, then it must be separable and so $e = 1$. Thus $\beta \in F$, that is, $\alpha = \beta^p \in F^p$ and so $F \subseteq F^p$. Since the reverse inclusion is manifest, we have $F = F^p$. Then 2) follows from Lemma 3.1.3.

Now assume that 2) holds. Then Lemma 3.1.3 implies that $F^p = F$. Suppose that $p(x) \in F[x]$ is irreducible. If $p(x)$ is not separable, then

$$p(x) = q(x^p) = \sum q_i (x^p)^i = \sum a_i^p (x^i)^p = (\sum a_i x^i)^p$$

contradicting the fact that $p(x)$ is irreducible. Hence, every irreducible polynomial is separable and so F is perfect. Thus, 2) implies 1). Since the Frobenius map is a monomorphism, statement 2), which says that σ_{p^k} is surjective, is equivalent to statement 3).□

We can now present an example of a nonperfect field.

Example 3.4.2 Let p be a prime. Since $\mathbb{Z}_p^p = \mathbb{Z}_p$, the field $\mathbb{Z}_p$ is perfect. However, if t is an independent variable, then the field $\mathbb{Z}_p(t)$ of all rational functions over $\mathbb{Z}_p$ is not perfect. We leave proof to the reader.□

While it is true that any algebraic extension of a perfect field is perfect, not all subfields of a perfect field need be perfect.

Theorem 3.4.4
1) If $F < E$ is algebraic and F is perfect then E is perfect.
2) If $F < E$ is finite and E is perfect then F is perfect.
Proof. Part 1) follows from Theorem 3.4.1 and the fact that every algebraic extension of E is an algebraic extension of F.

For part 2), let $\operatorname{char}(F) = p \neq 0$ and suppose first that $F < E$ is simple. Thus, $E = F(\alpha)$ is perfect and α is algebraic over F. Since $F(\alpha)$ is perfect, we have $F(\alpha) = [F(\alpha)]^p = F^p(\alpha^p)$. Consider the tower

$$F^p < F < F(\alpha) = F^p(\alpha^p)$$

If $p(x) = \sum a_i x^i$ is the minimal polynomial of α over F, then

$$0 = [\sum a_i \alpha^i]^p = \sum a_i^p \alpha^{pi}$$

and so $[F^p(\alpha^p) : F^p] \leq [F(\alpha) : F]$. It follows that in the tower above, $[F : F^p] = 1$, that is, $F = F^p$, whence F is perfect. Since $F < E$ is finitely generated by algebraic elements, the result follows by repetition of the previous argument. □

Note that we cannot drop the finiteness condition in part 2) of the previous theorem since, for example, $F < \overline{F}$ is algebraic and $\overline{F}$ is perfect even if F is not.

3.5 Pure Inseparability

The antithesis of a separable element is a *purely inseparable* element.

Definition *An element α algebraic over F is* **purely inseparable** *over F if its minimal polynomial* $\min(\alpha, F)$ *has the form $(x - \alpha)^n$ for some $n \geq 1$. An algebraic extension $F < E$ is* **purely inseparable** *if every element of E is purely inseparable over F.* □

It is clear that for a purely inseparable element α, the following are equivalent: (1) α is separable, (2) $n = 1$ and (3) $\alpha \in F$. In particular, for extensions of fields of characteristic 0 or finite fields, there are no “interesting" purely inseparable elements.

For $\alpha \notin F$, since the coefficient of x^{n-1} in $\min(\alpha, F)$ is $-n\alpha$, it follows that n must be a multiple of $p = \mathrm{char}(F)$, that is,

$$\min(\alpha, F) = (x - \alpha)^{mp^k}$$

But $\min(\alpha, F) = q(x^{p^d})$, where $q(x)$ is separable and d is the radical exponent of α over F. Hence, $k \geq d$ and we can write

$$\min(\alpha, F) = (x^{p^d} - \alpha^{p^d})^{mp^{k-d}}$$

which implies that $q(x) = (x - \alpha^{p^d})^{mp^{k-d}}$, which is separable if and only if $m = 1$ and $k = d$. Thus,

$$\min(\alpha, F) = (x - \alpha)^{p^d}$$

where d is the radical exponent of α over F.

Example 3.5.1 Let $\mathrm{char}(F) = 2$. If t is transcendental over F, then t is purely inseparable over $F(t^2)$, since its minimal polynomial over $F(t^2)$ is $x^2 - t^2 = (x - t)^2$. □

Example 3.5.2 Here we present an example of an element that is neither separable nor purely inseparable over a field F. Let $\mathrm{char}(F) = p$ and let $\alpha \in F$ be nonzero. Let t be transcendental over F and let

$$s = \frac{t^{p^2}}{t^p + \alpha}$$

According to Theorem 2.4.5 $F(s) < F(t)$ is algebraic and has degree equal to p^2. Since t is a root of the monic polynomial

$$p(x) = x^{p^2} - sx^p - s\alpha$$

of degree x^{p^2} over $F(s)$, this must be the minimal polynomial for t over $F(s)$. Since $p(x) = q(x^p)$, we deduce that t is not separable over $F(s)$. On the other hand, if t were purely inseparable over $F(s)$, we would have

$$x^{p^2} - sx^p - s\alpha = (x - t)^{p^2} = x^{p^2} - t^{p^2}$$

which would imply that $s = 0$, which is not the case. Hence, t is neither separable nor purely inseparable over $F(s)$. □

Definition *Let $F < E$ be finite. Since $[E : F]_s \mid [E : F]$, we may write*

$$[E : F] = [E : F]_s [E : F]_i$$

where $[E : F]_i$ is the **inseparable degree** *or* **degree of inseparability** *of E over F.* □

Note that while the separable degree is defined for infinite extensions, the inseparable degree is defined only for finite extensions.

Theorem 3.5.1 *Let* $F < E$ *be a finite extension with* $\operatorname{char}(F) = p \neq 0$.
1) *If* $F < K < E$ *then* $[E : F]_i = [E : K]_i[K : F]_i$.
2) $F < E$ *is separable if and only if* $[E : F]_i = 1$.
3) *If* $\alpha \in E$ *then* $[F(\alpha) : F]_i = p^d$, *where* d *is the radical exponent of* α.
4) $\alpha \in E$ *is purely inseparable if and only if* $[F(\alpha) : F]_s = 1$, *or equivalently,* $[F(\alpha) : F]_i = [F(\alpha) : F]$
5) $[E : F]_i$ *is a power of* p.
Proof. The first three statements are clear. Part 4) follows from the fact that α is purely inseparable if and only if its minimal polynomial has only one distinct root. But this is equivalent to saying that $\hom_F(F(\alpha), \overline{F})$ has cardinality 1. Part 5) follows from the fact that $F < E$ is finitely generated and the inseparable degree is multiplicative. We leave the details to the reader. □

We next characterize purely inseparable elements.

Theorem 3.5.2 (Purely inseparable elements) *Let* $\operatorname{char}(F) = p \neq 0$. *Let* α *be algebraic over* F, *with radical exponent* d *and let* $p(x) = \min(\alpha, F)$. *The following are equivalent.*
1) α *is purely inseparable over* F.
2) $F < F(\alpha)$ *is a purely inseparable extension*
3) $\alpha^{p^k} \in F$ *for some* $k \geq 0$.
Furthermore, d *is the smallest nonnegative integer for which* $\alpha^{p^d} \in F$.
Proof. If 1) holds and $\beta \in F(\alpha)$, then in the tower $F < F(\beta) < F(\alpha)$ the inseparable degree of the full extension is equal to the degree, and so the same holds for the lower step. Hence, β is purely inseparable over F and 2) holds. Clearly, 2) implies 1).

If 1) holds, then $\min(\alpha, F) = x^{p^d} - \alpha^{p^d}$ and so $\alpha^{p^d} \in F$, which implies 3). If 3) holds, then $\min(\alpha, F) \mid x^{p^k} - \alpha^{p^k}$ and, as we have seen, α is purely inseparable.□

Note that part 3) of the previous theorem, which can be written $F(\alpha^{p^d}) = F$, is the "antithesis" of the corresponding result $F(\alpha^{p^d}) = F(\alpha)$ for α separable.

The following result is the analogue of Theorem 3.2.4.

Theorem 3.5.3 (Purely inseparable extensions) *Let* $F < E$ *be algebraic. The following are equivalent.*
1) E *is* **purely inseparably generated**; *that is, generated by purely inseparable elements.*
2) $F < E$ *is* **degreewise purely inseparable**, *that is,* $[E : F]_s = 1$.

3) $F < E$ is a purely inseparable extension.

Proof. To prove that 1) implies 2), suppose that $E = F(I)$, where all elements of I are purely inseparable over F. Any embedding $\sigma: E \hookrightarrow L$ over F is uniquely determined by its values on the elements of I. But if $\alpha \in I$ then $\sigma\alpha$ is a root of the minimal polynomial $\min(\alpha, F)$ and so $\sigma\alpha = \alpha$. Hence σ must be the identity and $[E : F]_s = 1$.

To show that 2) implies 3), let $\alpha \in E$ and suppose that β is a root of $\min(\alpha, F)$ in $\overline{F}$. Then the identity on F can be extended to an embedding $\sigma: E \hookrightarrow \overline{F}$, for which $\sigma\alpha = \beta$. Since $[E : F]_s = 1$, we must have $\sigma = \iota$ and so $\beta = \alpha$. Thus, $\min(\alpha, F)$ has only one distinct root in $\overline{F}$ and so α is purely inseparable. It is clear that 3) implies 1).□

Purely Inseparable Extensions Are Distinguished

We can now show that the class of purely inseparable extensions is distinguished.

Theorem 3.5.4 *The class of purely inseparable extensions is distinguished. It is also closed under the taking of arbitrary composites.*

Proof. Let $F < K < E$. Since pure inseparability is equivalent to degreewise pure inseparability and $[E : F]_s = 1$ if and only if $[E : K]_s = 1$ and $[K : F]_s = 1$, it is clear that the tower property holds. For lifting, suppose that $F < E$ is purely inseparable and $F < K$. Since every element of E is purely inseparable over F, it is also purely inseparable over the larger field K. Hence $EK = K(E)$ is purely inseparably generated and therefore purely inseparable. We leave proof of the last statement to the reader. □

*3.6 Separable and Purely Inseparable Closures

Let $F < E$. Recall that the algebraic closure of F in E is the set A of all elements of E that are algebraic over F. The fact that A is a field is a consequence of the fact that an extension that is generated by algebraic elements is algebraic, since if $\alpha, \beta \in A$ then $F(\alpha, \beta) \subseteq A$ and so $\alpha \pm \beta, \alpha\beta$ and $\alpha^{-1} \in A$.

We can do exactly the same analysis for separable and purely inseparable elements. To wit, if $\alpha, \beta \in E$ are separable over F, then $F(\alpha, \beta)$ is separable over F. It follows that $\alpha \pm \beta, \alpha\beta$, and α^{-1} are separable over F. Hence, the set of all elements of E that are separable over F is a subfield of E. A similar statement holds for purely inseparable elements.

Definition *Let $F < E$. The field*

$$
\begin{aligned}
F_E^{\mathrm{sc}} &= \{\alpha \in E \mid \alpha \text{ separable over } F\} \\
&= \{\alpha \in E \mid F(\alpha^{p^k}) = F(\alpha) \text{ for some } k \geq 1\}
\end{aligned}
$$

is called the **separable closure** *of* F *in* E*. The field*

$$\begin{aligned} F_E^{\text{ic}} &= \{\alpha \in E \mid \alpha \textit{ is purely inseparable over } F\} \\ &= \{\alpha \in E \mid \alpha^{p^k} \in F \textit{ for some } k \geq 0\} \end{aligned}$$

is called the **purely inseparable closure** *of* F *in* E*. When the context is clear, we will drop the subscript and write* F^{sc} *and* F^{ic}*.*□

The separable closure allows us to decompose an arbitrary algebraic extension into separable and purely inseparable parts.

Theorem 3.6.1 *Let* $F < E$ *be algebraic.*

1) *In the tower* $F < F^{\text{sc}} < E$ *the first step is separable and the second step is purely inseparable.*
2) *Any automorphism* σ *of* E *over* F *is uniquely determined by its restriction to* F^{sc}*.*

Proof. For part 1), if $\alpha \in E \setminus F$ has radical exponent d, then α^{p^d} has a separable minimal polynomial and is therefore in F^{sc}. Thus, Theorem 3.5.2 implies that α is purely inseparable over F^{sc}. We leave proof of part 2) to the reader.□

Corollary 3.6.2 *Let* $F < E$ *be finite. Then* $[E : F]_s = [F^{\text{sc}} : F]$ *and* $[E : F]_i = [E : F^{\text{sc}}]$*.* □

Perfect Closures

Let $\text{char}(F) = p \neq 0$ and let $\overline{F}$ be an algebraic closure of F. Suppose that $F < P < \overline{F}$, where P is perfect. What can we say about P?

If $\alpha \notin P$, then α is separable over P and therefore cannot be purely inseparable over P. In other words, the purely inseparable closure F^{ic} is contained in P.

On the other hand, we claim that F^{ic} is perfect. For if $\alpha \in F^{\text{ic}}$, then $\alpha^p \in F$. Now, the polynomial $p(x) = x^p - \alpha$ has a root β in some extension and so $\alpha = \beta^p$. But then $\beta^{p^2} = \alpha^p \in F$ and so $\beta \in F^{\text{ic}}$. It follows that $\alpha = \beta^p \in [F^{\text{ic}}]^p$ and so $F^{\text{ic}} = [F^{\text{ic}}]^p$, that is, F^{ic} is perfect.

Thus, we have shown that the purely inseparable closure of F in $\overline{F}$ is the smallest intermediate field $F < F^{\text{ic}} < \overline{F}$ that is perfect. This field is also called the **perfect closure** of F in $\overline{F}$.

More on Separable and Inseparable Closures

The remainder of this section is somewhat more technical and may be omitted upon first reading.

Part 1 of Theorem 3.6.1 shows that any algebraic extension can be decomposed into a separable extension followed by a purely inseparable extension. In general, the reverse is not possible. Although $F < F^{\mathrm{ic}}$ is purely inseparable, the elements of $E \setminus F^{\mathrm{ic}}$ need not be separable over F^{ic}; they are simply not purely inseparable over F. However, it is not hard to see when $F^{\mathrm{ic}} < E$ is separable.

Theorem 3.6.3 *Let $F < E$ be algebraic. Then $F^{\mathrm{ic}} < E$ is separable if and only if $E = F^{\mathrm{sc}}F^{\mathrm{ic}}$.*

Proof. If $F^{\mathrm{ic}} < E$ is separable then so is the lifting $F^{\mathrm{sc}}F^{\mathrm{ic}} < E$. But since $F^{\mathrm{sc}} < E$ is purely inseparable, so is the lifting $F^{\mathrm{sc}}F^{\mathrm{ic}} < E$. Thus $E = F^{\mathrm{sc}}F^{\mathrm{ic}}$. Conversely, if $E = F^{\mathrm{sc}}F^{\mathrm{ic}}$ then $F^{\mathrm{ic}} < F^{\mathrm{sc}}F^{\mathrm{ic}}$, being a lifting of a separable extension $F < F^{\mathrm{sc}}$, is also separable. □

We can do better than the previous theorem when $F < E$ is a normal extension, which includes the case $E = \overline{F}$. Let $G = \mathrm{Aut}_F(E)$ be the set of all automorphisms of E over F. Since $F \lhd E$, G is also the set of all embeddings of E into $\overline{F}$ over F. We define the **fixed field** of G in E by

$$F(G) = \{\alpha \in E \mid \sigma\alpha = \alpha \text{ for all } \sigma \in G\}$$

Theorem 3.6.4 *Let $F \lhd E$. Let $G = \mathrm{Aut}_F(E)$ and let $F(G)$ be the fixed field of G in E. Then $F(G) = F^{\mathrm{ic}}$. Furthermore, in the tower $F < F^{\mathrm{ic}} < E$, the first step is purely inseparable and the second step is separable.*

Proof. Let $\alpha \in F(G)$. If $\beta \in \overline{F}$ is a root of $p(x) = \min(\alpha, F)$ then there exists an embedding $\sigma: E \to \overline{F}$ over F for which $\sigma\alpha = \beta$. But $\sigma\alpha = \alpha$ and so $\beta = \alpha$. Hence $\min(\alpha, F)$ has only one root and so $\alpha \in F^{\mathrm{ic}}$. On the other hand, if $\alpha \in F^{\mathrm{ic}}$ then any $\sigma \in G$ must map α to itself, since it must map α to a root of $\min(\alpha, F)$. Hence $\alpha \in F(G)$. This proves that $F(G) = F^{\mathrm{ic}}$.

Now let $\alpha \in E$ and $p(x) = \min(\alpha, F(G))$. Let $q(x) = \prod(x - r_i)$ where $R = \{r_1, \dots, r_n\}$ is the set of *distinct* roots of $p(x)$ in E. Since any $\sigma \in G$ is a permutation of R, we deduce that $q^\sigma(x) = q(x)$ and so the coefficients of $q(x)$ lie in $F(G)$. Hence $q(x) = p(x)$ and α is separable over $F(G)$. □

Corollary 3.6.5 *If $F \lhd E$ then $F^{\mathrm{ic}} < E$ is separable and $E = F^{\mathrm{sc}}F^{\mathrm{ic}}$.* □

Let us conclude this section with a characterization of simple algebraic extensions. If $E = F(\alpha)$ is a simple algebraic extension of F and if d is the radical exponent of α, we have seen that $p^d = [E : F]_i$ is the *smallest* nonnegative power of p such that α^{p^d} is separable over F, or equivalently, such that $E^{p^d} \subseteq F^{\mathrm{sc}}$. It turns out that this property actually characterizes simple algebraic extensions. Before proving this, we give an example where this property fails to hold.

Example 3.6.1 Let u and v be transcendental over K with $\mathrm{char}(K) = p \neq 0$. Let $E = K(u, v)$ and $F = K(u^p, v^p)$. It is easily seen that $F < E$ is purely inseparable with $[E:F]_i = p^2$. However, $\alpha \in E$ implies $\alpha^p \in F$ and so $E^p \subseteq F$. $\square$

Theorem 3.6.6 *Let $F < E$ be a finite extension with $[E:F]_i = p^d$. Then $F < E$ is simple if and only if d is the smallest nonnegative integer for which $E^{p^d} \subseteq F^{\mathrm{sc}}$.*

Proof. We have seen that if $F < E$ is simple then d is the smallest such nonnegative integer. For the converse, note first that if F is a finite field then so is E, implying that E^* is cyclic and so $F < E$ is simple. Let us assume that F is an infinite field and look at the second step in the tower $F < F^{\mathrm{sc}} < E$. This step is purely inseparable. Since $F^{\mathrm{sc}} < E$ is finite, we have

$$E = F^{\mathrm{sc}}(\beta_1, \ldots, \beta_n)$$

If for some $k < d$, we have $\beta_i^{p^k} \in F^{\mathrm{sc}}$ for all i, then $E^{p^k} \subseteq F^{\mathrm{sc}}$, contrary to hypothesis. Hence one of the β_i's, say β, satisfies

$$\beta^{p^d} \in F^{\mathrm{sc}}, \quad \beta^{p^k} \notin F^{\mathrm{sc}} \text{ for } k < d$$

It follows that

$$[F^{\mathrm{sc}}(\beta) : F^{\mathrm{sc}}]_i = p^d = [E:F]_i \geq [E:F^{\mathrm{sc}}]_i$$

Since $F^{\mathrm{sc}}(\beta) < E$, we have $[F^{\mathrm{sc}}(\beta) : F^{\mathrm{sc}}]_i = [E:F^{\mathrm{sc}}]_i$ and since the extensions involved are purely inseparable, we get $[F^{\mathrm{sc}}(\beta) : F^{\mathrm{sc}}] = [E:F^{\mathrm{sc}}]$. Hence, $E = F^{\mathrm{sc}}(\beta)$.

Our tower now has the form $F < F^{\mathrm{sc}} < F^{\mathrm{sc}}(\beta)$ where β is purely inseparable over F^{sc}. In addition, $F < F^{\mathrm{sc}}$ is finite and separable and therefore simple. Thus there exists $\alpha \in F^{\mathrm{sc}}$ such that $F^{\mathrm{sc}} = F(\alpha)$ and the tower takes the form $F < F(\alpha) < F(\alpha, \beta)$ where α is separable over F and β is purely inseparable over $F(\alpha)$. By Theorem 3.3.4, the extension $F < F(\alpha, \beta)$ is simple. $\square$

Note that Theorem 3.6.6 implies that the extension $F < E$ of Example 3.6.1 is not simple.

Exercises

1. Find an infinite number of primitive elements for $\mathbb{Q} < \mathbb{Q}(i, 2^{1/3})$.
2. A **biquadratic** extension is an extension of degree 4 of the form $F < F(\alpha, \beta)$ where α and β have degree 2 over F. Find all the proper intermediate fields of a biquadratic extension.
3. Show that all algebraically closed fields are perfect.
4. If t is transcendental over F and $\mathrm{char}(F) = p \neq 0$, then $F(t)$ is not perfect.
5. If $\mathrm{char}(F) = p \neq 0$ and F is not perfect, show that $F \neq F^{\mathrm{ic}}$.
6. Let α be algebraic over F, where $\mathrm{char}(F) = p \neq 0$ and let d be the radical exponent of α. Show that α^{p^k} is separable over F if and only if $k \geq d$.

7. Let p and q be distinct primes. Then $\mathbb{Q} < \mathbb{Q}(\sqrt{p}, \sqrt{q})$ is finite and separable and therefore simple. Describe an infinite class of primitive elements for this extension. Find the minimal polynomial for each primitive element.
8. Let $E = F(\alpha_1, \ldots, \alpha_n)$ be separable over an infinite field F. Prove that there is an infinite number of n-tuples $(a_1, \ldots, a_n) \in F^n$ for which $E = F(a_1\alpha_1 + \cdots + a_n\alpha_n)$.
9. Show that the class of purely inseparable extensions is closed under the taking of arbitrary composites.
10. Prove that for $F < E$ finite, $E^{[E:F]_i} \subseteq F^{\mathrm{sc}}$.
11. If $F < E$ is algebraic prove that any automorphism σ of E over F is uniquely determined by its restriction to F^{sc}.
12. Show that lifting an extension by a purely inseparable extension does not affect the separable degree. That is, show that if $F < E$ is algebraic and $F < P$ is purely inseparable then $[EP : P]_s = [E : F]_s$.
13. Let $F < S$ be finite separable and $F < P$ be finite purely inseparable. Prove that $P < SP$ is separable and $[SP : P] = [S : F]$. In fact, if B is a basis for S over F, prove that it is also a basis for SP over P.
14. Show that if $F < E$ is finite and $F < S$ is finite separable then $[ES : S]_i = [E : F]_i$.
15. Let $F < E$ be a finite extension and let $\alpha \in E$ be algebraic over F. Let H be the set of embeddings of E into $\overline{E}$ over F. The elements of H permute the roots of $p(x) = \min(\alpha, F)$. Let β be a root of $p(x)$. Show that

$$|\{\sigma \in H \mid \sigma\alpha = \beta\}| = [E : F(\alpha)]_s$$

Hence, the multiset $\{\sigma\alpha \mid \sigma \in H\}$ contains $[E : F(\alpha)]_s$ copies of each root of $p(x)$.
16. Let $F < E$ be a finite extension that is not separable. Show that for each $n \geq 1$ there exists a subfield E_n of E for which $E_n < E$ is purely inseparable and $[E : E_n]_i = p^n$.
17. Prove that if $F \neq \mathrm{pcl}(F)$ then the extension $F < \mathrm{pcl}(F)$ is infinite.

Chapter 4

Algebraic Independence

In this chapter, we discuss the structure of an arbitrary field extension $F < E$. We will see that for any extension $F < E$, there exists an intermediate field $F < F(S) < E$ whose upper step $F(S) < E$ is algebraic and whose lower step $F < F(S)$ is *purely transcendental*, that is, there is no nontrivial polynomial dependency (over F) among the elements of S, and so these elements act as "independent variables" over F. Thus, $F(S)$ is the field of all rational functions in these variables.

4.1 Dependence Relations

The reader is no doubt familiar with the notion and basic properties of linear independence of vectors, such as the fact that all bases for a vector space have the same cardinality. Independence is a common theme, which applies in the present context as well. However, here we are interested in *algebraic* independence, rather than *linear* independence. Briefly, a field element $\alpha \in F$ is *algebraically independent* of a subset $S \subseteq F$ if there is no nonconstant polynomial $p(x)$, with coefficients in $F(S)$, for which α is a root. Put another way, α is algebraically dependent on S if α is algebraic over $F(S)$.

Many of the common properties of linear independence, such as dependence (spanning sets) and bases, have counterparts in the theory of algebraic independence. However, these properties depend only on the most general properties of independence, so it is more "cost effective" to explore these properties in their most general setting, which is the goal of this section.

Definition *Let X be a nonempty set and let $\Delta \subseteq X \times \mathcal{P}(X)$ be a binary relation from X to the power set of X. We write $x \prec S$ (read: x is* **dependent on** *S) for $(x,S) \in \Delta$ and $S \prec T$ when $s \prec T$ for all $s \in S$. Then Δ is a* **dependence relation** *if it satisfies the following properties, for all S, T and $U \in \mathcal{P}(X)$:*

1) **(reflexivity)**

$$S \prec S$$

2) **(compactness)**

$$x \prec S \Rightarrow x \prec S_0 \textit{ for some finite subset } S_0 \textit{ of } S$$

3) **(transitivity)**

$$S \prec T, T \prec U \Rightarrow S \prec U$$

4) **(Steinitz exchange axiom)**

$$x \prec S, x \not\prec S \setminus \{s\} \Rightarrow s \prec (S \setminus \{s\}) \cup \{x\}$$

If $x \not\prec S$ we say that x is **independent of** *S.* □

Definition *A subset $S \subseteq X$ is* **dependent** *if there is an $s \in S$ for which $s \prec S \setminus \{s\}$. Otherwise, S is* **independent**. *(The empty set is independent.)* □

The reader should have no trouble supplying a proof for the following lemma.

Lemma 4.1.1
1) *If $S \prec T$ then $S \prec T'$ for any superset T' of T.*
2) *Any superset of a dependent set is dependent.*
3) *Any subset of an independent set is independent.*
4) *If S is a dependent set, then some finite subset S_0 of S is dependent. Equivalently, if every finite subset of T is independent, then T is independent.* □

Theorem 4.1.2 *If S is independent and $x \not\prec S$ then $S \cup \{x\}$ is independent.*
Proof. Let $s \in S$. If $s \prec (S \cup \{x\}) \setminus \{s\}$ then since $s \not\prec S \setminus \{s\}$, the exchange axiom implies that $x \prec S$, a contradiction. Hence $s \not\prec (S \cup \{x\}) \setminus \{s\}$. Furthermore, by hypothesis $x \not\prec S = (S \cup \{x\}) \setminus \{x\}$. Thus, $S \cup \{x\}$ is independent. □

Definition *A set $B \subseteq X$ is called a* **base** *if B is independent and $X \prec B$.* □

Theorem 4.1.3 Let X be a nonempty set with a dependence relation $\prec$.
1) *$B \subseteq X$ is a base for X if and only if it is a maximal independent set in X.*
2) *$B \subseteq X$ is a base for X if and only if B is minimal with respect to the property that $X \prec B$.*
3) *Let $A \subseteq S \subseteq X$, where A is an independent set (possibly empty) and $X \prec S$. Then there is a base B for X such that $A \subseteq B \subseteq S$.*

Proof. For part 1), assume B is a base. Then B is independent. If $x \in X \setminus B$ then $x \prec B$ implies that $B \cup \{x\}$ is not independent, that is, B is maximal independent. For the converse, if B is a maximal independent set and $x \not\prec B$

then $B \cup \{x\}$ is independent, which is not the case. Hence, $X \prec B$ and B is a base.

For part 2), if B is a base, then $X \prec B$. Suppose that some proper subset B_0 of B satisfies $X \prec B_0$. If $b \in B \setminus B_0$ then $b \prec B_0 \prec B \setminus \{b\}$, contradicting the independence of B. Hence B is minimal. Conversely, suppose that B is minimal with respect to the condition $X \prec B$. If B is dependent then $X \prec B \prec B \setminus \{b\}$ for some $b \in B$, a contradiction to the minimality of B. Hence B is independent and a base for X.

For part 3), we apply Zorn's lemma. The set $\mathcal{S}$ of all independent sets B in X satisfying $A \subseteq B \subseteq S$ is nonempty, since $A \in \mathcal{S}$. Order $\mathcal{S}$ by set inclusion. If $\mathcal{C} = \{C_i\}$ is a chain in $\mathcal{S}$, then the compactness property implies that the union $\bigcup C_i$ is an independent set, which also lies in $\mathcal{S}$. Hence, Zorn's lemma implies the existence of a maximal element $C \in \mathcal{S}$, that is, C is independent, $A \subseteq C \subseteq S$ and C is maximal with respect to these two properties. This maximality implies that $S \prec C$ and so $X \prec S \prec C$, which implies that C is a base. $\square$

To prove that any two bases for X have the same cardinality, we require a lemma, which says that we can remove a particular element from a dependent set and still have a dependent set.

Lemma 4.1.4 *Let S be a finite dependent set and let $A \subseteq S$ be an independent subset of S. Then there exists $\alpha \in S \setminus A$ for which $S \prec S \setminus \{\alpha\}$.*

Proof. The idea is simply to choose α from a maximal independent set containing A. In particular, among all subsets of $S \setminus A$, choose a maximal one B for which $A \cup B$ is independent. Then B is a proper (perhaps empty) subset of $S \setminus A$. If $\alpha \in S \setminus (A \cup B)$ then $\alpha \prec A \cup B \prec S \setminus \{\alpha\}$ and so $S \prec S \setminus \{\alpha\}$.$\square$

Theorem 4.1.5

1) If B is a finite set for which $X \prec B$ and if C is independent in X then $|C| \le |B|$.

2) Any two bases for a set X have the same cardinality.

Proof. For part 1), let $B = \{b_1, \ldots, b_m\}$. Choose $c_1 \in C$. The set $C_1 = \{c_1, b_1, \ldots, b_m\}$ satisfies the conditions of the previous lemma (with $A = \{c_1\}$) and so, after renumbering the b_i's if necessary, we deduce that

$$X \prec C_1 \prec \{c_1, b_1, \ldots, b_{m-1}\}$$

For any $c_2 \in C \setminus \{c_1\}$, the set $C_2 = \{c_1, c_2, b_1, \ldots, b_{m-1}\}$ satisfies the conditions of the lemma (with $A = \{c_1, c_2\}$) and so, again after possible renumbering, we get

$$X \prec C_2 \prec \{c_1, c_2, b_1, \ldots, b_{m-2}\}$$

Continuing this process, we must exhaust all of the elements of C before running out of elements of B, for if not, then a proper subset C' of C would have the property that $X \prec C'$, in contradiction to the independence of C. Hence, $|C| \le |B|$. Note that this also shows that C is finite. If B and C are bases, we may apply the argument with the roles of B and C reversed to get $|B| = |C|$.

Let us now assume that $B = \{b_i \mid i \in I\}$ and C are both infinite bases. Thus, $|B| = |I|$. For each $c \in C$, we have $c \prec B$ and so there is a finite subset $I_c \subseteq I$ such that $c \prec \{b_i \mid i \in I_c\}$. This gives a map $c \to I_c$ from C to the set of finite subsets of the index set I. Moreover,

$$I = \bigcup_{c \in C} I_c$$

for if $j \in I \setminus \bigcup I_c$ then, for any $c \in C$, we have

$$c \prec \{b_i \mid i \in I_c\} \prec B \setminus \{b_j\}$$

and so $b_j \prec C \prec B \setminus \{b_j\}$, which contradicts the independence of B. Hence,

$$|B| = |I| = \left| \bigcup_{c \in C} I_c \right| \le \aleph_0 |C| = |C|$$

Reversing the roles of B and C shows that $|B| = |C|$. □

4.2 Algebraic Dependence

Now that we have the basic theory of dependence, we can return to the subject matter of this book: fields. We recall a definition.

Definition *Let $F < E$. An element $t \in E$ is* **transcendental** *over F if t is not algebraic over F, that is, if there is no nonzero polynomial $p(x) \in F[x]$ such that $p(t) = 0$.* □

Recall that if t is transcendental over F then $F(t)$ is the field of all rational functions in the variable t, over the field F.

Definition *Let $F < E$ and let $S \subseteq E$. An element $\alpha \in E$ is* **algebraically dependent on** S **over** F, *written $\alpha \prec S$, if α is algebraic over $F(S)$. If α is not algebraically dependent on S over F, that is, if α is transcendental over $F(S)$ then α is said to be* **algebraically independent of** S **over** F *and we write $\alpha \not\prec S$.* □

Note that the relation $\prec$ depends on F, so we really should write $\prec_F$. However, we will not change the base field F so there should be no confusion in abbreviating the notation.

The condition $\alpha \prec S$ is equivalent to stating that $F(S) < F(S,\alpha)$ is algebraic. Thus, if $F < E$ and $A \subseteq E$, then $A \prec S$ if and only if $F(S) < F(S,\alpha)$ is algebraic for all $\alpha \in A$. But the class of algebraic extensions is closed under arbitrary composites and so this is equivalent to $F(S) < F(S,A)$ being algebraic. In short, $A \prec S$ if and only if A is algebraic over $F(S)$.

Now let us show that algebraic dependence is a dependence relation.

Theorem 4.2.1 *Algebraic dependence is a dependence relation.*
Proof. Since any $s \in S$ is algebraic over $F(S)$, we have reflexivity: $S \prec S$. To show compactness, let $\alpha \prec S$ and let $C \subseteq F(S)$ be the set of coefficients of $\min(\alpha, F(S))$. Since each $c \in C$ is a rational function over F in a finite number of elements of S, there is a finite subset S_0 of S for which $C \subseteq F(S_0)$. Hence α is algebraic over $F(S_0)$, that is, $\alpha \prec S_0$.

For transitivity, suppose that $\alpha \prec S$ and $S \prec T$. Then the tower

$$F(T) < F(T \cup S) < F(T \cup S, \alpha)$$

is algebraic and so α is algebraic over $F(T)$, that is, $\alpha \prec T$.

Finally, we verify the exchange axiom. Suppose that $\alpha \prec S$ and $\alpha \not\prec S \setminus \{s\}$. Then there is a finite set $S_1 \subseteq S$ for which $\alpha \prec S_1$ and $\alpha \not\prec S_1 \setminus \{s\}$. Let $S_0 = \{s\} \cup S_1$. Then $\alpha \prec S_0$ and $\alpha \not\prec S_0 \setminus \{s\}$. Our goal is to show that $s \prec S \setminus \{s\} \cup \{\alpha\}$, which will follow if we show that $s \prec S_0 \setminus \{s\} \cup \{\alpha\}$.

Note that $\{s\}$ is independent, for if s is algebraic over F then S_0 is algebraic over $S_0 \setminus \{s\}$ and so the tower $F(S_0 \setminus \{s\}) < F(S_0) < F(S_0, \alpha)$ is algebraic, in contradiction to $\alpha \not\prec S_0 \setminus \{s\}$. Thus, according to Lemma 4.1.4, we may remove elements of $S_0 \setminus \{s\}$ until the remaining set is independent, and yet α is still algebraic over this set. Hence, we may assume that S_0 is algebraically independent. Write $S_1 = \{s_1, \ldots, s_m\}$.

If $p(x) = \min(\alpha, F(S_0))$, then

$$p(x) = x^d + \sum_{i=0}^{d-1} \frac{f_i(s_1, \ldots, s_m, s)}{g_i(s_1, \ldots, s_m, s)} x^i$$

where $f_i(y_1, \ldots, y_m, y)$ and $g_i(y_1, \ldots, y_m, y)$ are polynomials.

Multiplying by the (nonzero) product

$$h(s_1, \ldots, s_m, s) = \prod g_i(s_1, \ldots, s_m, s)$$

of the denominators gives

$$h(s_1,\ldots,s_m,s)p(x) = h(s_1,\ldots,s_m,s)x^d + \sum_{i=0}^{d-1} h_i(s_1,\ldots,s_m,s)x^i$$

where $h(y_1,\ldots,y_m,y)$ and

$$h_i(y_1,\ldots,y_m,y) = h(y_1,\ldots,y_m,y)f_i(y_1,\ldots,y_m,y)$$

are polynomials over $F(S_1)$ and $h(s_1,\ldots,s_m,s) \neq 0$. Setting $x = \alpha$ gives

$$h(s_1,\ldots,s_m,s)\alpha^d + \sum_{i=0}^{d-1} h_i(s_1,\ldots,s_m,s)\alpha^i = 0$$

Now, if the polynomials $h(y_1,\ldots,y_m,y)$ and $h_i(y_1,\ldots,y_m,y)$ are constant with respect to y, then we have

$$h(s_1,\ldots,s_m,0)\alpha^d + \sum_{i=0}^{d-1} h_i(s_1,\ldots,s_m,0)\alpha^i = 0$$

in contradiction to $\alpha \not\prec S_0 \setminus \{s\} = S_1$. Hence, the polynomial

$$h(s_1,\ldots,s_m,y)\alpha^d + \sum_{i=0}^{d-1} h_i(s_1,\ldots,s_m,y)\alpha^i$$

is a nonconstant polynomial in y over $F(S_1,\alpha)$ satisfied by s, whence

$$s \prec S_0 \setminus \{s\} \cup \{\alpha\}$$

as desired.□

We may now take advantage of the results derived for dependence relations.

Definition *Let $F < E$.*

1) *A subset $S \subseteq E$ is* **algebraically dependent over** *F if there exists $s \in S$ that is algebraic over $F(S \setminus \{s\})$, that is, for which $F(S \setminus \{s\}) < F(S)$ is algebraic.*
2) *A subset $S \subseteq E$ is* **algebraically independent over** *F if s is transcendental over $F(S \setminus \{s\})$ for all $s \in S$. (The empty set is algebraically independent over F.)* □

Note that if $F < F(S)$ is algebraic, then certainly S is algebraically dependent over F, since *every* $s \in S$ is algebraic over F, let alone over $F(S \setminus \{s\})$. The converse, of course, is not true. For example, if t is transcendental over F then the set $S = \{t, 2t\}$ is algebraically dependent. In fact, t is algebraic over $F(2t)$ and $2t$ is algebraic over $F(t)$. However, $F < F(t, 2t) = F(t)$ is far from being algebraic.

Lemma 4.2.2

1) *Any superset of an algebraically dependent set is algebraically dependent.*
2) *Any subset of an algebraically independent set is algebraically independent.*□

Theorem 4.2.3 *If S is algebraically independent over F and α is transcendental over $F(S)$ then $S \cup \{\alpha\}$ is algebraically independent over F.* □

Algebraic Dependence and Polynomial Relationships

A subset S of a vector space is linearly dependent if there is a nontrivial *linear* relationship among the vectors of S. A similar statement holds in the present context.

Definition *Let $F < E$. A subset $S \subseteq E$ has a* **nontrivial polynomial relationship** *over F if there is a nonzero polynomial $p(x_1, \ldots, x_n)$ over F for which $p(s_1, \ldots, s_n) = 0$, for distinct $s_i \in S$. This is equivalent to saying that some $s \in S$ is algebraic over the* ring $F[S \setminus \{s\}]$ *of polynomials in $S \setminus \{s\}$.*□

To see that the two statements in the definition are equivalent, suppose that $p(s_1, \ldots, s_n) = 0$ for distinct $s_i \in S$, where $p(x_1, \ldots, x_n)$ is a nonzero polynomial over F. If $n = 1$, then this simply says that $p(s_1) = 0$ if and only if s_1 is algebraic over F. For $n > 1$, we may assume that $s_2, \ldots, s_n$ do not enjoy a similar polynomial dependency and hence that

$$p(x_1, \ldots, x_n) = \sum_{i=0}^{d} p_i(x_2, \ldots, x_n) x_1^i$$

where $p_d(x_2, \ldots, x_n) \neq 0$ and $p_d(s_2, \ldots, s_n) \neq 0$. Then the nonzero polynomial

$$p(x) = \sum_{i=0}^{d} p_i(s_2, \ldots, s_n) x^i$$

satisfies $p(s_1) = 0$, showing that s_1 is algebraic over $F[S \setminus \{s_1\}]$.

Now, to say that S is algebraically dependent is to say that s is algebraic over $F(S \setminus \{s\})$ for some $s \in S$. This is to say that s is algebraic over the field of *rational functions* in $S \setminus \{s\}$. But this is equivalent to saying that s is algebraic over the ring of polynomials in $S \setminus \{s\}$. One direction is clear, since a polynomial is a rational function. On the other hand, if s satisfies a polynomial $p(x)$ of degree $d > 0$ over $F(S \setminus \{s\})$ then $p(x)$ has the form

$$p(x) = \sum_{i=0}^{d} \frac{p_i(s_1, \ldots, s_m)}{q_i(s_1, \ldots, s_m)} x^i$$

where $p_d(s_1, \ldots, s_m) \neq 0$. Multiplying by the product P of the denominators gives a polynomial satisfied by s and whose leading coefficient is not zero.

We have proved the following.

Theorem 4.2.4 *Let $F < E$. A subset S of E is algebraically dependent over F if and only if there is a nontrivial polynomial relationship in S.* □

4.3 Transcendence Bases

We can now define an analogue of a (linear) basis for a vector space.

Definition *Let $F < E$. A* **transcendence basis** *for E over F is a subset $B \subseteq E$ that is algebraically independent over F and for which $F(B) < E$ is algebraic.* □

Since algebraic dependence is a dependence relation, we immediately get the following two results.

Theorem 4.3.1 *Let $F < E$. A subset $B \subseteq E$ is a transcendence basis for E over F if and only if it satisfies either one of the following.*
1) *B is a maximal algebraically independent subset of E over F.*
2) *B is a minimal set satisfying $E \prec B$, that is, B is minimal for the property that $F(B) < E$ is algebraic.* □

Theorem 4.3.2 *Let $F < E$.*
1) *Any two transcendence bases for E over F have the same cardinality, called the* **transcendence degree** *of E over F and denoted by $[E : F]_t$.*
2) *Suppose $F \subseteq A \subseteq S \subseteq E$ where A is algebraically independent over F and $F(S) < E$ is algebraic. Then there exists a transcendence basis B for E over F satisfying $A \subseteq B \subseteq S$. In particular, $[E : F]_t \leq |S|$.* □

While the vector space dimension is multiplicative over a tower of fields, the transcendence degree is additive, as we see in the next theorem.

Theorem 4.3.3 *Let $F < K < E$.*
1) *If $S \subseteq K$ is algebraically independent over F and $T \subseteq E$ is algebraically independent over K then $S \cup T$ is algebraically independent over F.*
2) *If S is a transcendence basis for K over F and T is a transcendence basis for E over K then $S \cup T$ is a transcendence basis for E over F.*
3) *Transcendence degree is additive, that is,*

$$[E : F]_t = [E : K]_t + [K : F]_t$$

Proof. For part 1), consider a polynomial dependence of $S \cup T$ over F, that is, a polynomial

$$p(x_1, \ldots, x_n, y_1, \ldots, y_m) \in F[x_1, \ldots, x_n, y_1, \ldots, y_m]$$

for which $p(s_1, \ldots, s_n, t_1, \ldots, t_m) = 0$, where $s_i \in S$ are distinct and $t_i \in T$ are distinct. Write

$$p(x_1,\ldots,x_n,y_1,\ldots,y_m)=\sum_{e_1,\ldots,e_m}\left(\sum_{f_1,\ldots,f_n}a_{f_1,\ldots,f_n}x_1^{f_1},\ldots,x_n^{f_n}\right)y_1^{e_1}\cdots y_m^{e_m}$$

where $a_{f_1,\ldots,f_n}\in F$ and where the monomials $y_1^{e_1}\cdots y_m^{e_m}$ are distinct and, for each such monomial, the monomials $x_1^{f_1},\ldots,x_n^{f_n}$ are distinct. Consider the polynomial

$$\begin{aligned}q(y_1,\ldots,y_m)&=p(s_1,\ldots,s_n,y_1,\ldots,y_m)\\&=\sum_{e_1,\ldots,e_m}\left(\sum_{f_1,\ldots,f_n}a_{f_1,\ldots,f_n}s_1^{f_1},\ldots,s_n^{f_n}\right)y_1^{e_1}\cdots y_m^{e_m}\end{aligned}$$

over K. Since T is algebraically independent over K, it follows that

$$\sum_{f_1,\ldots,f_n}a_{f_1,\ldots,f_n}s_1^{f_1},\ldots,s_n^{f_n}=0$$

However, S is algebraically independent over F and so $a_{f_1,\ldots,f_n}=0$. Thus, p is the zero polynomial and $S\cup T$ is algebraically independent over F.

For part 2), we know by part 1) that $S\cup T$ is algebraically independent over F. Also, since $F(S)<K$ and $K(T)<E$ are algebraic, each step in the tower $F(S\cup T)<K(T)<E$ is algebraic and so $F(S\cup T)<E$ is algebraic. Hence, $S\cup T$ is a transcendence basis for F over E. Part 3) follows directly from part 2). □

Purely Transcendental Extensions

When one speaks of the field of rational functions $F(x_1,\ldots,x_n)$ in the "independent" variables $x_1,\ldots,x_n$, one is really saying that the set $B=\{x_1,\ldots,x_n\}$ is algebraically independent over F and that $E=F(B)$. We have a name for such an extension.

Definition *An extension $F<E$ is said to be* **purely transcendental** *if $E=F(B)$ for some transcendence basis B for E over F.* □

We remark that if E is purely transcendental over F then $E=F(B)$ for *some* transcendence basis B, but not all transcendence bases C for E over F need satisfy $E\prec C$. The reader is asked to supply an example in the exercises.

The following is an example of an extension that is neither algebraic nor purely transcendental.

Example 4.3.1 Let $n\geq 3$ and let F be a field with $\operatorname{char}(F)\not| n$. Let u be transcendental over F, let v be a root of $p(x)=x^n+u^n-1$ in some splitting field and let $E = F(u, v)$. Clearly, E is not algebraic over F. We contend that E is also not purely transcendental over F. Since $\{u\}$ is algebraically independent and $F(u) < F(u, v)$ is algebraic, the set $B = \{u\}$ is a transcendence basis for E over F and so $[E : F]_t = 1$. If E were purely transcendental over F there would exist a transcendental element t over F for which $F(t) = F(u, v)$. Let us show that this is not possible.

If $F(t) = F(u, v)$ then

$$u = \frac{a(t)}{b(t)} \quad \text{and} \quad v = \frac{c(t)}{d(t)}$$

where $a(t), b(t), c(t)$ and $d(t)$ are polynomials over F. Hence

$$\frac{a^n(t)}{b^n(t)} + \frac{c^n(t)}{d^n(t)} = 1$$

or

$$[a(t)d(t)]^n + [b(t)c(t)]^n = [b(t)d(t)]^n$$

This can be written

$$f^n(t) + g^n(t) = h^n(t)$$

for nonconstant polynomials $f(t), g(t)$ and $h(t)$, which we may assume to be pairwise relatively prime. Let us assume that $\deg(f(t)) \le \deg(g(t))$, in which case $\deg(h(t)) \le \deg(g(t))$. We now divide by $h^n(t)$ and take the derivative with respect to t to get (after some simplification)

$$f^{n-1}[f'h - fh'] + g^{n-1}[g'h - gh'] = 0$$

Since f and g are relatively prime, we deduce that $g^{n-1} \mid f'h - fh'$. But this implies

$$(n-1)\deg(g) \le \deg(fh) - 1 = \deg(f) + \deg(h) - 1 \le 2\deg(g) - 1$$

which is not possible for $n \ge 3$. Hence, $F < F(u, v)$ is not purely transcendental. $\square$

Purely transcendental extensions $F < E$ are 100% transcendental, that is, every element of $E \setminus F$ is transcendental over F.

Theorem 4.3.4 *A purely transcendental extension $F < E$ is 100% transcendental, that is, any $\alpha \in E \setminus F$ is transcendental over F.*
Proof. Let B be a transcendence basis for E over F. Since $E = F(B)$, it follows that $\alpha \in F(t_1, \ldots, t_n)$ for some finite set $\{t_1, \ldots, t_n\} \subseteq B$, and we can assume that $\alpha \notin F(t_1, \ldots, t_{n-1})$. Letting $K = F(t_1, \ldots, t_{n-1})$, we have $\alpha \in E \setminus K$ where $E = K(t_n)$ is a simple transcendental extension of K.

Hence, Theorem 2.4.5 implies that α is transcendental over K, and therefore also over F.□

The following result will prepare the way to finishing the proof (promised in Chapter 2) that the class of finitely generated extensions is distinguished.

Theorem 4.3.5 *Let $F < K < E$ and suppose that $F < K$ is algebraic. If $T \subseteq E$ is algebraically independent over F, then T is also algebraically independent over K. In other words, T remains algebraically independent over any algebraic extension of the base field.*

Proof. We have the picture shown in Figure 4.3.1

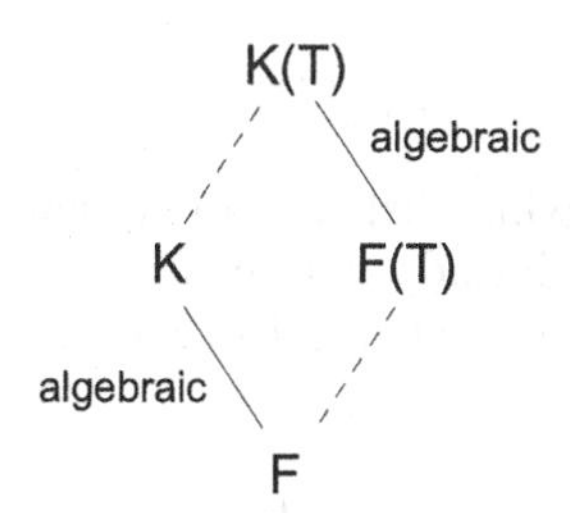

Figure 4.3.1

Since $F < K$ is algebraic , so is the lifting $F(T) < K(T)$. Now,

$$[K(T) : K]_t + [K : F]_t = [K(T) : F] = [K(T) : F(T)]_t + [F(T) : F]_t$$

and so

$$[K(T) : K]_t = [F(T) : F]_t = |T|$$

which shows that T must be a transcendence basis for $K(T)$ over K.

For an alternative proof, if T is not algebraically independent over K, there exists $t \in T$ that is algebraic over $K(T \setminus \{t\})$. Since $F < K$ is algebraic, the lifting $F(T \setminus \{t\}) < K(T \setminus \{t\})$ is also algebraic, and so the tower

$$F(T \setminus \{t\}) < K(T \setminus \{t\}) < K(T \setminus \{t\})(t) = K(T)$$

is algebraic, whence t is algebraic over $F(T \setminus \{t\})$, in contradiction to the algebraic independence of T over F. □

Finitely Generated Extensions Are Distinguished

We are now in a position to finish the proof that the class of finitely generated extensions is distinguished. Note how much more involved this task is than showing that finite or algebraic extensions are distinguished.

Theorem 4.3.6 *Let $F < K < E$. If E is finitely generated over F then K is also finitely generated over F. Thus, the set of finitely generated extensions is distinguished.*

Proof. Let $S = \{s_1, \ldots, s_k\}$ be a transcendence basis for K over F. Then the second step in the tower $F < F(S) < K < E$ is algebraic and E is finitely generated over $F(S)$. Hence, if we can prove the theorem for algebraic intermediate fields, we will know that K is finitely generated over $F(S)$ and therefore also over F, since S is a finite set.

Thus, we may assume that $F < K < E$ with $F < K$ algebraic and show that $[K : F]$ is finite. Let $T = \{t_1, \ldots, t_n\}$ be a transcendence basis for E over F. Our plan is to show that

$$[K : F] \leq [E : F(T)]$$

(see Figure 4.3.2) by showing that any finite subset of K that is linearly independent over F is also linearly independent over $F(T)$, as a subset of E. Since $F(T) < E$ is finitely generated and algebraic, $[E : F(T)]$ is finite and the proof will be complete.

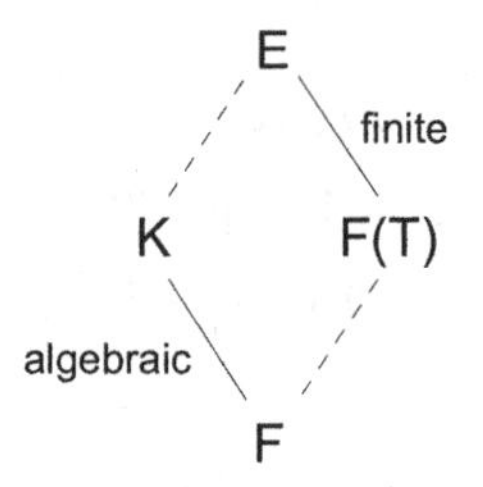

Figure 4.3.2

Let $A = \{\kappa_1, \ldots, \kappa_m\} \subseteq K$ be linearly independent over F and suppose that

$$\sum_i r_i(t_1, \ldots, t_n)\kappa_i = 0$$

where $r_i(t_1, \ldots, t_n) \in F(T)$. We wish to show that $r_i(t_1, \ldots, t_n) = 0$.

By clearing denominators if necessary, we may assume that each $r_i(t_1, \ldots, t_n)$ is a polynomial over F. Collecting terms involving like powers of the t_i's gives

$$\sum_{e_1, \ldots, e_n} \left(\sum_i a_{e_1, \ldots, e_n, i} \kappa_i \right) t_1^{e_1} \cdots t_n^{e_n} = 0$$

where $a_{e_1, \ldots, e_n, i} \in F$ is the coefficient of $t_1^{e_1} \cdots t_n^{e_n}$ in $r_i(t_1, \ldots, t_n)$. Since Theorem 4.3.5 implies that T is algebraically independent over K, it follows that T does not satisfy any polynomial relationships over K and so

$$\sum_i a_{e_1,\ldots,e_n,i} \kappa_i = 0$$

Then the linear independence of A over F gives

$$a_{e_1,\ldots,e_n,i} = 0$$

and so $r_i(t_1, \ldots, t_n) = 0$ for all i. This shows that A is linearly independent over $F(T)$, as desired. □

*4.4 Simple Transcendental Extensions

The class of purely transcendental extensions is much less well behaved than the class of algebraic extensions. For example, let t be transcendental over F. Then in the tower $F < F(t^2) < F(t)$, the extension $F < F(t)$ is purely transcendental (and simple) but the second step $F(t^2) < F(t)$ is not transcendental at all.

In addition, if $F < E$ is purely transcendental and $F < K < E$, it does not necessarily follow that the first step $F < K$ is purely transcendental. However, this is true for simple transcendental extensions. The proof of this simple statement illustrates some of the apparent complexities in dealing with transcendental extensions.

Theorem 4.4.1 (Luroth's Theorem) *Let t be transcendental over F. If $F < K < F(t)$ and $K \neq F$ then $K = F(s)$ for some $s \in F(t)$.*

Proof. Let us recall a few facts from Theorem 2.4.5. Since $K \neq F$, Theorem 2.4.5 implies that for any $s \in K \setminus F$, the tower $F(s) < K < F(t)$ is algebraic. Theorem 2.4.5 also implies that if $s = f(t)/g(t) \in K \setminus F$ where f and g are relatively prime polynomials over F, then

$$[F(t) : F(s)] = \max(\deg(f(x)), \deg(g(x)))$$

Now, we want to find an $s \in K \setminus F$ for which $[F(t) : F(s)] = [F(t) : K]$, showing that $K = F(s)$. Of course, $[F(t) : F(s)] \geq [F(t) : K]$.

Let

$$p(x) = \min(t, K) = x^n + \frac{a_1(t)}{b_1(t)} x^{n-1} + \cdots + \frac{a_n(t)}{b_n(t)}$$

where $a_i(t), b_i(t) \in F(t)$ are relatively prime. Since t is not algebraic over F, not all of the coefficients of $p(x)$ can lie in F. We will show that for any coefficient $a_k(t)/b_k(t) \neq F$, we may take

$$s = \frac{a_k(t)}{b_k(t)}$$

To this end, consider the polynomial

$$h(x) = a_k(x) - \frac{a_k(t)}{b_k(t)} b_k(x) \in K[x]$$

Since $s \notin F$, we have $h(x) \neq 0$. But $h(t) = 0$ and so $p(x) \mid h(x)$ over K. In other words, there exists $q(x) \in K[x]$ such that

$$a_k(x) - \frac{a_k(t)}{b_k(t)} b_k(x) = q(x)p(x)$$

or

$$a_k(x)b_k(t) - a_k(t)b_k(x) = b_k(t)q(x)p(x)$$

Multiplying both sides of this by

$$r(t) = b_1(t)\cdots b_n(t)$$

gives

$$r(t)[a_k(x)b_k(t) - a_k(t)b_k(x)] = b_k(t)q(x)r(t)p(x) \tag{4.4.1}$$

where

$$r(t)p(x) = b_1(t)\cdots b_n(t)x^n + \sum_{i=1}^{n} [b_1(t)\cdots b_{i-1}(t)a_i(t)b_{i+1}(t)\cdots b_n(t)]x^{n-i}$$

Now let $g(t)$ be the greatest common divisor of the coefficients on the right-hand side of this equation. Since $g(t)$ divides the first coefficient $b_1(t)\cdots b_n(t)$, it must be relatively prime to each $a_k(x)$ and so

$$g(t) \mid b_1(t)\cdots b_{k-1}(t)b_{k+1}(t)\cdots b_n(t)$$

for all k. Factoring out $g(t)$ gives

$$r(t)p(x) = g(t)p'(t,x)$$

where $p'(t,x) \in F[t,x]$ is *primitive* in t, that is, $p'(x,t)$ is not divisible by any nonconstant polynomial in t.

Note, however, that for each k, the polynomial $a_k(t)$ appears in the coefficent of x^{n-k}. Also, for each k, we have

$$b_k(t) \,\Big|\, \frac{b_1(t)\cdots b_n(t)}{g(t)}$$

It follows that the degree of $p'(t,x)$ with respect to t satisfies

$$t\text{-deg}(p'(t,x)) \geq \max(\deg(a_k(t)), \deg(b_k(t))) = [F(t) : F(s)] \tag{4.4.2}$$

Thus, (4.4.1) can be written

$$r(t)[a_k(x)b_k(t) - a_k(t)b_k(x)] = b_k(t)q(x)g(t)p'(t,x) \qquad (4.4.3)$$

Next, we multiply both sides of (4.4.3) by a polynomial $u(t)$ that will clear all of the denominators of $q(x)$, giving

$$u(t)r(t)[a_k(x)b_k(t) - a_k(t)b_k(x)] = b_k(t)q'(t,x)p'(t,x)$$

where $p'(t,x), q'(t,x) \in F[t,x]$. Since $p'(t,x)$ is t-primitive, we must have $u(t)r(t) \mid b_k(t)q'(t,x)$ and so the other factor $a_k(x)b_k(t) - a_k(t)b_k(x)$ must divide $p'(t,x)$, that is, there exists a polynomial $q''(t,x) \in F[t,x]$ for which

$$a_k(x)b_k(t) - a_k(t)b_k(x) = q''(t,x)p'(t,x) \qquad (4.4.4)$$

Now, the t-degree of the left-hand side of this equation is at most

$$\max(\deg(a_k(t)), \deg(b_k(t))) = [F(t) : F(s)]$$

and by (4.4.2), the t-degree of the right-hand side is at least $[F(t) : F(s)]$. Hence, the t-degree of either side of (4.4.4) is $[F(t) : F(s)]$ and (4.4.2) implies that $t\text{-deg}(q''(t,x)) = 0$, that is,

$$a_k(x)b_k(t) - a_k(t)b_k(x) = q''(x)p'(t,x) \qquad (4.4.5)$$

where $q''(x) \in F[x]$. Since the right side of (4.4.5) is not divisible by any nonconstant polynomial in t, neither is the left side. But the left side is symmetric in x and t, so it cannot be divisible by any nonconstant polynomial in x either. Hence, $q''(x)p'(t,x)$ is not divisible by any nonconstant polynomial in x, implying that $q''(x) \in F$, that is,

$$a_k(x)b_k(t) - a_k(t)b_k(x) = q''p'(t,x) \qquad (4.4.6)$$

where $q'' \in F$. Finally, since the x-degree and t-degree of the left side of (4.4.6) agree, this is also true of the right side. Hence by (4.4.2),

$$n = x\text{-deg}(p'(t,x)) = t\text{-deg}(p'(t,x)) \geq [F(t) : F(s)] \geq n$$

Thus, $[F(t) : F(s)] = n$, and the proof is complete. □

It can be shown that Luroth's theorem does not extend beyond simple transcendental extensions, but a further discussion of this topic would go beyond the intended scope of this book.

The Automorphims of a Simple Transcendental Extension

We conclude with a description of all F-automorphisms of a simple transcendental extension $F(t)$. Let $GL_n(F)$ denote the *general linear group*, that is, the group of all nonsingular $n \times n$ matrices over F. The proof, which is left as an exercise, provides a nice application of Theorem 2.4.5.

Theorem 4.4.2 *Let $F < F(t)$ be a simple transcendental extension and let $Aut_F(F(t))$ denote the group of all automorphisms of $F(t)$ over F.*

1) *For each $A = \begin{bmatrix} a & b \\ c & d \end{bmatrix} \in GL_2(F)$ there is a unique $\sigma_A \in Aut_F(F(t))$ for which*

$$\sigma_A(t) = \frac{at+b}{ct+d}$$

Moreover, all automorphisms of $F(t)$ over F have the form σ_A for some $A \in GL_2(F)$.

2) *If $A, B \in GL_2(F)$, then*

$$\sigma_{AB} = \sigma_A \sigma_B \quad \text{and} \quad \sigma_{A^{-1}} = \sigma_A^{-1}$$

Also, $\sigma_A = \sigma_B$ if and only if AB^{-1} is a nonzero scalar matrix. In other words, the map $\tau\colon GL_2(F) \to Aut_F(F(t))$ defined by $\tau A = \sigma_A$ is an epimorphism with kernel equal to the group of all nonzero scalar matrices in $GL_2(F)$.□

Exercises

1. Find an example of a purely transcendental extension $F < E$ with two transcendence bases B and C such that $E = F(B)$ but $F(C)$ is a proper subfield of E.
2. Let $F < E$ and $F < K$. Show that $[EK : K]_t \le [E : F]_t$.
3. Let $F < K < E$ and let $T \in E \setminus K$. Show that $[E(T) : F(T)]_t \le [K : F]_t$ with equality if T is algebraically independent over K or algebraic over F.
4. Use the results of the previous exercise to show that if $F < K < E$ and $F < L < E$ then $[KL : F]_t \le [K : F]_t + [L : F]_t$.
5. Let F be a field of characteristic $\ne 2$ and let u be transcendental over F. Suppose that $u^2 + v^2 = 1$. Show that $F(u, v)$ is a purely transcendental extension by showing that $F(u, v) = F(w)$ where $w = (1 + v)/u$.
6. Show that the extension $\mathbb{Q} < \mathbb{Q}(t, \sqrt{2t})$, where t is transcendental over $\mathbb{Q}$, is not purely transcendental.
7. Let $F < K < E$ and suppose that $S \subseteq E$ is algebraically independent over K. Prove that $F(S) < K(S)$ is algebraic if and only if $F < K$ is algebraic.
8. Prove that the transcendence degree of $\mathbb{R}$ over $\mathbb{Q}$ is uncountable.
9. a) Show that the only automorphism of $\mathbb{R}$ is the identity.
   b) Show that the only automorphisms of $\mathbb{C}$ over $\mathbb{R}$ are the identity and complex conjugation.
   c) Show that there are infinitely many automorphisms of $\mathbb{C}$ over $\mathbb{Q}$.
10. (An extension of Luroth's theorem) Suppose that $F < E$ is purely transcendental. Show that any simple extension of F contained in E (but not equal to F) is transcendental over F.

11. Prove part 1) of Theorem 4.3.5 by contradiction as follows. Suppose that $S \cup T$ is algebraically dependent over F. Then there exists an $\alpha \in S \cup T$ that is algebraic over $F(S_0 \cup T_0)$ for some finite sets $S_0 \subseteq S$ and $T_0 \subseteq T$ not containing α, and we may assume that no proper subset T_1 of T_0 has the property that α is algebraic over $F(S_0 \cup T_1)$. Prove that $\alpha \neq T$. Prove that T_0 is not empty. If $t \in T_0$, prove that t is algebraic over $F(S_0 \cup T_0 \cup \{\alpha\} \setminus \{t\})$. Complete the proof from here.
12. Prove Theorem 4.4.2.

Part II—Galois Theory

Chapter 5
Galois Theory I: An Historical Perspective

Galois theory sits atop a structure of work began about 4000 years ago on the question of how to solve polynomial equations algebraically *by radicals*, that is, how to solve equations of the form

$$a_n x^n + a_{n-1} x^{n-1} + \cdots + a_0 = 0$$

by applying the four basic arithmetical operations (addition, subtraction, multiplication and division), and the taking of roots, to the coefficients of the equation and to other "known" quantities (such as elements of the base field).

More specifically, a polynomial equation $p(x) = 0$ is *solvable by radicals* if there is a tower of fields

$$F < R_1 < R_2 < \cdots < R_n$$

where R_n contains a splitting field for $p(x)$ (and hence a full set of roots of $p(x)$) and where each field in the tower is obtained by adjoining some root of an element of the previous field, that is,

$$R_k = R_{k-1}(\alpha_k)$$

where $\alpha_k^{n_k} \in R_{k-1}$.

In this chapter, we will review this structure of work from its beginnings in Babylonia through the work of Galois. In subsequent chapters, we will set down the modern version of the theory that has become known as Galois theory.

5.1 The Quadratic Equation

Archeological findings indicate that as early as about 2000 B.C., the Babylonians (Mesopotamians) had an algorithm for finding two numbers a and b whose sum $s = a + b$ and product $p = ab$ were known. The algorithm is

1) Take half of s.
2) Square the result.

3) Subtract p.
4) Take the square root of this result.
5) Add half of s.

This results in one of the values a and b: the other is easily obtained. This algorithm can be expressed in modern notation by the formula

$$a = \sqrt{(s/2)^2 - p} + s/2$$

for solving the system of equations

$$\begin{aligned} a + b &= s \\ ab &= p \end{aligned}$$

The solutions to this system are solutions to the quadratic equation

$$x^2 - sx + p = 0$$

Thus, except for one issue, it can be said that the Babylonians knew the quadratic formula, but in algorithmic form.

The one issue is that the Babylonians had no notion of negative numbers! Indeed, they developed a *separate* algorithm to compute the numbers a and b whose *difference* and product were known. This is the solution to the system

$$\begin{aligned} a - b &= s \\ ab &= p \end{aligned}$$

whose solutions satisfy the quadratic equation

$$x^2 - sx - p = 0$$

Unfortunately, the origin of the Babylonian algorithms appears lost to antiquity. No texts uncovered from that period indicate who or how the algorithm was developed.

5.2 The Cubic and Quartic Equations

In the 3500 years or so between the apparent achievement of the Babylonians and the mid-Renaissance period of the 1500's, not much happened in Europe of a mathematical nature. However, during the Middle Ages (that is, prior to the Renaissance, which began in the late thirteenth century), the Europeans did learn about algebra from the Arabs and began to devise a new mathematical symbolism, which opened the way for the dramatic advancements of the mid-Renaissance period.

In particular, solutions to the general cubic and quartic equations were discovered. As to the cubic, we have the following excerpt from Girolamo Cardano's *Ars Magna* (1545). (Cardano was a highly educated and skilled physician, natural philosopher, mathematician and astrologer.)

> In our own days Scipione del Ferro of Bologna has solved the case of the cube and first power equal to a constant, a very elegant and admirable accomplishment. Since this art surpasses all human subtlety and the perspicuity of mortal talent, and is a truly celestial gift and a very clear test of the capacity of men's minds, whoever applies himself to it will believe that there is nothing that he cannot understand. In emulation of him, my friend Niccolò Tartaglia of Brescia, wanting not to be outdone, solved the same case when he got into a contest with his [Scipione's] pupil, Antonio Maria Fior, and, moved by my many entreaties, gave it to me.

The solution of the quartic equation was discovered by one of Cardano's students, Ludivico Ferrari, and published by Cardano. Let us briefly review these solutions in modern notation.

Solving the Cubic

1) An arbitrary monic cubic polynomial $x^3 + bx^2 + cx + d$ can be put in the form

$$p(x) = x^3 + px + q$$

by replacing x by $x - b/3$.

2) Introduce the variables u and v and set $x = u - v$. Then $p(x)$ has the form

$$p(x) = u^3 - 3u^2v + 3uv^2 - v^3 + pu - pv + q$$

or, equivalently,

$$p(x) = u^3 - v^3 + (u - v)(-3uv + p) + q$$

3) If $3uv = p$, then we get

$$p(x) = u^3 - v^3 + q$$

Thus, a solution to the pair of equations

$$\begin{aligned} 3uv &= p \\ u^3 - v^3 + q &= 0 \end{aligned}$$

provides a solution $u - v$ to the original cubic equation. Multiplying the second equation by $27u^3$ and using the fact that $27u^3v^3 = p^3$ gives

$$27u^6 - p^3 + 27u^3q = 0$$

which is a quadratic equation in u^3. If α is a cube root of a solution to this quadratic, then $v = p/3\alpha$, so that $\alpha - p/3\alpha$ is a root of the original cubic.

Solving the Quartic

1) An arbitrary monic quartic equation can be put in the form

$$x^4 + px^2 + qx + r = 0$$

2) Introducing a variable u, we have

$$(x^2 + u)^2 = x^4 + 2ux^2 + u^2$$

Using the quartic equation from 1) to replace x^4 on the right, we have

$$(x^2 + u)^2 = (2u - p)x^2 - qu + (u^2 - r)$$

3) If the right side of this equation can be put in the form $A(x + B)^2$, then we can take square roots. This happens if the quadratic on the right has a single root, which happens if its discriminant is 0, that is, if

$$q^2 - 4(2u - p)(u^2 - r) = 0$$

which is a cubic in u, and can therefore be solved, as described earlier.

4) Once u is found, we have $A = 2u - p$ and $B = -q/2A$, and so our quartic is

$$(x^2 + u)^2 = (2u - p)\left(x - \frac{q}{2(2u - p)}\right)^2$$

Hence,

$$x^2 + u = \sqrt{2u - p}\left(x - \frac{q}{2(2u - p)}\right)$$

which can be solved for a solution x of the original quartic.

5.3 Higher-Degree Equations

Naturally, solutions to the arbitrary cubic and quartic equations led to a search for methods of solution to higher-degree equations, but in vain. It was not until the 1820s, some 300 years later, in the work of Ruffini, Abel and then Galois, that it was shown that no solution similar to those of the cubic and quartic equations could be found, since none exists.

Specifically, for any $n \geq 5$, there is no algebraic formula, involving only the four basic arithmetic operations and the taking of roots, that gives the solutions to any polynomial equation of degree n. In fact, there are individual quintic (and higher–degree) equations whose solutions are not obtainable by these means. Thus, not only is there no *general* formula, but there are cases in which there is no *specific* formula.

5.4 Newton's Contribution: Symmetric Polynomials

It was not until the accomplishments of Vandermonde and, to a larger extent, Lagrange, in the period around 1770, that a deeper understanding of the work that led to the solutions of the cubic and quartic equations was revealed. However, even these fine mathematicians were unable to take the leap made by Abel and Galois a few decades later.

The cornerstone of the work of Vandermonde and Lagrange is the work of Isaac Newton on symmetric polynomials. We will go into precise detail at the appropriate time in a subsequent chapter, but here is an overview of Newton's contribution in this area.

The Generic Polynomial

If $t_1, \ldots, t_n$ are independent variables, the polynomial

$$g(x) = \prod_{i=1}^{n} (x - t_i)$$

is referred to as a **generic polynomial** of degree n. (Galois would have referred to this as a polynomial with "literal" coefficients.) Since the roots $t_1, \ldots, t_n$ of the generic polynomial $g(x)$ are independent, this polynomial is, in some sense, the most "general" polynomial of degree n and facts we learn about $g(x)$ often apply to all polynomials.

It can be shown by induction that the generic polynomial can be written in the form

$$g(x) = x^n - s_1 x^{n-1} + \cdots + (-1)^n s_n$$

where the coefficients are given by

$$s_1 = \sum_i t_i, \quad s_2 = \sum_{i<j} t_i t_j, \quad s_3 = \sum_{i<j<k} t_i t_j t_k \quad , \ldots, \quad s_n = \prod_{i=1}^{n} t_i$$

These polynomials are called the **elementary symmetric polynomials** in the variables t_i.

Thus, except for sign, the coefficients of $g(x)$ are the elementary symmetric polynomials of the roots of $g(x)$. Moreover, since this holds for the generic polynomial, it is clear that it holds for *all* polynomials.

Symmetric Polynomials

Intuitively, a polynomial $p(t_1, \ldots, t_n)$ in the variables $t_1, \ldots, t_n$ is **symmetric** if it remains unchanged when we permute the variables. More carefully, $p(t_1, \ldots, t_n)$ is symmetric if

$$p(t_{\sigma(1)},\ldots,t_{\sigma(n)}) = p(t_1,\ldots,t_n)$$

for any permutation σ of $\{1,\ldots,n\}$.

Of course, each elementary symmetric polynomial s_k, that is, each coefficient of $g(x)$, is a symmetric polynomial of the roots t_i, in this sense. It follows that *any* polynomial (symmetric or otherwise) in the coefficients of $g(x)$ is a symmetric polynomial of the roots t_i. For instance,

$$2s_1^2 + s_2 = 2\left(\sum_i t_i\right)^2 + \sum_{i<j} t_i t_j$$

is unchanged by a permutation of the t_i's.

Isaac Newton realized, sometime in the late 1600s, that a kind of converse to this holds: Any *symmetric* polynomial in the roots of $p(x)$ is a polynomial in the coefficients of $p(x)$. Let us state this theorem, known as *Newton's theorem*, first without reference to roots.

Newton's Theorem

1) *A polynomial $p(t_1,\ldots,t_n)$ is symmetric if and only if it is a polynomial in the elementary symmetric functions $s_1,\ldots,s_n$, that is,*

$$p(t_1,\ldots,t_n) = q(s_1,\ldots,s_n)$$

Moreover, if $p(t_1,\ldots,t_n)$ has integer coefficients, then so does $q(s_1,\ldots,s_n)$.

2) *Let $p(x)$ be a polynomial. Then the set of symmetric polynomials in the roots of $p(x)$ is the same as the set of polynomials in the coefficients of $p(x)$. In particular, any symmetric polynomial in the roots of $p(x)$ belongs to the same field as the coefficients, so if $p(x)$ is a polynomial over $\mathbb{Q}$, then any symmetric polynomial in the roots of $p(x)$ belongs to $\mathbb{Q}$. Also, if $p(x)$ has integer coefficients, then any symmetric polynomial in the roots of $p(x)$ is an integer.*□

The proof of Newton's theorem will be given in a later chapter. However, it should be noted that the proof is in the form of an *algorithm* (however impractical) for finding the polynomial q.

How can this be used to advantage in the present context? The answer is both simple and profound: When trying to find the roots of a polynomial $p(x)$, we can assume not only that the coefficients of $p(x)$ are known (obviously), but also that any symmetric polynomial in the roots of $p(x)$ is known! The reason is that an algorithm is known for computing this symmetric polynomial of the roots that requires knowledge of the coefficients of the polynomial only (and of other known quantities, such as rational numbers).

For instance, if $p(x)=x^2+bx+c$ has roots r and s, then not only are $r+s$ and rs known, being the coefficients of $p(x)$, up to sign, but we may also assume that expressions such as $r^3+s^3+2r^2s^2-3$ are known. More to the point, we cannot assume that $r-s$ is known, but we *can* assume that $(r-s)^2$ is known and so we may write

$$\begin{aligned} r+s&=\alpha \\ (r-s)^2&=\beta \end{aligned}$$

where α and β are known. Hence, $r-s=\sqrt{\beta}$. Adding this to the first equation gives $2r=\alpha+\sqrt{\beta}$, or

$$r=\frac{1}{2}\left(\alpha+\sqrt{\beta}\right)$$

Of course, $r+s=-b$ and $(r-s)^2=(r+s)^2-4rs=b^2-4c$ and so this becomes the well-known quadratic formula

$$r=\frac{1}{2}\left(-b+\sqrt{b^2-4c}\right)$$

(Note that there is another solution to $(r-s)^2=\beta$, which gives the other root.)

We are very close here to the work of Vandermonde and Lagrange.

5.5 Vandermonde

How can we apply the previous analysis to the cubic equation? The previous solution to the quadratic can be expressed as

$$r_1=\frac{1}{2}\left(r_1+r_2+\sqrt{(r_1-r_2)^2}\right)$$

where the solutions are r_1 and r_2. Now let r_1,r_2 and r_3 be solutions to a cubic equation. Again, the sum $r_1+r_2+r_3$ is known, being symmetric in the roots. As to the analogue of the difference, note that the coefficients $+1$ and -1 of r_1-r_2 are the two roots of the equation $x^2-1=0$, that is, they are the *square roots of unity*.

In general, the complex **nth roots of unity** are the roots (in the complex field) of the equation

$$x^n-1=0$$

As we will see in a later chapter, this equation has n distinct complex roots, which we denote by U_n. The set U_n is a cyclic group under multiplication. Any generator of U_n is called a **primitive** nth root of unity. The set of primitive nth roots of unity is denoted by Ω_n. Note that if $\omega\in\Omega_n$, then

$$1+\omega+\omega^2+\cdots+\omega^{n-1}=\frac{1-\omega^n}{1-\omega}=0$$

This fact is used many times.

Now, for the analogue of the difference r_1-r_2, we require the two expressions

$$\begin{aligned} t_1 &= r_1+\omega r_2+\omega^2 r_3 \\ t_2 &= r_1+\omega^2 r_2+\omega r_3 \end{aligned}$$

where ω is a primitive cube root of unity. Then

$$r_1=\frac{1}{3}\left[(r_1+r_2+r_3)+(r_1+\omega r_2+\omega^2 r_3)+(r_1+\omega^2 r_2+\omega r_3)\right]$$

Now, the expressions t_1 and t_2 are not symmetric in the roots, so we cannot conclude *directly* from Newton's theorem that they are known. However, the previous expression can be written in the form

$$r_1=\frac{1}{3}\left[(r_1+r_2+r_3)+\sqrt[3]{(r_1+\omega r_2+\omega^2 r_3)^3}+\sqrt[3]{(r_1+\omega^2 r_2+\omega r_3)^3}\right]$$

and while the expressions

$$\begin{aligned} u &= (r_1+\omega r_2+\omega^2 r_3)^3 \\ v &= (r_1+\omega^2 r_2+\omega r_3)^3 \end{aligned}$$

are also not symmetric in the roots, the expressions $u+v$ and uv are symmetric.

To see this, first note that interchanging r_2 and r_3 has the effect of interchanging u and v, thus preserving both $u+v$ and uv. Also, the cyclic permutation $\sigma=(r_1r_2r_3)$, which sends r_1 to r_2, r_2 to r_3 and r_3 to r_1, actually fixes both u and v. For example,

$$\begin{aligned} \sigma u &= \sigma[(r_1+\omega r_2+\omega^2 r_3)^3] \\ &= (r_2+\omega r_3+\omega^2 r_1)^3 \\ &= [\omega^2(\omega r_2+\omega^2 r_3+r_1)]^3 \\ &= \omega^6 u \\ &= u \end{aligned}$$

Thus, both $u+v$ and uv are known quantities, from which we can compute u and v using the *quadratic* formula. It follows that the root

$$r_1=\frac{1}{3}\left[(r_1+r_2+r_3)+\sqrt[3]{u}+\sqrt[3]{v}\right]$$

is known. Note that there are three possible values for each cube root in this expression, leading to nine possible value of r_1, of which exactly 3 are roots of the cubic. Of course, it is a simple matter (in theory) to determine which of the nine candidates are roots. Thus, the solution to the general cubic equation is reduced to solving a quadratic equation and to the taking of cube roots.

This analysis of the cubic equation is the work of Vandermonde, who presented it to the Paris Academy in 1770, along with a similar analysis of the quartic and some additional work on higher–degree polynomial equations. However, Vandermonde appears not to have pursued this work beyond this point.

Perhaps we can find one reason in the fact that Lagrange's major (over 200 page) treatise *Réflexions sur la Résolution Algébrique des Equations*, which included similar but independent work in more depth on this subject, was published a few months after Vandermonde's presentation, while Vandermonde had to wait until 1774 to see his work published by the Paris Academy!

5.6 Lagrange

In his *Réflexions,* Lagrange gives a thorough treatment of the quadratic, cubic and quartic equations. His approach is essentially the same as Vandermonde's, but with a somewhat different perspective. He also addresses some issues that Vandermonde did not.

The Cubic Equation

Lagrange also considers the expression

$$t(x_1, x_2, x_3) = x_1 + \omega x_2 + \omega^2 x_3$$

but looks directly at all six quantities obtained from this expression by substituting the roots r_1, r_2 and r_3:

$$\begin{aligned}
t_1 &= r_1 + \omega r_2 + \omega^2 r_3 \\
t_2 &= r_1 + \omega r_3 + \omega^2 r_2 \\
t_3 &= r_2 + \omega r_1 + \omega^2 r_3 = \omega t_2 \\
t_4 &= r_2 + \omega r_3 + \omega^2 r_1 = \omega^2 t_1 \\
t_5 &= r_3 + \omega r_1 + \omega^2 r_2 = \omega t_1 \\
t_6 &= r_3 + \omega r_2 + \omega^2 r_1 = \omega^2 t_2
\end{aligned}$$

The roots of $p(x)$ are given in terms of the t_i's and other *known* quantities by

$$\begin{aligned}
r_1 &= \frac{1}{3}[(r_1 + r_2 + r_3) + t_1 + t_2] \\
r_2 &= \frac{1}{3}[(r_1 + r_2 + r_3) + t_3 + t_4] \\
r_3 &= \frac{1}{3}[(r_1 + r_2 + r_3) + t_5 + t_6]
\end{aligned}$$

Note that, in the notation of the previous section, $t_1^3 = u$ and $t_2^3 = v$.

Now, permuting the roots in any of the t_i's results in another t_i and so the coefficients of the 6th degree polynomial

$$f(x) = (x - t_1)\cdots(x - t_6)$$

are symmetric in the t_i's and therefore also symmetric in the roots r_1, r_2 and r_3, and are therefore known quantities. Lagrange called the equation $f(x) = 0$ the **resolvent equation** and the solutions t_i to this equation **resolvents**.

Lagrange observed that although the resolvent equation is of degree 6, it is also a *quadratic* equation in x^3, due to the relationships among the t_i's. In particular, $f(x)$ can be expressed in terms of t_1 and t_2 only:

$$\begin{aligned} f(x) &= (x - t_1)(x - t_2)(x - \omega t_2)(x - \omega^2 t_1)(x - \omega t_1)(x - \omega^2 t_2) \\ &= (x - t_1)(x - \omega t_1)(x - \omega^2 t_1)(x - t_2)(x - \omega t_2)(x - \omega^2 t_2) \\ &= (x^3 - t_1^3)(x^3 - t_2^3) \\ &= x^6 - (t_1^3 + t_2^3)x^3 + t_1^3 t_2^3 \end{aligned}$$

Thus, the resolvent equation is easily solved for the six resolvents t_i, using the quadratic formula, followed by the taking of cube roots—the same operations required by Vandermonde's approach. It is then a matter of determining which roots correspond to t_1 and t_2.

Lagrange addresses (or avoids) the latter issue by observing that if t is any resolvent, we can assume, by renaming the roots r_1, r_2 and r_3, that $t = t_1$. Then since it is easily checked that the product $t' = t_1 t_2$ is symmetric in r_1, r_2 and r_3 and therefore known, the three roots of $p(x)$ are given by

$$\begin{aligned} r_i &= \frac{1}{3}[(r_1 + r_2 + r_3) + t + t'/t] \\ r_j &= \frac{1}{3}\left[(r_1 + r_2 + r_3) + \omega t + \omega^2 t'/t\right] \\ r_k &= \frac{1}{3}\left[(r_1 + r_2 + r_3) + \omega^2 t + \omega t'/t\right] \end{aligned}$$

Thus, the solutions to the cubic are expressed in terms of any resolvent.

The important points to note here are that

1) Each resolvent t is an expression (polynomial) in the roots of $p(x)$ and other *known* quantities.
2) Conversely, the roots of $p(x)$ can be expressed in terms of a single resolvent and other *known* quantities.
3) Each resolvent can be determined *in a tractable way*, in this case by solving a quadratic equation and taking cube roots.

The Quartic Equation

Lagrange and Vandermonde each employed their similar lines of analysis with success for quartic equations. For a quartic $p(x)$, the resolvent expression is

$$\begin{aligned} t_\omega(x_1, x_2, x_3, x_4) &= x_1 + \omega x_2 + \omega^2 x_3 + \omega^3 x_4 \\ &= x_1 + i x_2 - x_3 - i x_4 \end{aligned}$$

where the x_i's represent the roots of $p(x)$ and where $\omega = i$ is a primitive 4th root of unity. It follows that there are $4! = 24$ distinct resolvents, satisfying a resolvent equation of degree 24. By analogy with the cubic case, one root of the quartic $p(x)$ is given by

$$\begin{aligned} r_1 = \frac{1}{4}[(r_1 + r_2 + r_3 + r_4) &+ t_\omega(r_1, r_2, r_3, r_4) \\ &+ t_\omega(r_1, r_3, r_4, r_2) + t_\omega(r_1, r_4, r_2, r_3)] \end{aligned}$$

since r_2, r_3 and r_4 each appear in all three of the last positions in $t_\omega(x_1, x_2, x_3, x_4)$ and so have coefficient $1 + \omega + \omega^2 + \omega^3 = 0$.

It is possible to proceed in a manner analogous to the cubic case, but Lagrange and Vandermonde both observed that a simplification is possible for the quartic. In particular, unlike the case of the cubic (and the quintic), where the degrees are prime, in the case of a quartic, there is a *nonprimitive* 4th root of unity other than 1, namely, $\beta = -1$.

The resolvent expression with respect to β,

$$t_\beta(x_1, x_2, x_3, x_4) = x_1 - x_2 + x_3 - x_4$$

has only $24/2!2! = 6$ distinct resolvents, which have the form $\pm t_1, \pm t_2$ and $\pm t_3$. Moreover, the roots of $p(x)$ are given by

$$\begin{aligned} r_1 &= \frac{1}{4}[(r_1 + r_2 + r_3 + r_4) + t_1 + t_2 + t_3] \\ r_2 &= \frac{1}{4}[(r_1 + r_2 + r_3 + r_4) - t_1 + t_2 - t_3] \\ r_3 &= \frac{1}{4}[(r_1 + r_2 + r_3 + r_4) + t_1 - t_2 - t_3] \\ r_4 &= \frac{1}{4}[(r_1 + r_2 + r_3 + r_4) - t_1 - t_2 + t_3] \end{aligned}$$

Since the resolvent polynomial in this case is

$$\begin{aligned} f(x) &= (x - t_1)^4 (x + t_1)^4 (x - t_2)^4 (x + t_2)^4 (x - t_3)^4 (x + t_3)^4 \\ &= \left[(x^2 - t_1^2)(x^2 - t_2^2)(x^2 - t_3^2)\right]^4 \\ &= [g(x^2)]^4 \end{aligned}$$

the resolvent equation, whose coefficients are known, can be solved by solving a known *cubic* equation $g(x) = 0$. This gives solutions t_1^2, t_2^2 and t_3^2, leaving only an ambiguity of sign in determining the resolvents t_1, t_2 and t_3. Lagrange addressed the issue of how to choose the correct sign, but Vandermonde simply left the issue to one of trial and error.

The Quintic Equation

The case of the 5th degree equation stymied both mathematicians, and for good reason. The Lagrange resolvent equation has degree 120 and is a 24th degree equation in x^5. It seems that both mathematicians doubted that their lines of analysis would continue to be fruitful. The somewhat ad hoc trick used for the quartic will not work for the quintic, and it is clear that the Lagrange–Vandermonde resolvent approach is simply running out of steam.

This is essentially where Lagrange (and Vandermonde) left the situation in his *Réflexions*.

5.7 Gauss

We need to say a word about roots of unity with respect to solvability by radicals. It is an obvious fact that since we allow the taking of roots in constructing a tower

$$F < R_1 < R_2 < \cdots < R_n$$

that shows that $p(x) = 0$ is solvable by radicals, then every equation of the form $x^n - 1 = 0$ is solvable by radicals, that is, the nth roots of unity are obtainable by taking—well—roots. This is not a very useful statement.

Note, however, that if α is an nth root of unity, then α is a root of the polynomial

$$x^{n-1} + x^{n-2} + \cdots + x + 1 = 0$$

which has degree $n - 1$. It would be much more interesting (and useful) to know that α could be obtained by adjoining roots whose degree is at most $n - 1$, that is, various kth roots, where $k \leq n - 1$.

This was Gauss's contribution, published in 1801 in his *Disquisitiones Arithmeticae*, when he was only 24 years old. We should mention that while Gauss is considered by many to be perhaps the greatest mathematician of all time, in this particular case, the ideas that Gauss used appear not to have originated with him. Moreover, Gauss seems to leave a gap in his proof, so one could argue that this was not really a completely Gaussian affair. Let us briefly outline Gauss's approach, which uses Lagrange resolvents.

First, it is not hard to show that if $n = ab$, where a and b are relatively prime, then every primitive nth root of unity is the product of a primitive ath root of unity and a primitive bth root of unity. In symbols,

$$\Omega_n = \Omega_a \Omega_b$$

Moreover, since $\Omega_n^{n/d} = \Omega_d$ (proof postponed until a later chapter), where $d \mid n$, we have

$$\Omega_{p^k} = \Omega_p^{1/p^{k-1}}$$

and so we need to prove the result only for pth roots of unity, where p is a prime.

A primitive pth root of unity α is a solution to the polynomial equation

$$g(x) = x^{p-1} + x^{p-2} + \cdots + x + 1 = 0$$

whose solutions are

$$R = \{\alpha, \alpha^2, \ldots, \alpha^{p-1}\}$$

These are all primitive pth roots of unity, since p is prime. Note that the exponents of α constitute the cyclic group Z_p^* of nonzero elements of the field $\mathbb{Z}_p$. Any generator a of this group is called a **primitive root modulo p**. For any such a, we have

$$R = \{\alpha, \alpha^a, \alpha^{a^2} \ldots, \alpha^{a^{p-1}}\} \tag{5.7.1}$$

Now, since the equation $g(x) = 0$ has degree $p - 1$, a Lagrange resolvent for this equation requires a primitive $(p - 1)$st root of unity β, and the resolvent expression is

$$t = x_1 + \beta x_2 + \beta^2 x_3 + \cdots + \beta^{p-2} x_{p-1}$$

where, as usual, a resolvent is obtained by substituting the roots of $g(x)$ for the x_i's.

The key idea (which may have been in part Vandermonde's) is to choose a resolvent in a specific way. In particular, the roots are chosen in the order given by a primitive root modulo p, as shown in (5.7.1). Hence, the resolvent is

$$t(\alpha, \beta) = \alpha + \beta\alpha^a + \beta^2\alpha^{a^2} + \cdots + \beta^{p-2}\alpha^{a^{p-2}}$$

Note that for any $k \geq 1$,

$$y_k = \beta^k + (\beta^k)^2 + \cdots + (\beta^k)^{p-1} = -1 + \frac{1 - \beta^{kp}}{1 - \beta^k} = 0$$

Accordingly, if we take the sum

$$z = t(\alpha, \beta) + t(\alpha, \beta^2) + \cdots + t(\alpha, \beta^{p-1})$$

the coefficient of α^{a^k} will be $y_k = 0$, for all $k \geq 1$. Also, the coefficient of α is

$p-1$ and so $z=(p-1)\alpha$, that is,

$$\alpha = \frac{1}{p-1}\sum_{k=1}^{p-1} t(\alpha,\beta^k) = \frac{1}{p-1}\sum_{k=1}^{p-1} \sqrt[p-1]{[t(\alpha,\beta^k)]^{p-1}}$$

Thus, if it can be shown that the expressions

$$s(\alpha,\beta^k) = [t(\alpha,\beta^k)]^{p-1}$$

under the radical signs are known, then α will be known, at least up to determining the correct $(p-1)$st roots. This is where the order of the roots in the resolvent $t(\alpha,\beta)$ is important. (Actually, the issue of which roots to take can be mitigated considerably, but we will not go into the details here.)

The "hard part" is thus to show that the expressions $s(\alpha,\beta^k)$ are known. Since we can assume that β is known (being a smaller primitive root of unity), it suffices to show that $s(\alpha,\beta^k)$ does not depend on α. This is done using a result whose origin is somewhat obscure. Gauss apparently used the result without proof at one point and then later gave an incomplete proof. In any case, it is not entirely clear whether Gauss possessed a complete proof of this result, which can be stated as follows.

Theorem 5.7.1 *Let α be a primitive pth root of unity and let β be a primitive $(p-1)$st root of unity. Then the powers*

$$\alpha, \alpha^2, \ldots, \alpha^{p-1}$$

are linearly independent over $\mathbb{Q}(\beta)$.

Proof. We need the following additional facts about roots of unity, whose proofs will be given in a later chapter.

1) If ω is a primitive nth root of unity, then $[\mathbb{Q}(\omega):\mathbb{Q}] = \phi(n)$, where ϕ is the Euler phi function, that is, $\phi(n)$ is the number of positive integers less than n and relatively prime to n.
2) If a and b are relatively prime, then $\phi(ab) = \phi(a)\phi(b)$.
3) If p is a prime, then $\phi(p) = p-1$.

Consider the tower

$$\mathbb{Q} < \mathbb{Q}(\beta) < \mathbb{Q}(\alpha,\beta)$$

The lower step has degree $\phi(p-1)$ and the upper step, being a lifting of $\mathbb{Q} < \mathbb{Q}(\alpha)$, has degree $d \le \phi(p) = p-1$. Consider also the tower

$$\mathbb{Q} < \mathbb{Q}(\alpha\beta) < \mathbb{Q}(\alpha,\beta)$$

The lower step has degree $\phi(p(p-1)) = (p-1)\phi(p-1)$ and if the upper step has degree e, then

$$e(p-1)\phi(p-1) = [\mathbb{Q}(\alpha,\beta) : \mathbb{Q}] = d\phi(p-1)$$

Hence, $e(p-1) = d \le p-1$, which implies that $e = 1$ and $d = p-1$, that is,

$$[\mathbb{Q}(\beta)(\alpha) : \mathbb{Q}(\beta)] = p-1$$

Hence, the set

$$\{1 = \alpha^{p-1}, \alpha, \ldots, \alpha^{p-2}\}$$

is a basis for $\mathbb{Q}(\beta)(\alpha)$ over $\mathbb{Q}(\beta)$.□

Now let us look at how this result can be used to show that

$$s(\alpha, \beta^k) = [t(\alpha, \beta^k)]^{p-1}$$

does not depend on α. If we replace α by α^a (recall that a is a primitive root modulo p), we have

$$t(\alpha^a, \beta^k) = \sum_{i=0}^{p-2} \beta^{ki}\alpha^{a^{i+1}} = \sum_{i=1}^{p-1} \beta^{k(i-1)}\alpha^{a^i} = \beta^{-k}\sum_{i=1}^{p-1} \beta^{ki}\alpha^{a^i} = \beta^{-k}t(\alpha, \beta^k)$$

It follows that

$$s(\alpha^a, \beta^k) = [t(\alpha^a, \beta^k)]^{p-1} = [\beta^{-k}t(\alpha, \beta^k)]^{p-1} = [t(\alpha, \beta^k)]^{p-1} = s(\alpha, \beta^k)$$

In other words, $s(\alpha, \beta^k)$ is invariant under the replacement $\alpha \mapsto \alpha^a$.

Now, $s(\alpha, \beta^k)$ is a polynomial in α and β. Collecting powers of α (which are linearly independent by Theorem 5.7.1) gives

$$s(\alpha, \beta^k) = q_0(\beta) + q_1(\beta)\alpha + q_2(\beta)\alpha^a + \cdots + q_{p-1}(\beta)\alpha^{a^{p-1}}$$

Then the invariance under $\alpha \mapsto \alpha^a$ implies that

$$\begin{aligned} q_0(\beta) &+ q_1(\beta)\alpha + q_2(\beta)\alpha^a + \cdots + q_{p-1}(\beta)\alpha^{a^{p-1}} \\ &= q_0(\beta) + q_1(\beta)\alpha^a + q_2(\beta)\alpha^{a^2} + \cdots + q_{p-1}(\beta)\alpha^{a^p} \\ &= q_0(\beta) + q_{p-1}(\beta)\alpha + q_1(\beta)\alpha^a + q_2(\beta)\alpha^{a^2} + \cdots + q_{p-2}(\beta)\alpha^{a^{p-1}} \end{aligned}$$

Equating coefficients of the linearly independent powers of α gives

$$\begin{aligned} q_1(\beta) &= q_{p-1}(\beta) \\ q_2(\beta) &= q_1(\beta) \\ q_3(\beta) &= q_2(\beta) \\ &\vdots \\ q_{p-1}(\beta) &= q_{p-2}(\beta) \end{aligned}$$

and so the polynomial expressions $q_i(\beta)$, for $i > 0$, are equal. Hence,

$$s(\alpha, \beta^k) = q_0(\beta) + q_1(\beta)[\alpha + \alpha^a + \cdots + \alpha^{a^{p-1}}] = q_0(\beta) - q_1(\beta)$$

which is independent of α, as desired.

Thus, we have shown that a primitive pth root of unity α can be expressed in terms of a primitive $(p-1)$st root of unity β, using only root of degree at most $p-1$. An induction completes the proof that any nth root of unity can be expressed by taking roots of degree at most $n-1$.

As a very simple illustration, let us compute a primitive cube root of unity α. We begin with a primitive square root of unity $\beta = -1$ and form the expressions

$$t(\alpha, \beta) = \alpha - \alpha^2$$
$$t(\alpha, \beta^2) = \alpha + \alpha^2$$

Then since $1 + \alpha + \alpha^2 = 0$, we have

$$t(\alpha, \beta)^2 = (\alpha - \alpha^2)^2 = \alpha^2 - 2\alpha^3 + \alpha^4 = -3$$
$$t(\alpha, \beta^2)^2 = (\alpha + \alpha^2)^2 = \alpha^2 + 2\alpha^3 + \alpha^4 = 1$$

Thus,

$$\alpha = \frac{1}{2}\left[\sqrt{t(\alpha, \beta)^2} + \sqrt{t(\alpha, \beta^2)^2}\right] = \frac{1}{2}\left[\pm\sqrt{-3} \pm 1\right]$$

and we need only choose the correct combination of signs.

5.8 Back to Lagrange

As we have remarked, Lagrange's (and Vandermonde's) resolvent has three properties:

1) Each resolvent t is a polynomial in the roots of $p(x)$ and other *known* quantities, including perhaps the nth roots of unity.
2) Conversely, the roots of $p(x)$ can be expressed in terms of a single resolvent and other *known* quantities.
3) Each resolvent can be determined in a tractable way.

Lagrange doubted that it would be possible to find a resolvent that could be determined in a tractable way for the quintic, let alone for higher–degree polynomials. On the other hand, he did spend considerable effort considering "resolvents" that satisfy only 1) and 2). In fact, the following theorem of Lagrange, and its corollary, is a cornerstone of Galois theory. The version we present here appears in Edwards, and is from Lagrange's *Réflexions*, Article 104.

Theorem 5.8.1 *If t and y are any two functions [polynomials] in the roots $x', x'', x''', \ldots$ of $x^\mu + mx^{\mu-1} + nx^{\mu-2} + [px^{\mu-3} +]\cdots$ and if these functions* *are such that every permutation of the roots $x', x'', x''', \ldots$ which changes y also changes t, one can, generally speaking, express y rationally in terms of t and $m, n, p, \ldots,$ so that when one knows a value of t one will also know immediately the corresponding value of y; we say* generally speaking *because if the known value of t is a double or triple or higher root of the equation for t then the corresponding value of y will depend on an equation of degree 2 or 3 or higher with coefficients that are rational in t and $m, n, p, \ldots$.*

If we think of t as a known polynomial of the roots, then this theorem states that under the conditions of the theorem, the value of y, which could simply be a root of $p(x)$, is expressible as a known function of t. Lagrange's theorem has the following corollary (in slightly more modern notation).

Corollary 5.8.2 *Suppose that $p(x)$ has distinct roots, say, $\alpha_1, \ldots, \alpha_n$. If there exists a polynomial $t(x_1, x_2, \ldots, x_n)$ with the property that the $n!$ values*

$$t(\alpha_{\sigma 1}, \alpha_{\sigma 2}, \ldots, \alpha_{\sigma n})$$

are distinct, that is, if $t(\alpha_1, \ldots, \alpha_n)$ is changed by every permutation of the roots, then any polynomial $y(\alpha_1, \ldots, \alpha_n)$ in the roots, including the roots themselves, is a known rational expression in $t(\alpha_1, \ldots, \alpha_n)$. $\square$

We will be able to rephrase this in more modern terms in a later chapter. For the curious, it is as follows: If $p(x)$ is separable over F, with splitting field K and Galois group G and if $t \in K$ has the property that $\sigma t \neq t$ for all $\sigma \in G$, then $G_{F(t)}(K) = \{\iota\}$ and so taking fixed fields gives $F(t) = K$, that is, every polynomial in the roots of $p(x)$ is a polynomial in t.

A polynomial t as described in the previous corollary is a "resolvent" in the sense that it satisfies the first two conditions of a Lagrange resolvent: t is a known function of the (unknown) roots and the roots are a known function of t. Any t with these properties is called a **Galois resolvent**, because Galois was the first to recognize that such a resolvent always exists (provided that $p(x)$ has no multiple roots). He was also the first to realize the importance of such resolvents.

We can describe Galois resolvents in more modern terms as follows. Let $E = F(\alpha_1, \ldots, \alpha_n)$ be a splitting field for $p(x)$ over F. We may assume that F is the field of "known" quantities. Then $t \in E$ is a Galois resolvent if and only if $F(\alpha_1, \ldots, \alpha_n) = F(t)$, that is, if and only if t is a primitive element of E.

Now we see that the existence of Galois resolvents follows from the Theorem of the Primitive Element. Assuming that $p(x)$ has no multiple roots—an assumption that Galois also made—the fact that $F < E$ is finite and separable implies that it is simple.

5.9 Galois

It is not hard to place the work of Evariste Galois in time, since he was born in 1811 and died only 21 years later, of a gunshot wound, in 1832. However, it is much harder to describe the importance of his work, which sparked the foundations of modern algebra. (Of course, Cauchy, Cayley, Lagrange, Vandermonde, Newton, Gauss and others had a hand in the foundations of algebra as well.)

Galois realized that while a (Galois) resolvent might not be able to provide the actual values of the roots of a polynomial, it does lead the way to a beautiful theory, now called Galois theory that, among other things, shows that there are no *Lagrange* resolvents for polynomials of degree 5 or greater.

In his 1831 *Memoir on the Conditions for Solvability of Equations by Radicals*, Galois states a result akin to the corollary of Lagrange given above, without mention of either Lagrange or his theorem (although he had read Lagrange as a student). Moreover, Galois' proof is, to say the least, sketchy. In fact, when Poisson read Galois' memoir, as submitted for publication to the Paris Academy of Sciences, Poisson remarked

> "We have made every effort to understand Mr. Galois' proof. His arguments are not clear enough, nor developed enough, for us to be able to judge their correctness. . . ."

Galois' paper was rejected for publication.

In his memoir of 1831, Galois proved the following result (Proposition VIII):

> "For an equation of prime degree, which has no commensurable divisors, to be solvable by radicals, it is necessary and sufficient that all roots be rational functions of any two of them."

In more modern language, this theorem says that if $f(x)$ is irreducible and separable of prime degree p, then the equation $f(x) = 0$ is solvable by radicals if and only if $F[\alpha, \beta]$ is a splitting field for $f(x)$, for *any* two roots α and β of $f(x)$. Since, for example, any quintic polynomial with exactly two nonreal roots fails to meet this condition, it cannot be solvable by radicals. This theorem is covered in detail in the chapter on solvable extensions.

Galois and Groups

Galois' great achievement was not the actual result that polynomial equations of degree 5 and higher have no general algebraic solution. Indeed, even the formulas for cubic and quartic equations are not of much practical use. Galois' great achievement lies in the *path* he took to prove this result, in particular, his discovery and application of the notion of a "Galois-style" group, described below.

While on the subject of groups, it cannot be said that Galois discovered in its entirety the modern notion of a group. As we will see, Galois dealt only with sets of permutations and stated only that these sets must be closed under composition (although not in these words). The other properties of the definition of a modern group: associativity, identity and inverses, were not mentioned explicitly by Galois. (Perhaps he thought them too obvious for explicit mention.)

When Galois' work was finally published in 1846, the theory of finite permutation groups had already been formalized by Cauchy, who likewise required only closure under product, but who clearly recognized the importance of the other axioms by introducing notations for the identity and for inverses.

Cayley (1854) was the first to consider the possibility of more abstract groups, and the need to axiomatize associativity. He also axiomatized the identity property, but still assumed that each group was a finite set, and so had no need to axiomatize inverses (only the validity of cancellation). It was not until 1883 that Dyck, in studying the relationship between groups and geometry, made explicit mention of inverses.

It is also interesting to note that Cayley's famous theorem of group theory, to the effect that every group is isomorphic to a permutation group, completes a full circle back to Galois (at least for finite groups)!

Galois-Style Groups

Galois' version of a group is as follows (although the terminology is not necessarily that of Galois). Consider a table in which each row contains an ordered arrangement of a set S of distinct symbols (such as the roots of a polynomial), for example

$$\begin{array}{ccccc} a & b & c & d & e \\ c & a & b & d & e \\ b & c & a & d & e \\ a & b & c & e & d \\ c & a & b & e & d \\ b & c & a & e & d \end{array}$$

Then each pair of rows defines a permutation of S, that is, a bijective function on S. Galois considered tables of ordered arrangements with the property that the set A_i of permutations that transform any given row r_i into the other rows (or into itself) is the same for all rows r_i, that is, $A_i = A_j$ for all i, j. Let us refer to this type of table, or list of ordered arrangements, as a **Galois-style group**.

It is not hard to show that a list of arrangements is a Galois–style group if and only if the corresponding set A ($= A_i$) of permutations is a subgroup of the group of all permutations of the set S, that is, if and only if A is a *permutation group*, in the modern sense.

To see this, let the permutation that transforms row r_i to row r_j be $\pi_{i,j}$. Then Galois' assumption is that the sets $A_i = \{\pi_{i,1}, \ldots, \pi_{i,n}\}$ are the same for all i. This implies that for each i, u and j, there is a v for which $\pi_{i,u} = \pi_{j,v}$. Hence,

$$\pi_{i,u}\pi_{i,j} = \pi_{j,v}\pi_{i,j} = \pi_{i,v} \in A_i$$

and so A_i is closed under composition. It is also closed under inverses, since for any $\pi_{i,j} \in A_i$, it is true that $\pi_{i,j}^{-1} = \pi_{j,i} \in A_i$. Finally, the identity is in A_i, since it is the substitution associated to the pair of rows (r_1, r_1).

Conversely, if A_1 is a permutation group, then since

$$\pi_{i,j} = \pi_{1,j}\pi_{i,1} = \pi_{1,j}(\pi_{1,i})^{-1} \in A_1$$

it follows that $A_i = A_1$ for all i.

Galois appears not to be entirely clear about a precise meaning of the term group, but for the most part, he uses the term for what we are calling a Galois–style group. Galois also worked with subgroups and recognized the importance of what we now call normal subgroups, although his "definition" is quite different from what we would see today.

The Galois Group

For a modern mathematician, the Galois group of a polynomial $p(x)$ over a field F is defined in terms of a splitting field. Galois and his predecessors talked about the "roots" of a polynomial without regard to considerations of their existence (much as our students do today) and it was not until Kronecker came upon the scene, several decades later, that the issue of existence was explicitly addressed.

In any case, the modern definition of the Galois group of a polynomial $p(x)$ over F is the group $G_F(E)$ of all automorphisms of a splitting field E of $p(x)$ over F that fix F pointwise, in symbols

$$G_F(E) = \mathrm{Aut}_F(E)$$

Galois would have defined the Galois-style group of a polynomial $p(x)$, with distinct roots, essentially as follows (but in different terms). Let $E = F(t)$ be a splitting field for $p(x)$. Let $p_t(x)$ be the minimal polynomial of t over F and let

$$R = \{t = t_1, \ldots, t_d\}$$

be the conjugates of t, that is, the roots of $p_t(x)$. Note that since $p(x)$ is assumed to have only simple roots, the extension $F < E$ is separable and so $p_t(x)$ is separable, that is, $p_t(x)$ has distinct roots. Also, since $F < E$ is normal, $p_t(x)$ splits over E and so $[E : F] = d$.

Let $\alpha_1, \ldots, \alpha_n$ be the roots of $p(x)$. Each root α_i is a polynomial $f_i(t)$ in the primitive element t (that is, the Galois resolvent). Consider the list of arrangements

$$\begin{array}{cccc} f_1(t_1) & f_2(t_1) & \cdots & f_n(t_1) \\ f_1(t_2) & f_2(t_2) & \cdots & f_n(t_2) \\ & \vdots & & \\ f_1(t_d) & f_2(t_d) & \cdots & f_n(t_d) \end{array} \tag{5.9.1}$$

the first row of which is just the set of roots of $p(x)$. We claim (as did Galois, in a different way) that this is a Galois-style group.

To see this, we make the following observations:

1) Since $F < E$ is normal,
$$G_F(E) = \hom_F(E, \overline{F})$$
where $F < E < \overline{F}$.
2) According to Theorem 2.8.3, for each i, there is a $\sigma_i \in \hom_F(E, \overline{F}) = G_F(E)$ that maps t to t_i. Furthermore, each element of $G_F(E)$ is uniquely determined by its value on t. Hence,
$$|G_F(E)| = [E : F] = d$$

Thus, letting $G_F(E) = \{\sigma_1 = \iota, \sigma_2, \ldots, \sigma_d\}$, we can rewrite the previous list of arrangements as

$$\begin{array}{cccc} f_1(t) & f_2(t) & \cdots & f_n(t) \\ f_1(\sigma_2 t) & f_2(\sigma_2 t) & \cdots & f_n(\sigma_2 t) \\ & \vdots & & \\ f_1(\sigma_d t) & f_2(\sigma_d t) & \cdots & f_n(\sigma_d t) \end{array}$$

or

$$\begin{array}{cccc} f_1(t) & f_2(t) & \cdots & f_n(t) \\ \sigma_2 f_1(t) & \sigma_2 f_2(t) & \cdots & \sigma_2 f_n(t) \\ & \vdots & & \\ \sigma_d f_1(t) & \sigma_d f_2(t) & \cdots & \sigma_d f_n(t) \end{array}$$

Therefore, in the notation of Galois-style groups used earlier, $A_1 = G_F(E)$ and so this list does indeed represent a Galois-style group.

Of course, Galois did not prove that his list (5.9.1) is a Galois-style group in the same way we have done. His first task is to show that each row of (5.9.1) is a permutation of the first row.

The first step is to show that all of the elements of the table are roots of $p(x)$, that is, that $p(f_i(t_j)) = 0$ for all i and j. For this, Galois considers the polynomials $p(f_i(x))$. Since

$$p(f_i(t_1)) = p(\alpha_i) = 0$$

it follows that the polynomial $p(f_i(x))$ and the *irreducible* polynomial $p_t(x)$ have a common root t_1. Galois knew that this implies that $p_t(x) \mid p(f_i(x))$. Hence, every root of $p_t(x)$ is a root of $p(f_i(x))$, that is, $p(f_i(t_j)) = 0$ for all i and j, as desired.

Then Galois reasoned that if two elements $f_i(t_k)$ and $f_j(t_k)$ of the same row, where $i \neq j$, are equal, then the polynomial $f_i(x) - f_j(x)$ has root t_k and so, as above,

$$p_t(x) \mid f_i(x) - f_j(x)$$

which implies that all conjugates t_u are roots of $f_i(x) - f_j(x)$. In particular, $f_i(t_1) = f_j(t_1)$. But these are roots from the first row of (5.9.1), which are distinct and so $i = j$, a contradiction.

For more details on Galois' approach to these issues, we refer to the reader to Edwards.

Solvability by Radicals

So let us recap: Galois developed the notion of a Galois resolvent, that is, a primitive element of a splitting field of $p(x)$ and showed that Galois resolvents always exist. He then used this notion to develop the concept of the Galois-style Galois group of $p(x)$. The stage is now set for his most famous result, namely, that the roots of a 5th or higher degree polynomial equation are not always solvable by radicals. Galois' approach was to consider the conditions imposed on the Galois group of a polynomial by the requirement that the polynomial equation be solvable by radicals. Here is a brief sketch.

Note that since the roots of unity can be considered as known quantities (obtainable by the taking of roots), once a single root α of a quantity is known, all other roots of that quantity, being of the form $\omega\alpha$ where ω is a root of unity, are also known.

Since if $n = ab$, then

$$\sqrt[n]{r} = \sqrt[a]{\sqrt[b]{r}}$$

it follows that an extension $F < F(\alpha)$ obtained by adjoining a single nth root α can be decomposed into a tower in which each step is obtained by adjoining a *prime* root of an element. Hence, a polynomial equation $p(x) = 0$ is solvable by radicals if and only if a splitting field E for $p(x)$ over F can be "captured" within a finite tower of fields

$$F < F(\alpha_1) < F(\alpha_1, \alpha_2) < \cdots < F(\alpha_1, \ldots, \alpha_m) \tag{5.9.2}$$

where each α_i is a p_ith root (p_i a prime) of some element in the previous field of the tower. Moreover, we may assume that the required roots of unity appear as necessary, in particular, we may assume that if $K < K(\beta^{1/p})$ is a step in the tower (5.9.2), then K contains the pth roots of unity.

Now let us examine, as Galois did, the Galois groups $G_{F(\alpha_1,\ldots,\alpha_k)}(E)$. It is clear from the definition that they form a nonincreasing sequence

$$G_F(E) > G_{F(\alpha_1)}(E) > G_{F(\alpha_1,\alpha_2)}(E) > \cdots > G_{F(\alpha_1,\ldots,\alpha_m)}(E) \tag{5.9.3}$$

Moreover, if $E < F(\alpha_1, \ldots, \alpha_m)$ then, since the taking of Galois groups reverses inclusion, we have

$$G_{F(\alpha_1,\ldots,\alpha_m)}(E) < G_E(E) = \{\iota\}$$

that is, $G_{F(\alpha_1,\ldots,\alpha_m)}(E) = \{\iota\}$.

Galois studied the properties of the sequence (5.9.3). In particular, he showed that each group in (5.9.3) is a *normal* subgroup of its predecessor, and has prime index in its predecessor. A sequence of subgroups in which each group is normal in its immediate parent is called a **normal series**, and if the indices are prime, then the top group, which in Galois' case is $G_F(E)$, is called **solvable**.

Galois proved that if $p(x) = 0$ is solvable by radicals, then its Galois group $G_F(E)$ is solvable. He also proved the converse.

Galois used his remarkable theory in his *Memoir on the Conditions for Solvability of Equations by Radicals* of 1831 (but not published until 1846), to show that the general equation of degree 5 or larger is not solvable by radicals. It is worth noting that Ruffini, in 1799, offered the first "proof" that the 5th degree equation is not solvable by radicals. However, his proof was not completely convincing and a complete proof was given by Abel in 1826. Nevertheless, Galois' achievement is not diminished by these facts.

5.10 A Very Brief Look at the Life of Galois

Evariste Galois life was, to say the least, very short and very controversial. Of course, it would not be the subject of such legend today were it not for his remarkable discoveries, which spanned only a few short years.

Galois was born on October 25, 1811, near Paris. Apparently, Galois was recognized at an early age as a brilliant student with some bizarre and rebellious tendencies.

In 1828, at the age of 17, Galois attempted to enter the prestigious École Polytechnique, but failed the entrance exams, so he remained at the royal school of Louis-le-Grand, where he studied advanced mathematics. His teacher urged Galois to publish his first paper, which appeared on April 1, 1829.

After this, things started to go very badly for Galois. An article that Galois sent to the Academy of Sciences was given to Cauchy, who lost it. (Apparently, Cauchy had a tendency to lose papers; he had already lost a paper by Abel.) On April 2, 1829, Galois' father committed suicide.

Galois once again tried to enter the École Polytechnique, but again failed under some rather controversial circumstances. So he entered the École Normale, considered to be on a much lower level than the École Polytechnique. While at the École Normale, Galois wrote up his research and entered it for the Grand Prize in Mathematics of the Academy of Sciences. The work was given to Fourier for consideration, who took it home, but promptly died, and the manuscript appears now to be lost.

Galois possessed very strong political opinions. On July 14, 1831, he was arrested during a political demonstration, and condemned to six months in prison. In May 1832, Galois had a brief love affair with a young woman. He broke off the affair on May 14, and this appears to be the cause of a subsequent duel that proved fatal to Galois. Galois died on May 31, 1832.

On September 4, 1843, Liouville announced to the Academy of Sciences that he had discovered, in the papers of Galois, the theorem, from his 1831 Memoir, that we mentioned earlier concerning the solvability by radicals of a prime–degree equation, and referred to it with the words “as precise as it is deep.” However, he waited until 1846 to publish Galois' work.

In the 1850s, the complete texts of Galois' work became available to mathematicians, and it initiated a great deal of subsequent work by the likes of Betti, Kronecker, Dedekind, Cayley, Hermite, Jordan and others.

Now it is time that we left the past, and pursued Galois' theory from a modern perspective.

Chapter 6

Galois Theory II: The Theory

6.1 Galois Connections

The traditional Galois correspondence between intermediate fields of an extension and subgroups of the Galois group is one of the main themes of this book. We choose to approach this theme through a more general concept, however.

Definition *Let P and Q be partially ordered sets. A* **Galois connection** *on the pair (P,Q) is a pair (Π,Ω) of maps $\Pi\colon P\to Q$ and $\Omega\colon Q\to P$, where we write $\Pi(p)=p^*$ and $\Omega(q)=q'$, with the following properties:*

1) **(Order-reversing** *or* **antitone)** *For all $p,q\in P$ and $r,s\in Q$,*

$$p\le q\Rightarrow p^*\ge q^* \text{ and } r\le s\Rightarrow r'\ge s'$$

2) **(Extensive)** *For all $p\in P$, $q\in Q$,*

$$p\le p^{*\prime} \text{ and } q\le q^{\prime *}$$

□

Closure Operations

Lurking within a Galois connection we find two *closure operations*.

Definition *Let P be a partially ordered set. A map $p\to \mathrm{cl}(p)$ on P is an (algebraic)* **closure operation** *if the following properties hold for all $p,q\in P$:*

1) **(Extensive)**

$$p\le \mathrm{cl}(p)$$

2) **(Idempotent)**

$$\mathrm{cl}(\mathrm{cl}(p))=\mathrm{cl}(p)$$

3) **(Isotone)**

$$p \le q \Rightarrow \mathrm{cl}(p) \le \mathrm{cl}(q)$$

An element $p \in P$ *is said to be* **closed** *if* $\mathrm{cl}(p) = p$. *The set of all closed elements in* P *is denoted by* $\mathrm{Cl}(P)$.□

Theorem 6.1.1 *Let* (Π, Ω) *be a Galois connection on* (P, Q). *Then the maps*

$$p \to p^{*\prime} \quad \text{and} \quad q \to q^{\prime *}$$

are closure operations on P and Q, respectively, and we write $p^{*\prime} = \mathrm{cl}(p)$ *and* $q^{\prime *} = \mathrm{cl}(q)$. *Moreover,*

1) $p^{*\prime *} = p^*$, *that is,*

$$\mathrm{cl}(p^*) = \mathrm{cl}(p)^* = p^*$$

2) $q^{\prime *\prime} = q^\prime$, *that is,*

$$\mathrm{cl}(q^\prime) = \mathrm{cl}(q)^\prime = q^\prime$$

Proof. Since $p \le p^{*\prime}$, the order-reversing property of $*$ gives

$$p^{*\prime *} \le p^* \le (p^*)^{\prime *}$$

and so $p^{*\prime *} = p^*$, from which part 1) follows. Part 2) is similar. □

Theorem 6.1.2 *The maps* $\Pi: P \to \mathrm{Cl}(Q)$ *and* $\Omega: Q \to \mathrm{Cl}(P)$ *are surjective and the restricted maps* $\Pi: \mathrm{Cl}(P) \to \mathrm{Cl}(Q)$ *and* $\Omega: \mathrm{Cl}(Q) \to \mathrm{Cl}(P)$ *are inverse bijections.*

Proof. Since $\mathrm{cl}(p^*) = p^*$, we see that p^* is closed, that is, Π maps P into $\mathrm{Cl}(Q)$. Moreover, Π is surjective since if $q \in \mathrm{Cl}(Q)$, then $q = \mathrm{cl}(q) = (q^\prime)^*$. To see that Π is injective when restricted to closed elements, if $p, r \in \mathrm{Cl}(P)$ and $p^* = r^*$, then $p^{*\prime} = r^{*\prime}$, that is, $p = r$. Similar arguments apply to Ω. Finally, since

$$\Omega \circ \Pi(\mathrm{cl}(p)) = \mathrm{cl}(\mathrm{cl}(p)) = \mathrm{cl}(p)$$

we see that $\Omega \circ \Pi = \iota$ on $\mathrm{Cl}(P)$ and similarly, $\Pi \circ \Omega = \iota$ on $\mathrm{Cl}(Q)$.□

Theorem 6.1.3 *Let* (Π, Ω) *be a Galois connection on a pair* (P, Q) *of lattices.*

1) *If P is a complete lattice, then so is* $\mathrm{Cl}(P)$, *under the same meet as P. A similar statement holds for Q.*

2) **De Morgan's Laws** *hold in* $\mathrm{Cl}(P)$ *and* $\mathrm{Cl}(Q)$, *that is, for* $p, q \in \mathrm{Cl}(P)$ *and* $r, s \in \mathrm{Cl}(Q)$,

$$(p \wedge q)^* = p^* \vee q^*, \quad (p \vee q)^* = p^* \wedge q^*$$

and

$$(r \wedge s)^\prime = r^\prime \vee s^\prime, \quad (r \vee s)^\prime = r^\prime \wedge s^\prime$$

Proof. For part 1), we apply Theorem 0.1.1 to the subset $\mathrm{Cl}(P)$ of P. First, since $1 \in P$ has the property that $1 \geq \mathrm{cl}(1) \geq 1$, it follows that $1 \in \mathrm{Cl}(P)$. Suppose that $p_i \in \mathrm{Cl}(P)$. Then the meet $\bigwedge p_i$ exists in P and since $\bigwedge p_i \leq p_j$ for all j, we have

$$\mathrm{cl}(\bigwedge p_i) \leq \mathrm{cl}(p_j) = p_j$$

whence $\mathrm{cl}(\bigwedge p_i) \leq \bigwedge p_j$. Since the reverse inequality holds as well, equality holds and $\bigwedge p_i \in \mathrm{Cl}(P)$. It follows from Theorem 0.1.1 that $\mathrm{Cl}(P)$ is a complete lattice under meet in P. A similar argument can be made for Q.

For part 2), observe first that $p \wedge q \leq p$ and $p \wedge q \leq q$ imply that $(p \wedge q)^* \geq p^*$ and $(p \wedge q)^* \geq q^*$, whence $(p \wedge q)^* \geq p^* \vee q^*$. If $r \geq p^*$ and $r \geq q^*$ for $r \in \mathrm{Cl}(P)$ then $r' \leq p$ and $r' \leq q$, whence $r' \leq p \wedge q$. Thus, $r \geq (p \wedge q)^*$. It follows by definition of join that $(p \wedge q)^* = p^* \vee q^*$. The other parts of De Morgan's laws are proved similarly. □

Examples of Galois Connections

Our interest in Galois connections is the famous Galois correspondence between intermediate fields of a field extension and subgroups of the Galois group of an extension (to be defined later). However, let us take a look at some other examples of Galois connections.

Example 6.1.1 Let X and Y be nonempty sets and $P = \mathcal{P}(X)$ and $Q = \mathcal{P}(Y)$ be the corresponding power sets. Let $R \subseteq X \times Y$ be a relation on $X \times Y$. Then the maps

$$S \in \mathcal{P}(X) \mapsto S^* = \{y \in Y \mid (x, y) \in R \text{ for all } x \in S\}$$

and

$$T \in \mathcal{P}(Y) \mapsto T' = \{x \in X \mid (x, y) \in R \text{ for all } y \in T\}$$

form a Galois connection on $(\mathcal{P}(X), \mathcal{P}(Y))$.□

Example 6.1.2 Let $n > 1$ and let F be a field. Let $P = \mathcal{P}(F[x_1, \ldots, x_n])$ be the set of all subsets of polynomials over F in the variables $x_1, \ldots, x_n$. Let $Q = \mathcal{P}(F^n)$ be the set of all subsets of F^n, the set of all ordered n-tuples over F.

Let Π: $F[x_1, \ldots, x_n] \to \mathcal{P}(F^n)$ be defined by

$$\begin{aligned}\Pi(S) &= \text{Set of all common roots of the polynomials in } S \\ &= \{x \in \mathcal{P}(F^n) \mid p(x) = 0 \text{ for all } p \in S\}\end{aligned}$$

and let $\Omega: \mathcal{P}(F^n) \to F[x_1, \ldots, x_n]$ be defined by

$$\begin{aligned}\Omega(T) &= \text{Set of all polynomials whose root set includes } T \\ &= \{p \in F[x_1, \ldots, x_n] \mid p(t) = 0 \text{ for all } t \in T\}\end{aligned}$$

We leave it as an exercise to show that (Π, Ω) is a Galois connection on $(F[x_1, \ldots, x_n], \mathcal{P}(F^n))$.□

Top and Bottom Elements

In many examples of Galois connections, P and Q have both top and bottom elements.

A top element is closed, since $1_P \leq \text{cl}(1_P) \leq 1_P$ and similarly for 1_Q. Note also that a top element is the image of the corresponding bottom element (if it exists), for $1_P = \Omega[\Pi(p)]$ is the image of $\Pi(p)$ and since $0_Q \leq \Pi(p)$, the image of 0_Q must be at least as large as 1_P, and therefore equal to 1_P.

However, a bottom element need not be closed. Indeed, the smallest closed element of Q is $\Pi(1_P)$ and so 0_Q is closed if and only if $0_Q = \Pi(1_P)$, for example. In other words, a bottom element is closed if and only if it is the image of the corresponding top element.

Indexed Galois Connections

Let $\mathbb{Z}^+$ denote the set of positive integers. In the set $\mathbb{Z}^+ \cup \{\infty\}$, we observe some obvious understandings about ∞, in particular, $n \leq \infty$ for all $n \in \mathbb{Z}^+$, $\infty \leq \infty, n \cdot \infty = \infty$ for $n \in \mathbb{Z}^+$ and $\infty \leq k \leq \infty$ implies $k = \infty$.

Definition *A Galois connection* (Π, Ω) *on* (P, Q) *is* **indexed** *if*

a) *For each $p, q \in P$ with $p \leq q$, there exists a number $(q : p)_P \in \mathbb{Z}^+ \cup \{\infty\}$, called the* **degree**, *or* **index** *of q over p.*

b) *For each $r, s \in Q$ with $r \leq s$, there exists a number $(s : r)_Q \in \mathbb{Z}^+ \cup \{\infty\}$, called the* **degree**, *or* **index** *of s over r.*

We generally write $(q : p)$ without a subscript to denote the appropriate index. Moreover, the following properties must hold:

1) **(Degree is multiplicative)** *If $s_1, s_2, s_3 \in P$ or $s_1, s_2, s_3 \in Q$ then*

$$s_1 \leq s_2 \leq s_3 \Rightarrow (s_3 : s_1) = (s_3 : s_2)(s_2 : s_1)$$

2) **(Π and Ω are degree-nonincreasing)** *If $p, q \in P$ then*

$$p \leq q \Rightarrow (p^* : q^*) \leq (q : p)$$

If $r, s \in Q$ then

$$r \leq s \Rightarrow (r' : s') \leq (s : r)$$

3) **(Equality by degree)** *If $s,t \in P$ or $s,t \in Q$ then*

$$(s:t)=1 \Leftrightarrow s=t$$

If $(s:t)<\infty$, then s is said to be a **finite extension** *of t. If P has a top and bottom element then the* **index** *of P is* $\operatorname{index}(P)=(1_P:0_P)$, *and similarly for Q.*□

From now on, when we write $(q:p)$, it is with the tacit assumption that $p \le q$.

The importance of indexing is described in the next theorem. It says that if a Galois connection is indexed, then the connection preserves the index of closed elements and that any finite extension of a closed element is also closed.

Theorem 6.1.4 Let (Π,Ω) be an indexed Galois connection on (P,Q).

1) **(Degree-preserving on closed elements)** *If $p,r \in \operatorname{Cl}(P)$ and $p \le r$ then $(r:p)=(p^*:r^*)$. A similar statement holds for Q.*

2) **(Finite extensions of closed elements are closed)** *If $p \in \operatorname{Cl}(P)$ and $(r:p)<\infty$ then $r \in \operatorname{Cl}(P)$. In particular, if 0 is closed and $(1:0)$ is finite then all elements are closed. A similar statement holds for Q.*

Proof. For part 1), we have

$$(r:p) \ge (p^*:r^*) \ge (r^{*\prime}:p^{*\prime}) = (\operatorname{cl}(r):\operatorname{cl}(p)) = (r:p)$$

so equality holds throughout.

For part 2), if $p \in \operatorname{Cl}(P)$ and $(r:p)<\infty$ then

$$(r:p) \ge (p^*:r^*) \ge (\operatorname{cl}(r):p) = (\operatorname{cl}(r):r)(r:p)$$

and since $(r:p)<\infty$, we may cancel to get $(\operatorname{cl}(r):r)=1$, which shows that r is closed.□

Thus, in an indexed Galois connection, the maps are *degree-preserving*, order-reversing bijections between the collections of closed sets $\operatorname{Cl}(P)$ and $\operatorname{Cl}(Q)$.

A Simple Degree Argument

There is a situation in which a simple degree argument can show that an element is closed. Referring to Figure 6.1.1,

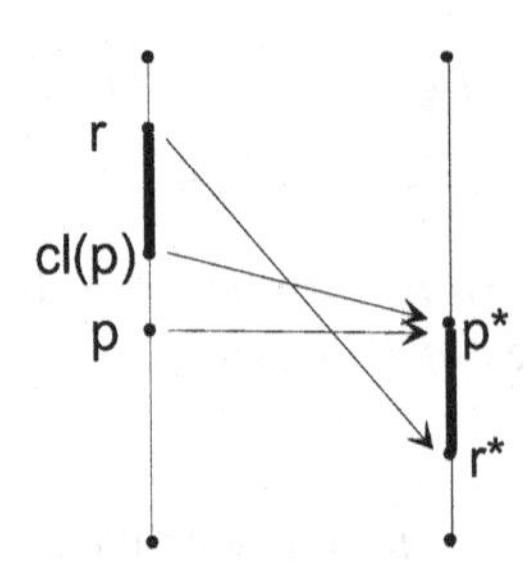

Figure 6.1.1

suppose that $(r : \mathrm{cl}(p)) < \infty$. Then r is closed and since $\mathrm{cl}(p)^* = p^*$, we have

$$(r : \mathrm{cl}(p)) = (p^* : r^*)$$

Now, if

$$(r : p) = (p^* : r^*)$$

then

$$(r : \mathrm{cl}(p)) = (r : p) = (r : \mathrm{cl}(p))(\mathrm{cl}(p) : p)$$

and so $(\mathrm{cl}(p) : p) = 1$, that is, $p = \mathrm{cl}(p)$ is closed.

Theorem 6.1.5 *If $r, p \in P$ and one of the following holds*
1) $(r : \mathrm{cl}(p)) < \infty$ *and* $(r : p) = (p^* : r^*)$
2) $\mathrm{cl}(p) \le r$ *and* $(r : p) = (p^* : r^*) < \infty$
then p is closed. In particular, for $r = 1_P$, if

$$(1_P : p) = (p^* : 0_Q) < \infty$$

then p is closed.□

When 0_Q is Closed

The following nonstandard definition will come in handy.

Definition *For a Galois connection on (P, Q), we say that P is* **completely closed** *if every element of P is closed, and similarly for Q. Also, the pair (P, Q) (or the connection) is* **completely closed** *if all elements of P and all elements of Q are closed.*□

We have remarked that the top elements 1_P and 1_Q, if they exist, are always closed, but the bottom elements 0_P and 0_Q need not be closed.

However, the most important example of a Galois connection, namely, the Galois correspondence of a field extension $F < E$, which is the subject of our investigations, has the property that 0_Q is closed. So let us assume that 0_Q is closed and see what we can deduce.

Since

$$\text{index}(P) = (1_P : 0_P) \geq (1_Q : 1_P^*) = (1_Q : 0_Q) = \text{index}(Q)$$

it follows that if P has finite index, then so does Q. Hence, if either P or Q has finite index, then Q is completely closed. Finally, if P has finite index and 0_P is also closed, then the connection is completely closed.

Theorem 6.1.6 (0_Q is closed) *Let (Π, Ω) be a Galois connection on (P, Q), where P and Q have top and bottom elements. Assume that 0_Q is closed. Then*

$$\text{index}(Q) \leq \text{index}(P)$$

Also,
1) If $\text{index}(Q) < \infty$ or $\text{index}(P) < \infty$, then Q is completely closed.
2) If $\text{index}(P) < \infty$ and 0_P is closed, then (P, Q) is completely closed.□

6.2 The Galois Correspondence

Now we describe the main theme of the rest of the book.

Definition *The* **Galois group** *of a field extension $F < E$, denoted by $G_F(E)$, is the group $\text{Aut}_F(E)$ of all automorphisms of E over F. The group $G_F(E)$ is also called the* **Galois group of E over F**.□

Note that when $F < E$ is algebraic,

$$G_F(E) = \text{Aut}_F(E) = \text{hom}_F(E, E)$$

and when $F < E$ is normal,

$$G_F(E) = \text{hom}_F(E, \overline{E})$$

Let $F < E$ and let $\mathcal{F}$ be the complete lattice of all intermediate fields of $F < E$, ordered by set inclusion. Let $\mathcal{G}$ be the complete lattice of all subgroups of the Galois group $G_F(E)$, ordered by set inclusion. We define two maps $\Pi: \mathcal{F} \to \mathcal{G}$ and $\Omega: \mathcal{G} \to \mathcal{F}$ by

$$\Pi(K) = G_K(E)$$

and

$$\Omega(H) = \text{fix}(H) = \{\alpha \in E \mid \sigma\alpha = \alpha \text{ for all } \sigma \in H\}$$

where $\text{fix}(H)$ is called the **fixed field** of H. These are pictured in Figure 6.2.1.

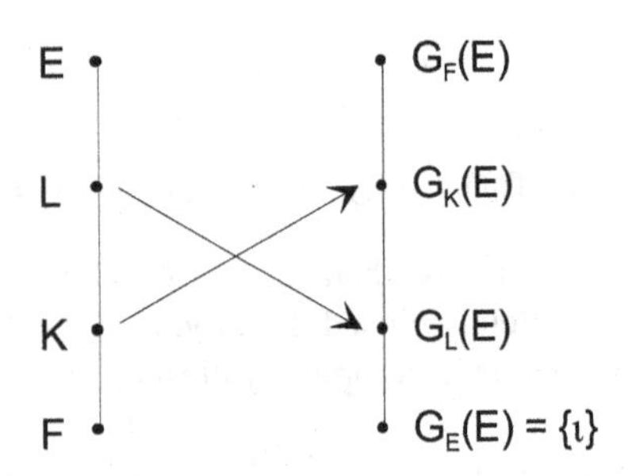

Figure 6.2.1–The Galois correspondence

Theorem 6.2.1 *Let $F < E$. The pair of maps*

$$(\Pi\colon K \mapsto G_K(E), \Omega\colon H \mapsto \text{fix}(H))$$

is a Galois connection on $(\mathcal{F}, \mathcal{G})$ called the **Galois correspondence** *of the extension $F < E$.*

Proof. It is clear from the definitions that both maps are order-reversing, that is,

$$K \subseteq J \Rightarrow G_J(E) \subseteq G_K(E)$$

and

$$H \subseteq I \Rightarrow \text{fix}(I) \subseteq \text{fix}(H)$$

Also, any element of K is fixed by every element of $G_K(E)$, that is,

$$K \subseteq \text{fix}(G_K(E))$$

Finally, any $\sigma \in J$ fixes every element in $\text{fix}(J)$, that is,

$$J \subseteq G_{\text{fix}(J)}(E)$$

□

Since $\mathcal{F}$ and $\mathcal{G}$ are complete lattices, Theorem 6.1.3 provides the following corollary.

Corollary 6.2.2 *The set $\text{Cl}(\mathcal{F})$ of closed intermediate fields and the set $\text{Cl}(\mathcal{G})$ of closed subgroups of $G_F(E)$ are complete lattices, where meet is intersection. In particular, the intersection of closed intermediate fields is closed and the intersection of closed subgroups is closed.* □

Note that both partially ordered sets $\mathcal{F}$ and $\mathcal{G}$ are topped and bottomed (as are all complete lattices). The top of $\mathcal{F}$ is E and the bottom is F. The top of $\mathcal{G}$ is $G_F(E)$ and the bottom of $\mathcal{G}$ is the trivial subgroup $\{\iota\}$. Also, the image of the top E is $G_E(E) = \{\iota\}$ and so the bottom of $\mathcal{G}$ is closed. Hence, three out of the four extreme elements are closed. We will spend much time discussing the issue of the closedness of the bottom element F.

The Plan

Now that we have established that the Galois correspondence is a Galois connection, our plan is as follows. First, we will show that the Galois correspondence is indexed, where $(K : L) = [K : L]$ is the degree of $F < E$ and $(H : J)$ is the index of the subgroup J in the group H.

Then we will describe the closed intermediate fields and the closed subgroups. The next step is to describe the connection between intermediate normal extensions and normal subgroups of the Galois group. (They don't call splitting fields normal extensions for nothing.) Finally, we describe the Galois group of a lifting and a composite.

The Galois Correspondence Is Indexed

We would like to show that the Galois correspondence of an extension $F < E$ is indexed, where $(K : L) = [K : L]$ is the degree of the extension $F < E$ and $(H : J)$ is the index of the subgroup J in the group H. We know that the degrees are multiplicative and that

$$[K : L] = 1 \Rightarrow K = L$$
$$(H : J) = 1 \Rightarrow H = J$$

The next theorem shows that the map $\Pi: K \mapsto G_K(E)$ is degree-nonincreasing. Recall that if $F < E$ is finite, then $[E : F]_s \le [E : F]$. When $F < E$ is infinite, this inequality still holds provided that we interpret it, not as an inequality of infinite cardinals, but simply as saying that $n \le \infty$ or $\infty \le \infty$.

Theorem 6.2.3 *For the tower $F < K < L < E$, we have*

$$(G_K(E) : G_L(E)) \le [L : K]_s \le [L : K]$$

as elements of $\mathbb{Z}^+ \cup \{\infty\}$.

Proof. Consider the function $\phi: G_K(E) \to \hom_K(L, E)$ that maps $\sigma \in G_K(E)$ to its restriction $\sigma|_L \in \hom_K(L, E)$. Then $\phi(\sigma) = \phi(\tau)$ if and only if σ and τ agree on L, that is, if and only if $\sigma G_L(E) = \tau G_L(E)$. Hence ϕ is constant on the cosets of $G_L(E)$ in $G_K(E)$ and so induces an injection on $G_K(E)/G_L(E)$, whence

$$(G_K(E) : G_L(E)) = |\mathrm{im}(\phi)| \le |\hom_K(L, E)| \le [L : K]_s$$

But as elements of $\mathbb{Z}^+ \cup \{\infty\}$, we have $[L : K]_s \le [L : K]$.□

Showing that $[\mathrm{fix}(J) : \mathrm{fix}(H)] \le (H : J)$ is a bit more difficult.

Theorem 6.2.4 *Let $F < E$ and let $J < H < G_F(E)$. Then*

$$[\mathrm{fix}(J) : \mathrm{fix}(H)] \le (H : J)$$

Proof. First, if $(H : J)$ is infinite, then there is nothing to prove, so let us assume that $(H : J) < \infty$, that is, $H/J = \{h_1 J, \ldots, h_m J\}$ is a finite set. Thus, $S = \{h_1, \ldots, h_m\}$ is a complete set of distinct coset representatives for H/J, and we may assume that $h_1 \in J$.

Let $E^{H/J}$ denote the set of all functions from H/J into E. Then $E^{H/J}$ is a vector space over E, where if $\sigma, \tau \in E^{H/J}$ and $a, b \in E$, then

$$(a\sigma + b\tau)(hJ) = a\sigma(hJ) + b\tau(hJ)$$

Moreover, since the functions $\epsilon_i \colon H/J \to E$ defined by $\epsilon_i(h_k J) = \delta_{i,k}$ form a basis for $E^{H/J}$ over E, we have

$$\dim(E^{H/J}) = |H/J| = (H : J)$$

Thus, we have two vector spaces: $\mathrm{fix}(J)$ is a vector space over $\mathrm{fix}(H)$ of dimension $[\mathrm{fix}(J) : \mathrm{fix}(H)]$ and $E^{H/J}$ is a vector space over E of dimension $(H : J)$. We wish to show that $\dim(\mathrm{fix}(J)) \le \dim(E^{H/J})$.

To do this, we will show that if $\alpha_1, \ldots, \alpha_n \in \mathrm{fix}(J)$ are linearly independent over $\mathrm{fix}(H)$, then the evaluation functions $\widehat{\alpha}_1, \ldots, \widehat{\alpha}_n \in E^{H/J}$, defined by

$$\widehat{\alpha}_k(h_i J) = h_i(\alpha_k)$$

are linearly independent over E. (In fact, the converse also holds.)

First, we must show that $\widehat{\alpha}_k$ is a well-defined function from H/J to E. If $h_1 J = h_2 J$ then $h_2^{-1} h_1 = j \in J$ and so

$$(h_2^{-1} h_1)(\alpha_k) = j(\alpha_k) = \alpha_k$$

which implies that $h_1(\alpha_k) = h_2(\alpha_k)$, that is, $\widehat{\alpha}_k(h_1 J) = \widehat{\alpha}_k(h_2 J)$. Hence, $\widehat{\alpha}_k$ is well-defined.

So assume that $\alpha_1, \ldots, \alpha_n \in \mathrm{fix}(J)$ are linearly independent over $\mathrm{fix}(H)$ and, by reindexing if necessary, let

$$e_1 \widehat{\alpha}_1 + \cdots + e_s \widehat{\alpha}_s = 0$$

be a nontrivial linear combination over E that is shortest among all nontrivial linear combinations equal to 0. Thus, $e_i \neq 0$ for all i. Dividing by e_s if necessary, we may also assume that $e_s = 1$. Thus

$$e_1 \widehat{\alpha}_1 + \cdots + e_{s-1} \widehat{\alpha}_{s-1} + \widehat{\alpha}_s = 0 \tag{6.2.1}$$

Then applying this to $h_k J$ gives

$$e_1 h_k(\alpha_1) + \cdots + e_{s-1} h_k(\alpha_{s-1}) + h_k(\alpha_s) = 0$$

for all $h_k \in S$. Since the α_i's are fixed by any element of J, and any $h \in H$ has the form $h = h_k j$ for some $j \in J$, we deduce that

$$e_1 h(\alpha_1) + \cdots + e_{s-1} h(\alpha_{s-1}) + h(\alpha_s) = 0 \tag{6.2.2}$$

for all $h \in H$. In particular, if $h = \iota$ then

$$e_1 \alpha_1 + \cdots + e_{s-1} \alpha_{s-1} + \alpha_s = 0 \tag{6.2.3}$$

which implies, owing to the independence of the α_i's over $\operatorname{fix}(H)$, that not all of the e_i's can lie in $\operatorname{fix}(H)$. Let us assume that $e_1 \notin \operatorname{fix}(H)$. Hence, there is a $\tau \in H$ for which $\tau e_1 \neq e_1$.

We can replace h by $\tau^{-1} h$ in (6.2.2) to get

$$e_1 \tau^{-1} h(\alpha_1) + \cdots + e_{s-1} \tau^{-1} h(\alpha_{s-1}) + \tau^{-1} h(\alpha_s) = 0$$

Applying τ gives

$$\tau(e_1)(h\alpha_1) + \cdots + \tau(e_{s-1})(h\alpha_{s-1}) + h\alpha_s = 0$$

for all $h \in H$ and so

$$(\tau e_1)\widehat{\alpha}_1 + \cdots + (\tau e_{s-1})\widehat{\alpha}_{s-1} + \widehat{\alpha}_s = 0$$

Finally, subtracting (6.2.1) from (6.2.3) gives

$$[(\tau e_1) - e_1]\widehat{\alpha}_1 + \cdots + [(\tau \widehat{\alpha}_{s-1}) - e_{s-1}]\widehat{\alpha}_{s-1} = 0$$

whose first coefficient is nonzero. But this is shorter than (6.2.1), a contradiction that completes the proof. □

Thus, the Galois correspondence of an algebraic extension $F < E$ is indexed. We can now summarize our results in a famous theorem.

Theorem 6.2.5 (Fundamental Theorem of Galois Theory Part 1: The correspondence) *The Galois correspondence* (Π, Ω) *of an extension* $F < E$ *is an indexed Galois connection and the bottom group* $\{\iota\}$ *is closed. It follows that the restrictions of* Π *and* Ω *to closed elements are order-reversing, degree-preserving inverse bijections as well as lattice anti-isomorphisms, that is, if* K_i *are closed intermediate fields and* H_i *are closed subgroups, then*

$$G_{\bigcap K_i}(E) = \bigvee G_{K_i}(E), \quad G_{\bigvee K_i}(E) = \bigcap G_{K_i}(E)$$

and

$$\operatorname{fix}\left(\bigcap H_i\right) = \bigvee \operatorname{fix}(H_i), \quad \operatorname{fix}\left(\bigvee H_i\right) = \bigcap \operatorname{fix}(H_i)$$ □

We should note that the joins in the previous theorem are joins in the corresponding lattices. Thus, for instance, $\bigvee G_{K_i}(E)$ is the smallest *closed* subgroup of $G_F(E)$ containing all of the subgroups $G_{K_i}(E)$, and this need not be the smallest subgroup of $G_F(E)$ containing these groups.

As a result of the closedness of $G_E(E) = \{i\}$, Theorem 6.1.6 gives the following.

Corollary 6.2.7 *Let* (Π, Ω) *be the Galois correspondence of* $F < E$. *Then*

$$|G_F(E)| \le [E:F]$$

Also,

1) *If* $|G_F(E)| < \infty$, *then* $\mathcal{G}$ *is completely closed.*
2) *If* $[E:F] < \infty$, *then* $\mathcal{G}$ *is completely closed.*
3) *If* $[E:F] < \infty$ *and* F *is closed, then* $\mathcal{F}$ *and* $\mathcal{G}$ *are completely closed.*□

6.3 Who's Closed?

We turn our attention to the question of which intermediate fields of an extension and which subgroups of the Galois group are closed.

We know on general principles that top elements are always closed. Thus, E and $G_F(E)$ are closed. Moreover, the bottom group $G_E(E) = \{\iota\}$ is also closed. We also know that any finite extension of a closed element is closed.

Now we require a definition.

Definition *A normal separable extension* $F < E$ *is called a* **Galois extension**, *or simply* **Galois**. □

The next theorem follows from the relevant properties of normal and separable extensions.

Theorem 6.3.1

1) **(Full extension Galois implies upper step Galois)** *Let* $F < K < E$. *If* $F < E$ *is Galois then the upper step* $K < E$ *is Galois.*
2) **(Closed under lifting)** *The class of Galois extensions is closed under lifting.*
3) **(Closed under arbitary composites and intersections)** *The class of Galois extensions is closed under arbitrary composites and intersections.* □

Let $F < E$ be algebraic. We wish to show that an intermediate field K is closed if and only if the extension $K < E$ is Galois.

First, suppose that K is closed and let $\alpha \in E \setminus K$. Then the finite extension $K(\alpha)$ of K is also closed and so

$$d = (G_K(E) : G_{K(\alpha)}(E)) = [K(\alpha) : K] < \infty$$

Let $S = \{\sigma_1, \ldots, \sigma_d\}$ be a complete system of distinct coset representatives for

$G_K(E)/G_{K(\alpha)}(E)$. Each element of S gives a distinct value on α, that is, a distinct root of $\min(\alpha, K)$, for if $\sigma_i\alpha = \sigma_j\alpha$, then $\sigma_j^{-1}\sigma_i \in G_{K(\alpha)}(E)$, which is not possible for $i \neq j$. Hence, the d roots of $\min(\alpha, K)$ are $\{\sigma_1\alpha, \ldots, \sigma_d\alpha\}$, which are distinct and lie in E. Thus, α is separable and $\min(\alpha, K)$ splits in E, implying that $K < E$ is a Galois extension.

For the converse, suppose that $K < E$ is Galois. If $\alpha \in \mathrm{cl}(K) = \mathrm{fix}(G_K(E))$, has minimal polynomial $p(x) = \min(\alpha, K)$, then $p(x)$ can have no roots other than α. For if β is a root of $p(x)$ in some extension, then there is an embedding $\sigma: E \hookrightarrow \overline{F}$ over K for which $\sigma\alpha = \beta$. But since $K \lhd E$, it follows that $\sigma \in G_K(E)$ and so $\beta = \sigma\alpha = \alpha$. Thus $p(x)$ has only one distinct root. Since $p(x)$ is separable, it must be linear, which implies that $\alpha \in K$. Thus, $\mathrm{cl}(K) = K$ and K is closed.

Let us summarize, with the help of Theorem 6.2.7.

Theorem 6.3.2 (Fundamental Theorem of Galois Theory Part 2: Who's closed?) *Let $F < E$ be algebraic and consider the Galois correspondence on $F < E$.*

1) **(Closed fields)** *The closed intermediate fields are precisely the fixed fields, that is, the fields of the form $\mathrm{fix}(H)$ for some $H \leq G_F(E)$.*
   a) *An intermediate field K is closed if and only if $K < E$ is Galois.*
   b) *Any extension of a closed intermediate field is closed. In particular, if F is closed, then $F < E$ is completely closed.*
   c) *If $F < \mathrm{cl}(K) < L < E$ and*

$$[L : K] = (G_K(E) : G_L(E)) < \infty$$

   then K is closed. In particular, if

$$[E : K] = |G_K(E)| < \infty$$

   then K is closed.
2) **(Closed groups)** *The closed subgroups of $G_F(E)$ are precisely the Galois groups of E, that is, the subgroups of the form $G_K(E)$, for some intermediate field K.*
   a) *Any finite extension of a closed subgroup is closed.*
   b) *$\{\iota\}$ is closed and so any* finite *subgroup of $G_F(E)$ is closed.*
   c) *When $F < E$ is finite, so is $G_F(E)$ and so $\{\iota\} < G_F(E)$ is completely closed.*
3) *If F is a finite Galois extension, then the correspondence is completely closed.* □

As the next example shows, in the general algebraic case, not all subgroups need be closed.

Example 6.3.1 For this example, we borrow from a later chapter the fact that for any prime power p^d, there exists a finite field $GF(p^d)$ of size p^d and $GF(p^d) < GF(p^r)$ if and only if $d \mid r$.

Referring to Figure 6.3.1, let $F = \mathbb{Z}_p = GF(p)$ and let $E = \overline{\mathbb{Z}_p}$. Since F is a finite field, it is perfect and so $F < E$ is separable. Since E is algebraically closed, $F \triangleleft E$. Hence $F < E$ is a Galois extension and therefore F is closed. The extension $F < E$ is not finite, however, since $[GF(p^k) : GF(p)] = k$ and $GF(p) < GF(p^k) < E$ for all $k \geq 1$.

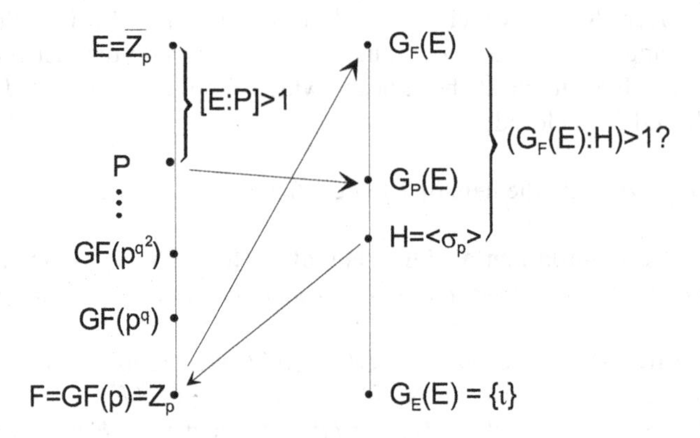

Figure 6.3.1

Let $H = \langle\sigma_p\rangle$ be the subgroup of $G_F(E)$ generated by the Frobenius map $\sigma_p\colon \alpha \to \alpha^p$. The fixed field $\mathrm{fix}(H)$ is the set of all $\alpha \in E$ for which $\alpha^p = \alpha$, in other words, the roots in E of the polynomial $p(x) = x^p - x$. But $p(x)$ has p roots in F and so $\mathrm{fix}(H) = F$. It follows that

$$\mathrm{cl}(H) = G_{\mathrm{fix}(H)}(E) = G_F(E)$$

Hence, all we need do is show that $H \neq G_F(E)$ to conclude that H is not closed. The key is that any $\mu \in H$ has the form $\mu = \sigma_p^k$ for some k and so the fixed set of μ is

$$\{\alpha \in E \mid \sigma_p^k \alpha = \alpha\} = \{\alpha \in E \mid \alpha^{p^k} = \alpha\} = GF(p^k)$$

which is a finite set. Thus, we need only show that there is an element of $G_F(E)$ that fixes infinitely many elements of E.

To this end, let q be a prime and consider the field

$$P = GF(p^q) \cup GF(p^{q^2}) \cup GF(p^{q^3}) \cup \cdots$$

Then P is a proper subfield of E, since it does not contain, for instance, the subfield $GF(p^{q+1})$. Hence $[E : P] > 1$ and since $P < E$ is Galois, the group $G_P(E)$ is not trivial. But if $\tau \in G_P(E)$, then τ fixes the infinite field P.□

Starting with a Field E and a Subgroup of $\operatorname{Aut}(E)$

The Galois correspondence begins with a field extension $F < E$ and the corresponding Galois group $G_F(E)$. Referring to Figure 6.3.2, we may also begin with a field E and a subgroup G of $\operatorname{Aut}(E)$. Then we can form the fixed field

$$\operatorname{fix}(G) = \{\alpha \in E \mid \sigma\alpha = \alpha \text{ for all } \sigma \in G\}$$

and consider the Galois correspondence of the extension $\operatorname{fix}(G) < E$, which we assume to be algebraic. The Galois group $G_{\operatorname{fix}(G)}(E)$ contains G, but the containment may be proper.

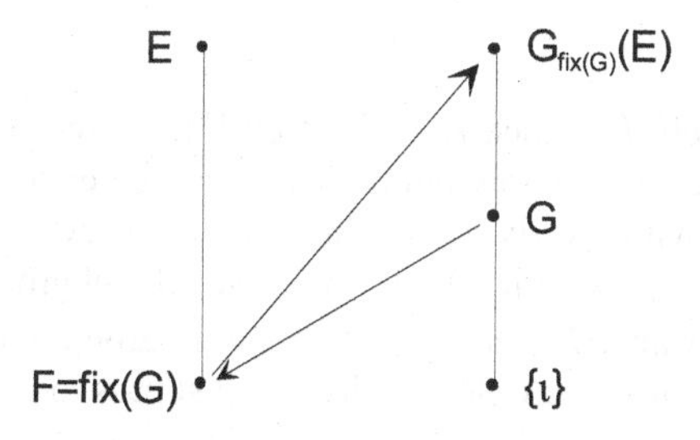

Figure 6.3.2–Starting with a field E and a subgroup G of $\operatorname{Aut}(E)$

Since $\operatorname{fix}(G) < E$ is algebraic and the base field $\operatorname{fix}(G)$ is closed, it follows that $\operatorname{fix}(G) < E$ is a Galois extension. Moreover, if $[E : \operatorname{fix}(G)] < \infty$, then the correspondence is completely closed (all intermediate fields and all subgroups are closed).

We emphasize that G may be a proper subgroup of its closure $G_{\operatorname{fix}(G)}(E)$, as in Example 6.3.1. However, this does not happen if G is finite, since finite subgroups are closed.

Theorem 6.3.3 *Let E be a field and let G be a group of automorphisms of E.*
1) *If $\operatorname{fix}(G) < E$ is algebraic, then it is Galois and all intermediate fields are closed.*
2) *If $\operatorname{fix}(G) < E$ is finite, then all intemediate fields and all subgroups are closed.*
3) *If G is closed (which happens if G is finite), then $G = G_{\operatorname{fix}(G)}(E)$ is the top group of the correspondence.* □

More on Closed Subgroups: Closure Points

Let $F < E$ be algebraic. The closure $\operatorname{cl}(H)$ of a subgroup H of the Galois group $G_F(E)$ can be characterized in a useful way. The following nonstandard definition will help.

Definition *Let $F < E$ be algebraic. Let H be a subgroup of the Galois group $G_F(E)$. A function $\tau\colon E \to E$ is a* **closure point** *of H if for any finite set $U \subseteq E$, we have $\tau|_U \in H|_U$, that is, τ agrees with some member of H on U. Let $\overline{H}$ denote the set of closure points of H.*$\square$

First, note that a closure point τ of H is a member of the Galois group $G_F(E)$, in fact, τ is in the closure of H, that is,

$$\overline{H} < G_{\mathrm{fix}(H)}(E) = \mathrm{cl}(H)$$

Indeed, $\tau \in \overline{H}$ is a homomorphism because it agrees with a homomorphism on any finite set in E and it fixes each element of $\mathrm{fix}(H)$ because every member of H fixes $\mathrm{fix}(H)$.

We claim that $\overline{H} = \mathrm{cl}(H)$. Since $H < \overline{H} < \mathrm{cl}(H)$, the result would follow if H were closed, but of course, it may not be. However, given any finite set $U \subseteq E$, we need only work with the *finite* extension $\mathrm{fix}(H) < K = \mathrm{fix}(H)(U)$, whose Galois group is $G_{\mathrm{fix}(H)}(\mathrm{fix}(H)(U))$. In this case, all subgroups are closed. The problem is that we want $H|_K$ to be in the Galois group and this requires that $\mathrm{fix}(H) < K$ be normal. No problem really: we just pass to a normal closure.

Consider the extension

$$\mathrm{fix}(H) < K = \mathrm{nc}(\mathrm{fix}(H)(U)/F)$$

which is finite, normal, contains U and has Galois group $G_{\mathrm{fix}(H)}(K)$. Since all subgroups are closed, $H|_K$ is a closed subgroup of the Galois group $G_{\mathrm{fix}(H)}(K)$.

Hence, in the Galois correspondence on $\mathrm{fix}(H) < K$, we have

$$H|_K = \overline{H|_K} = \mathrm{cl}(H|_K) = G_{\mathrm{fix}(H|_K)}(K)$$

It follows that any $\sigma \in G_{\mathrm{fix}(H|_K)}(K)$ agrees with a member of $H|_K$ on U. But if $\tau \in \mathrm{cl}(H) = G_{\mathrm{fix}(H)}(E)$, then

$$\tau|_K \in G_{\mathrm{fix}(H)}(E)|_K = G_{\mathrm{fix}(H|_K)}(K)$$

and so $\tau|_K$ agrees with a member of $H|_K$ on U, that is, τ agrees with a member of H on U. Thus, $\overline{H} = \mathrm{cl}(H)$, as desired.

Theorem 6.3.4 *Let $F < E$ be algebraic and let H be a subgroup of the Galois group $G_F(E)$. Then $\mathrm{cl}(H)$ is the set of closure points of H. More specifically, the following are equivalent:*

1) $\tau \in \mathrm{cl}(H)$

2) For any finite set $U \subseteq E$, we have $\tau|_U \in H|_U$.

Consequently, a subgroup H of $G_F(E)$ is closed if and only if it contains all of its closure points. In particular, any subgroup of the form $G_K(E)$ contains all of its closure points. □

*The Krull Topology

For those familiar with elementary topology, we can make this discussion a bit more topological.

We begin by extending the definition of closure point to apply to any set of functions in E^E, not just subgroups of the Galois group. In particular, a function $\tau \in E^E$ is a **closure point** of $S \subseteq E^E$ if for any finite set $U \subseteq E$, we have $\tau|_U \in S|_U$.

It is not hard to show that the operation $H \mapsto \overline{H}$ is an *algebraic* closure operation, in the sense defined earlier in the chapter. In addition, we have $\overline{\emptyset} = \emptyset$ and

$$\overline{H \cup K} = \overline{H} \cup \overline{K}$$

To see the latter, note that if $f \in \overline{H \cup K}$, then for any finite subset $X \subseteq E$, the function f agrees with an element of $H \cup K$ on X. But if $f \notin \overline{H}$, then there is a finite set $U \subseteq E$ for which f does not agree with any element of H on U. Similarly, if $f \notin \overline{K}$, then there is a finite set $V \subseteq E$ for which f does not agree with any element of K on V. However, $X = U \cup V$ is a finite set and so there must be some element $g \in H \cup K$ that agrees with f on $U \cup V$, and therefore on *both* U and V, that is, $f = g$ on U and $f = g$ on V. But $g \in H$ or $g \in K$, either one of which provides a contradiction.

It follows that the operation $H \mapsto \overline{H}$ is also a *topological* closure operation. Hence, the set of all complements of closed elements forms a toplology on E^E. This topology is actually quite famous.

Definition *Let E^E be the set of all functions from E into E. The* **finite topology** *$\mathcal{T}$ on E^E is defined by specifying as subbasis all sets of the form*

$$S_{u,v} = \{f\colon E \to E \mid fu = v\}$$

where u, $v \in E$. Thus, a basis for $\mathcal{T}$ consists of all sets of the form

$$\{f\colon E \to E \mid fu_1 = v_1, \ldots, fu_k = v_k\}$$

where u_i, $v_i \in E$. □

To show that the topology obtained from closure points is the finite topology, let S be any subset of E^E. If $f \in E^E$ is in the closure S^c of S under the finite topology, then any basis set that contains f also contains an element of S. It follows that for any finite set $U \subseteq E$, there is a $g \in S$ for which $f|_U = g|_U$, that is, $f|_U \in S|_U$. In other words, f is a closure point of S. Thus $S^c \subseteq \overline{H}$.

On the other hand, if $f \in \overline{S}$, then f agrees with some element of S on any finite set and so any basis element containing f must intersect S, showing that $f \in S^c$. Thus, $S^c = \overline{S}$.

Since the set of closed sets is the same in the topology of closure points and in the finite topology, these topologies are the same. Moreover, the Galois group $G_F(E)$ is closed in the sense of closure points and so it is closed in the finite topology. Thus, the induced (subspace) topologies are the same and, in view of Theorem 6.3.4, we can state the following.

Theorem 6.3.5 *Let $F < E$ be algebraic. Then the Galois group $G_F(E)$ is closed in the finite topology on E^E. Moreover, a subgroup $H < G_F(E)$ is closed in the Galois correspondence if and only if it is closed in the finite subspace topology on $G_F(E)$.*□

The subspace topology of the finite topology inherited by $G_F(E)$ is called the **Krull topology** on $G_F(E)$. We may phrase the previous theorem as follows: A subgroup of $G_F(E)$ is Galois-closed if and only if it is Krull-closed.

Note that we do *not* say that the set of Galois-closed subgroups of $G_F(E)$ is the set of closed sets for a topology. We say only that these closed subgroups are closed in the Krull topology. There are other subsets of $G_F(E)$ that are Krull-closed, for example, sets of the form $G_K(E) \cup G_L(E)$ which in general are not even groups.

6.4 Normal Subgroups and Normal Extensions

We now wish to discuss intermediate fields $F < K < E$ and their Galois groups $G_F(K)$. We begin with a result concerning the conjugates of a Galois group.

Definition *Let $F < K, L < E$. If there is a $\sigma \in G_F(E)$ for which $\sigma K = L$, then K and L are said to be* **conjugate**.□

Theorem 6.4.1

1) *If $F < K < E$, then for any $\sigma \in \hom_F(E, \overline{E})$,*

$$\sigma G_K(E)\sigma^{-1} = G_{\sigma K}(\sigma E)$$

2) *If $F \lhd K < E$, then for any $\sigma \in \hom_F(E, \overline{E})$,*

$$\sigma G_K(E)\sigma^{-1} = G_K(\sigma E)$$

3) If $F < K < E$ with $F \lhd E$, then for any $\sigma \in \hom_F(E, \overline{E})$,

$$\sigma G_K(E)\sigma^{-1} = G_{\sigma K}(E)$$

4) Let $F < K, L < E$, with $F < E$ Galois. Then K and L are conjugate if and only if the Galois groups $G_K(E)$ and $G_K(E)$ are conjugate.

Proof. For part 1), let $\tau \in G_{\sigma K}(\sigma E)$. Then $\sigma^{-1}\tau\sigma$ is an automorphism of E. Moreover, since τ fixes σK, we have for $\alpha \in K$,

$$\sigma^{-1}\tau\sigma\alpha = \sigma^{-1}\tau(\sigma\alpha) = \sigma^{-1}\sigma\alpha = \alpha$$

and so $\sigma^{-1}\tau\sigma \in G_K(E)$. Hence

$$G_{\sigma K}(\sigma E) \subseteq \sigma G_K(E)\sigma^{-1}$$

For the reverse inclusion, let $\mu = \sigma\tau\sigma^{-1}$, where $\tau \in G_K(E)$. Then μ is an automorphism of σE and if $\alpha \in K$, then $\tau\alpha = \alpha$ and so

$$\mu(\sigma\alpha) = \sigma\tau\sigma^{-1}(\sigma\alpha) = \sigma\tau\alpha = \sigma\alpha$$

which shows that $\mu \in G_{\sigma K}(\sigma E)$.

Part 2) follows from part 1), since when $F \lhd K$ then any $\sigma \in \hom_F(E, \overline{E})$ satisfies $\sigma K = K$. Part 3) is similar. For part 4), if $\sigma K = L$, then part 1) implies that

$$\sigma G_K(E)\sigma^{-1} = G_L(\sigma E) = G_L(E)$$

Conversely, if $G_L(E) = \sigma G_K(E)\sigma^{-1}$ then part 1) implies that $G_L(E) = G_{\sigma K}(E)$ and taking field fields gives $L = \sigma K$.$\square$

Now, $G_K(E)$ is normal in $G_F(E)$ if and only if

$$\sigma G_K(E)\sigma^{-1} = G_K(E)$$

for all $\sigma \in G_F(E)$. According to the previous theorem,

$$\sigma G_K(E)\sigma^{-1} = G_{\sigma K}(\sigma E) = G_{\sigma K}(E)$$

and so $G_K(E) \lhd G_F(E)$ if and only if

$$G_{\sigma K}(E) = G_K(E)$$

If $F \lhd K$, then $\sigma K = K$ and so $G_K(E) \lhd G_F(E)$. For the converse, if $G_K(E) \lhd G_F(E)$ then taking fixed fields gives

$$\sigma K < \text{cl}(\sigma K) < \text{cl}(K)$$

Thus, if K is closed, then $\sigma K < K$ for all $\sigma \in G_F(E)$ and if, in addition, $F \lhd E$, then $\sigma K < K$ for all $\sigma \in \hom_F(E, \overline{E})$, that is, $F < K$ is normal.

Note that when $F < K$ is normal, the restriction map $\phi: G_F(E) \to \hom_F(K, E)$ defined by

$$\phi(\sigma) = \sigma|_K$$

is a homomorphism, whose kernel is none other than the normal subgroup $G_K(E)$. Hence, the first isomorphism theorem of group theory shows that

$$\frac{G_F(E)}{G_K(E)} \hookrightarrow G_F(K)$$

Moreover, if $F < E$ is normal, then ϕ is surjective, since any $\sigma \in G_F(K)$ can be extended to an embedding of E into $\overline{F}$ over F, which must be an element of $G_F(E)$. Hence, if $F \triangleleft E$, then

$$\frac{G_F(E)}{G_K(E)} \approx G_F(K)$$

Now we are ready to summarize.

Theorem 6.4.2 (Fundamental Theorem of Galois Theory Part 3: Normality) *Let $F < K < E$. Let $\phi: G_F(E) \to \hom_F(K, E)$ be the restriction map*

$$\phi(\sigma) = \sigma|_K$$

1) *If $F \triangleleft K$ then $G_K(E) \triangleleft G_F(E)$ and ϕ induces an embedding*

$$\frac{G_F(E)}{G_K(E)} \hookrightarrow G_F(K)$$

which is an isomorphism if the full extension $F < E$ is normal.

2) *If $G_K(E) \triangleleft G_F(E)$ and in addition, $F \triangleleft E$ and K is closed (that is, $K < E$ is Galois), then $F \triangleleft K$ and ϕ induces an isomorphism*

$$\frac{G_F(E)}{G_K(E)} \approx G_F(K)$$

3) *If $F < E$ is Galois, then $F \triangleleft K$ if and only if $G_K(E) \triangleleft G_F(E)$.*□

An Example

Now that we have a complete picture of the Galois correspondence, let us consider a simple example: the Galois correspondence of a splitting field E for the polynomial $p(x) = x^4 - 2$ over $\mathbb{Q}$. Of course, $\mathbb{Q} < E$ is finite and Galois. Hence, the Galois correspondence is completely closed.

The roots of this polynomial are $r = \sqrt[4]{2}, -r, ri$ and $-ri$ and so any member of the Galois group G is a permutation of these roots. As to degree, we have

$$\mathbb{Q} < \mathbb{Q}(r) < \mathbb{Q}(i, r) = E$$

where the lower step has degree 4 since $p(x) = \min(r, \mathbb{Q})$, by Eisenstein's criterion. The upper step has degree at most 2, but cannot be 1 because $\mathbb{Q}(r) < \mathbb{R}$, which does not contain i. Hence, the upper step has degree 2 and $[E : \mathbb{Q}] = 8$.

One way to help find the Galois group is to look for an intermediate field N that is normal, because the elements of $G_{\mathbb{Q}}(N)$ are precisely the restrictions of the members of G. Since any extension of degree 2 is normal, we have $\mathbb{Q} \lhd \mathbb{Q}(i)$. The elements of $G_{\mathbb{Q}}(\mathbb{Q}(i))$ are the identity $\sigma = \iota$ and the map $\tau\colon i \mapsto -i$.

Since $[\mathbb{Q}(i, r) : \mathbb{Q}(i)] = 4$, we have $p(x) = \min(r, \mathbb{Q}(i))$ and so each of the automorphisms σ and τ can be extended to an element of G by sending r to any of the roots of $p(x)$. This gives

1) $\sigma_1\colon i \mapsto i, r \mapsto r$ $(\sigma_1 = \iota)$
2) $\sigma_2\colon i \mapsto i, r \mapsto -r$
3) $\sigma_3\colon i \mapsto i, r \mapsto ri$
4) $\sigma_4\colon i \mapsto i, r \mapsto -ri$

and

5) $\tau_1\colon i \mapsto -i, r \mapsto r$
6) $\tau_2\colon i \mapsto -i, r \mapsto -r$
7) $\tau_3\colon i \mapsto -i, r \mapsto ri$
8) $\tau_4\colon i \mapsto -i, r \mapsto -ri$

which constitute the 8 elements of G.

Could G be cyclic? Of course, one can tell this simply by checking G for an element of order 8. A more elegant way is the following: If G were cyclic, then all of its subgroups would be normal and so all of the intermediate fields would be normal extensions of $\mathbb{Q}$. But $\mathbb{Q} < \mathbb{Q}(r)$ is not normal, since $\mathbb{Q}(r)$ does not contain all of the roots of $p(x) = \min(r, \mathbb{Q})$.

Thus, all nonidentity elements $\sigma \in G$ have order 2 or 4, and this is determined by whether or not $\sigma^2 r = r$. In particular, σ_2 and all τ_i have order 2 and σ_3 and σ_4 have order 4. Thus, G has a normal cyclic subgroup $\langle \sigma_3 \rangle = \{1, \sigma_3, \sigma_3^2, \sigma_3^3\}$ where $\tau_1 \sigma_3 \tau_1 = \sigma_3^3$, and G is the dihedral group $\mathbb{D}_4$ of symmetries of the square.

All nontrivial subgroups of G have order 2 or 4. The subgroups of order 2 correspond to the elements of order 2:

1) $S_1 = \{\iota, \sigma_2\}$
2) $S_2 = \{\iota, \tau_1\}$
3) $S_3 = \{\iota, \tau_2\}$

4) $S_4 = \{\iota, \tau_3\}$
5) $S_5 = \{\iota, \tau_4\}$

The subgroups of order 4 are the cyclic subgroup $S_6 = \langle\sigma\rangle$ and the subgroups isomorphic to $\mathbb{Z}_2 \times \mathbb{Z}_2$. A computation shows that

6) $T_1 = \{1, \sigma_2, \sigma_3, \sigma_4\}$
7) $T_2 = \{1, \sigma_2, \tau_1, \tau_2\}$
8) $T_3 = \{1, \sigma_2, \tau_3, \tau_4\}$

The lattice of subgroups is shown in Figure 6.4.1.

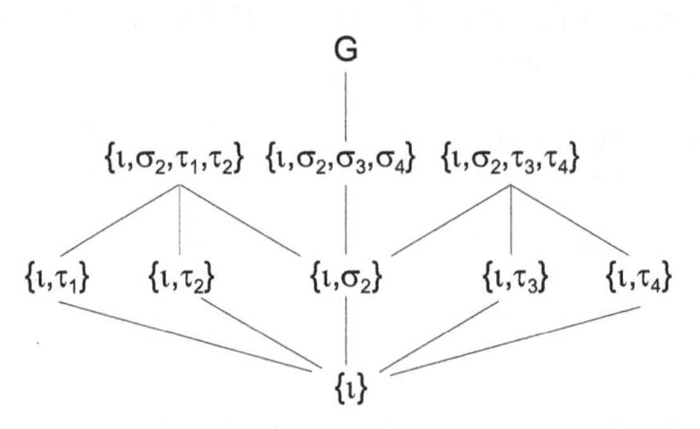

Figure 6.4.1

Of course, the lattice of intermediate (fixed) fields is a reflection of this. To compute fixed fields, we use the fact that $\{1, i\}$ is a basis for $\mathbb{Q}(i)$ over $\mathbb{Q}$ and $\{1, r, r^2, r^3\}$ is a basis for $\mathbb{Q}(i, r)$ over $\mathbb{Q}(i)$ and so the products form a basis for $\mathbb{Q}(i, r)$ over $\mathbb{Q}$. Hence, each $\sigma \in E$ has the form

$$\alpha = a + b_1 r + b_2 r^2 + b_3 r^3 + d_1 i + d_2 ir + d_3 ir^2 + d_4 ir^3$$

Thus for instance, $\alpha \in \text{fix}(\{\iota, \tau_1\})$ if and only if $\tau_1 \alpha = \alpha$, that is,

$$\begin{aligned} &a + b_1 r + b_2 r^2 + b_3 r^3 - d_1 i - d_2 ir - d_3 ir^2 - d_4 ir^3 \\ &\quad = a + b_1 r + b_2 r^2 + b_3 r^3 + d_1 i + d_2 ir + d_3 ir^2 + d_4 ir^3 \end{aligned}$$

Equating coefficients of the basis vectors gives $d_i = 0$ for all i. Thus,

$$\text{fix}(\{\iota, \tau_1\}) = \mathbb{Q}(r)$$

As another example, note that σ_2 fixes both i and $r^2 = \sqrt{2}$ and so

$$\text{fix}(\{\iota, \sigma_2\}) = \mathbb{Q}(\sqrt{2}, i) = \mathbb{Q}(\sqrt{2} + i)$$

(see Example 3.4.1). Moreover, $\{\iota, \sigma_2\}$ is a normal subgroup of G and so $\mathbb{Q} < \mathbb{Q}(\sqrt{2} + i)$ is a normal extension of degree 4. In fact, the roots of the polynomial $q(x) = x^4 - x^2 - 2$ are $\pm\sqrt{2}$ and $\pm i$ and so $\mathbb{Q}(\sqrt{2} + i)$ is a splitting field for this polynomial.

More generally, the normal subgroups of G correspond to the normal extensions of $\mathbb{Q}$. These subgroups are G, $\{\iota\}$, the subgroups of order 4 (index 2) and $\{\iota, \sigma_2\}$.

6.5 More on Galois Groups

We now examine the behavior of Galois groups under lifting and under composites. As usual, we assume that all composites mentioned are defined.

The Galois Group of a Lifting

Let $F < E$ be normal and let $F < K$. Any $\sigma \in G_K(EK)$, the Galois group of the lifting, is uniquely determined by what it does to E (since it fixes K) and so the restriction map $\sigma \mapsto \sigma|_E$ is an injection. Since $F < E$ is normal, it follows that $\sigma|_E \in G_F(E)$. But $\sigma|_E$ may fix more than F: It also fixes every element of E that is fixed by σ, that is,

$$\sigma|_E \in G_{E \cap \text{fix}(G_K(EK))}(E) = G_{E \cap \text{cl}(K)}(E)$$

Note also that the restriction map is a homomorphism, and hence an embedding of $G_K(EK)$ into $G_{E\cap \text{cl}(K)}(E)$. We will show that this embedding is actually an isomorphism and

$$G_K(EK) \approx G_{E\cap \text{cl}(K)}(E)$$

Note that if $F < E$ is Galois, then $K < EK$ is Galois and so K is closed, which simplifies the preceding to

$$G_K(EK) \approx G_{E\cap K}(E)$$

Theorem 6.5.1 (The Galois group of a lifting) *Let $F < E$ be normal and let $F < K$. The restriction map*

$$\phi: G_K(EK) \to G_{E\cap \text{cl}(K)}(E)$$

where $\text{cl}(K) = \text{fix}(G_K(EK))$, defined by $\phi\sigma = \sigma|_E$ is an isomorphism and

$$G_K(EK) \approx G_{E\cap \text{cl}(K)}(E)$$

Proof. We have already proved that ϕ is an embedding. It remains to show that ϕ is surjective. To avoid confusion, let us use the notation fix_E for the fixed field with respect to the Galois correspondence on $F < E$, and fix_{EK} for the fixed field with respect to the Galois correspondence on $K < EK$. Then

$$\begin{aligned}\text{fix}_E(\text{im}(\phi)) &= \{\alpha \in E \mid \tau\alpha = \alpha \text{ for all } \tau \in \text{im}(\phi)\}\\ &= \{\alpha \in E \mid (\sigma|_E)\alpha = \alpha \text{ for all } \sigma \in G_K(EK)\}\\ &= \{\alpha \in E \mid \sigma\alpha = \alpha \text{ for all } \sigma \in G_K(EK)\}\\ &= E \cap \text{fix}_{EK}(G_K(EK))\end{aligned}$$

Now, if we show that $\text{im}(\phi)$ is a closed subgroup with respect to the Galois correspondence on $F < E$, it follows by taking Galois groups (of E) that

$$\operatorname{im}(\phi) = G_{E\cap \operatorname{fix}(G_K(EK))}(E)$$

and thus ϕ is surjective, completing the proof. If $F < E$ is finite, then all subgroups of the Galois group $G_F(E)$ are closed, and we are finished.

When $F < E$ is not finite, we must work a bit harder. We show that $I = \operatorname{im}(\phi)$ is closed by showing that I contains all of its closure points. So suppose that $\tau \in \overline{I}$. To show that $\tau \in I$, we must find a $\sigma \in G_K(EK)$ for which $\sigma|_E = \tau$. But any σ in $G_K(EK)$ is completely determined by its action on E and so this completely determines σ, that is, if it exists. To this end, note that every $\alpha \in EK$ has the form

$$\alpha = \sum e_i k_i$$

where $e_i \in E$ and $k_i \in K$. Define a function $\sigma: EK \to EK$ by

$$\sigma\alpha = \sum (\tau e_i) k_i$$

To see that σ is well-defined, let

$$\alpha = \sum e_i' k_i'$$

where $e_i' \in E$ and $k_i' \in K$. Then since $\tau \in \overline{I}$, there exists a $\sigma' \in G_K(EK)$ that agrees with τ on the elements $\{e_i, e_i'\}$, and so

$$\sum (\tau e_i) k_i = \sum (\sigma' e_i) k_i = \sigma'\left(\sum e_i k_i\right) = \sigma'\left(\sum e_i' k_i'\right) = \sum (\tau e_i') k_i'$$

Thus, σ is well-defined. Clearly, σ fixes K and agrees with τ on E.

Next, we show that σ is a closure point of $G_K(EK)$. Then, since $G_K(EK)$ is closed, it will follow that $\sigma \in G_K(EK)$, and the proof will be complete.

First note that $\sigma = \tau$ on E. Since $\tau \in \overline{I}$, it agrees with some element of I on any finite set U. Hence, σ agrees with some element of I on any finite subset U of E, and so also with some element of $G_K(EK)$ on any finite subset U of E. But σ also fixes K and so agrees with *any* element of $G_K(EK)$ on K. Thus, σ agrees with some element of $G_K(EK)$ on any finite subset of $E \cup K$. But any finite subset $\{\alpha_1, \ldots, \alpha_n\}$ of EK has the form

$$\alpha_1 = \sum e_{1,i} k_i,\ \alpha_2 = \sum e_{2,i} k_i\ , \ldots, \alpha_n = \sum e_{n,i} k_i$$

where $\{e_{i,j}\} \cup \{k_i\}$ is a finite subset of $E \cup K$ and so σ agrees with some element of $G_K(EK)$ on $\{\alpha_1, \ldots, \alpha_n\}$. Thus, $\sigma \in \overline{G_K(EK)}$, as desired.□

For a Galois extension $F < E$, the previous theorem simplifies a bit.

Corollary 6.5.2 (The Galois group of a lifting) *The lifting $K < EK$ of a Galois extension $F < E$ by an arbitrary extension $F < K$ is Galois. Moreover, the restriction map $\phi: G_K(EK) \to G_{E\cap K}(E)$ defined by $\phi\sigma = \sigma|_E$ is an isomorphism and*

$$G_K(EK) \approx G_{E\cap K}(E)$$

Also,
1) $E \cap K = F$ *implies* $G_K(EK) \approx G_F(E)$.
2) *If* $F < E$ *is finite, then* $G_K(EK) \approx G_F(E)$ *implies* $E \cap K = F$.
Proof. We have proved all but the last two statements. Statement 1) is clear. As to statement 2), since all is finite, we have $G_{E\cap K}(E) = G_F(E)$ and the result follows by taking fixed fields. □

Corollary 6.5.2 yields a plethora of useful statements about degrees, all of which can be read from Figure 6.5.1. We leave details of the proof to the reader.

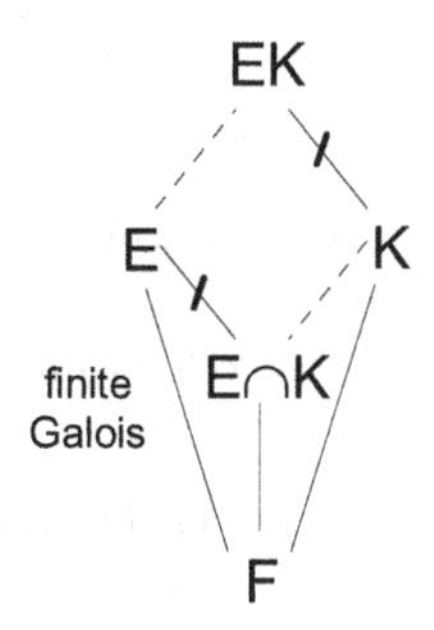

Figure 6.5.1

Corollary 6.5.3 *Suppose that $F < E$ is finite Galois and $F < K$. Then*
1) $[EK : K] = [E : E \cap K]$ *and so* $[EK : K] \mid [E : F]$.
If $F < K$ *is also finite then*
2) $[EK : F] = [E : E \cap K][K : F]$.
3) $[EK : F]$ *divides* $[E : F][K : F]$, *with equality if and only if* $E \cap K = F$.
More generally, if $F < E_i$ is finite Galois for $i = 1, \ldots, n-1$ and $F < E_n$ is finite then. letting $E_{i+1}\cdots E_n = F$ when $i = n$, we have
4) $[E_1\cdots E_n : F] = \prod_{i=1}^{n} [E_i : E_i \cap (E_{i+1}\cdots E_n)]$
5) $[E_1\cdots E_n : F] = \prod_{i=1}^{n} [E_i : F]$ *if and only if* $E_i \cap (E_{i+1}\cdots E_n) = F$ *for all* i.□

The Galois Group of a Composite

We now turn to the Galois group of a composite. Let $F \lhd E$ and $F \lhd K$. Then any $\sigma \in G_F(EK)$ is completely determined by its action on E and K, that is, by its restrictions $\sigma|_E$ and $\sigma|_K$, or put another way, by the element

$$(\sigma|_E, \sigma|_K) \in G_F(E) \times G_F(K)$$

Indeed, the map $\phi: G_F(EK) \to G_F(E) \times G_F(K)$ is an embedding of groups. Moreover, as we will see, in the finite case, if the fields enjoy a form of independence ($E \cap K = F$), then the embedding is an isomorphism.

The following theorem gives the general case.

Theorem 6.5.4 (The Galois group of a composite)

1) Let $\mathcal{F} = \{E_i \mid i \in I\}$ be a family of fields, with $F < E_i$ normal for all $i \in I$. Let $G = \prod G_F(E_i)$ be the direct product of the Galois groups $G_F(E_i)$ and let $\pi_i: G \to G_F(E_i)$ be projection onto the ith coordinate. Then the map

$$\phi: G_F(\bigvee E_i) \to \prod G_F(E_i)$$

defined by

$$\pi_i(\phi\sigma) = \sigma|_{E_i}$$

is an embedding of groups. Hence, $G_F(\bigvee E_i)$ is isomorphic to a subgroup of $\prod G_F(E_i)$.

2) If $\mathcal{F} = \{E_1, \ldots, E_n\}$ is a finite family of finite Galois extensions, then the map ϕ is surjective and

$$G_F(E_1 \vee \cdots \vee E_n) \approx G_F(E_1) \times \cdots \times G_F(E_n)$$

if and only if

$$E_i \cap (E_{i+1} \cdots E_n) = F$$

for all $i = 1, \ldots, n$.

Proof. Since $F \lhd E_i$, Theorem 6.4.1 implies that each individual restriction map

$$\phi_k = (\pi_k \circ \phi): \sigma \mapsto \sigma|_{E_k}$$

is a surjective homomorphism from $G_F(\bigvee E_i)$ onto $G_F(E_k)$, with kernel $G_{E_k}(\bigvee E_i)$. Hence, ϕ is a homomorphism from $G_F(\bigvee E_i)$ into $\prod G_F(E_i)$.

As to the kernel of ϕ, if $\phi(\sigma) = \iota$, then

$$\sigma|_{E_k} = \phi_k(\sigma) = \pi_k \phi(\sigma) = \pi_k \iota = \iota$$

and so $\sigma = \iota$ on each E_k, which implies that $\sigma = \iota$. Hence, $\ker(\phi) = \{\iota\}$ and ϕ is an embedding.

When $\mathcal{F}$ is a finite family of finite Galois extensions, all Galois groups are finite and all subgroups and intermediate fields are closed. Since ϕ is injective, we have

$$|\text{im}(\phi)| = \left|G_F\left(\bigvee E_i\right)\right| = \left[\bigvee E_i : F\right]$$

and also

$$\left|\prod G_F(E_i)\right| = \prod |G_F(E_i)| = \prod [E_i : F]$$

Hence ϕ is surjective if and only if $[\bigvee E_i : F] = \prod [E_i : F]$ and Corollary 6.5.3 gives the desired result. □

If $F < E$ is a finite Galois extension whose Galois group is a direct product $G_1 \times \cdots \times G_n$, then we may wish to find intermediate fields $F < E_i < E$ whose Galois groups (over F) are isomorphic to the individual factors G_i in the direct product.

Corollary 6.5.5 *Suppose that $F < E$ is a Galois extension with Galois group of the form*

$$G = G_F(E) = G_1 \times \cdots \times G_n$$

If

$$H_i = G_1 \times \cdots \times \{\iota\} \times \cdots \times G_n$$

where $\{\iota\}$ is in the ith coordinate and if

$$E_i = \text{fix}(H_i)$$

then
1) $F < E_i$ is Galois, with Galois group $G_F(E_i) \approx G_i$.
2) $E = E_1 \vee \cdots \vee E_n$.
3) $E_i \cap (E_{i+1} \cdots E_n) = F$ for all $i = 1, \ldots, n$.
Proof. Since $H_i \lhd G$, $E_i = \text{fix}(H_i)$ is closed and $F < E$ is normal, it follows from Theorem 6.4.2 that $F \lhd E_i$ and

$$G_F(E_i) \approx \frac{G_F(E)}{G_{E_i}(E)} = \frac{G}{H_i} \approx G_i$$

In addition, $F < \bigvee E_i$ is Galois and since

$$G_{\bigvee E_i}(E) = \bigcap G_{E_i}(E) = \bigcap H_i = \{\iota\} = G_E(E)$$

taking fixed fields gives $\bigvee E_i = E$. Hence,

$$G_F(\bigvee E_i) = G_F(E) = \prod G_i \approx \prod G_F(E_i)$$

and Theorem 6.5.4 implies that $E_i \cap (E_{i+1} \cdots E_n) = F$ for all $i = 1, \ldots, n$. □

The Galois Group of the Normal Closure

We next wish to consider the Galois group of a normal closure, which is a special composite of fields.

Theorem 6.5.6 *Let $F < E$ be separable.*
1) If

$$F \lhd K \lhd E < \mathrm{nc}(E/F)$$

then $G_K(\mathrm{nc}(E/F))$ is isomorphic to a subgroup of

$$\prod_{\sigma \in \hom_F(E,\overline{E})} G_K(\sigma E) = \prod_{\sigma \in \hom_F(E,\overline{E})} \sigma G_K(E)\sigma^{-1}$$

2) If, in addition to the conditions of part 1), $F < E$ is finite, then the direct product given above is a finite direct product.

Proof. Let $N = \mathrm{nc}(E/F) = \bigvee(\sigma E)$, the join being over all $\sigma \in \hom_F(E, \overline{E})$. Then

$$G_K(N) = G_K(\bigvee_{\sigma \in \hom_F(E,\overline{E})} (\sigma E))$$

Since $K < E$ is Galois, so is $K < \sigma E$ and Theorem 6.5.4 implies that $G_K(\bigvee \sigma E)$ is isomorphic to a subgroup of $\prod G_K(\sigma E)$. The rest of part 1) follows from Theorem 6.4.1. For the second statement, if $F < E$ is finite, then

$$\left|\hom_F(E, \overline{E})\right| = [E : F]_s \leq [E : F]$$

and so the direct sum is a finite sum.□

6.6 Abelian and Cyclic Extensions

Extensions are often named after their Galois groups. Here is a very important example.

Definition *A Galois extension $F < E$ is* **abelian** *if its Galois group $G_F(E)$ is abelian and* **cyclic** *if the Galois group $G_F(E)$ is cyclic.* □

The basic properties of abelian and cyclic extensions are given in the next theorem, whose proof is left as an exercise. Note that abelian and cyclic extensions are *not* (quite) distinguished.

Theorem 6.6.1
1) **(Composite of abelian is abelian)** *If $F < E_i$ are abelian, then $F < \bigvee E_i$ is abelian.*
2) **(Lifting of abelian/cyclic is abelian/cyclic)** *If $F < E$ is abelian (cyclic) and $F < K$, then $K < EK$ is abelian (cyclic).*

3) **(Steps in an abelian/cyclic tower are abelian/cyclic)** *If $F < K < E$ with $F < E$ abelian (cyclic), then $F < K$ and $K < E$ are abelian (cyclic).*□

Abelian and cyclic extensions fail to be distinguished because, and only because if the steps in a tower are abelian (cyclic), this does not imply that the full extension is abelian (cyclic). What does it imply?

Suppose that

$$F_1 < F_2 < \cdots < F_n$$

is a tower in which each step $F_i < F_{i+1}$ is abelian (cyclic). Taking Galois groups gives the series

$$\{1\} = G_{F_n}(F_n) < G_{F_{n-1}}(F_n) < \cdots < G_{F_1}(F_n)$$

Consider the subtower $F_i < F_{i+1} < F_n$. Since the lower step is normal, it follows from Theorem 6.4.2 that $G_{F_{i+1}}(F_n)$ is a normal subgroup of its parent $G_{F_i}(F_n)$ and that

$$\frac{G_{F_i}(F_n)}{G_{F_{i+1}}(F_n)} \hookrightarrow G_{F_i}(F_{i+1})$$

Since the latter is abelian (cyclic), so is the former. Thus,

$$\{1\} = G_{F_n}(F_n) \lhd G_{F_{n-1}}(F_n) \lhd \cdots \lhd G_{F_1}(F_n)$$

where each quotient group is abelian (cyclic). In the language of group theory, this series of subgroups is an **abelian series**. (When the groups are finite, the cyclic case and the abelian case are equivalent.) A group that has an abelian series is said to be **solvable**.

Theorem 6.6.2 *If*

$$F_1 < F_2 < \cdots < F_n$$

is a tower of fields in which each step $F_i < F_{i+1}$ is abelian, then the Galois group $G_{F_1}(F_n)$ is solvable.□

*6.7 Linear Disjointness

If $F < K$ and $F < L$ are finite extensions, the degree $[KL : F]$ provides a certain measure of the "independence" of the extensions. Assuming that $[K : F] \le [L : F]$, we have

$$[L : F] \le [KL : F] \le [K : F][L : F]$$

The "least" amount of independence occurs when $[KL : F] = [L : F]$, or equivalently, when $K < L$ and the "greatest" amount of independence occurs when

$$[KL:F]=[K:F][L:F] \tag{6.7.1}$$

We have seen (Corollary 6.5.3) that if one of the extensions is Galois, then (6.7.1) holds if and only if $K \cap L = F$. For finite extensions in general, we cannot make such a simple statement. However, we can express (6.7.1) in a variety of useful ways. For instance, we will show that (6.7.1) holds for arbitrary finite extensions if and only if whenever $\{\kappa_i\} \subseteq K$ is linearly independent over F and $\{\lambda_j\} \subseteq L$ is independent over F then $\{\kappa_i\lambda_j\}$ is also independent over F.

To explore the situation more fully (and for not necessarily finite extensions), it is convenient to employ tensor products. (All that is needed about tensor products is contained in Chapter 0.) The multiplication map $\sigma\colon K \times L \to KL$ defined by $\sigma(\kappa, \lambda) = \kappa\lambda$ is bilinear and so there exists a unique linear map $\phi\colon K \otimes L \to KL$ for which $\phi(\kappa \otimes \lambda) = \kappa\lambda$.

Note that the image of ϕ is the F-algebra $K[L] = L[K]$ of all elements of the form

$$\kappa_1\lambda_1 + \cdots + \kappa_n\lambda_n$$

for $\kappa_i \in K$ and $\lambda_i \in L$. Hence, if $F < K$ or $F < L$ is algebraic, say $F < L$ is algebraic, then $KL = K(L) = K[L]$ and so the map ϕ is surjective.

If F is a field, we use the term F**-independent** to mean linearly independent over F.

Theorem 6.7.1 *Let $F < E$ and suppose that K and L are intermediate fields. Then K and L are* **linearly disjoint** *over F if any of the following equivalent conditions holds.*

1) *The multiplication map $\phi\colon K \otimes L \to KL$ is injective.*
2) *If $\{\kappa_i\} \subseteq K$ is F-independent, then it is also L-independent.*
3) *If $\{\kappa_i\} \subseteq K$ and $\{\lambda_j\} \subseteq L$ are both F-independent, then $\{\kappa_i\lambda_j\}$ is also F-independent.*
4) *If $\{\kappa_i\}$ is a basis for K over F and $\{\lambda_j\}$ is a basis for L over F, then $\{\kappa_i\lambda_j\}$ is a basis for $K[L]$ over F.*
5) *There is a basis for K over F that is L-independent.*

Moreover,

6) *K and L are linearly disjoint if and only if K_0 and L_0 are linearly disjoint, for all* finite *extensions $F < K_0 < K$ and $F < L_0 < L$.*
7) *If K and L are linearly disjoint then*

$$K \cap L = F$$

Proof. $[1 \Rightarrow 2]$ Let $\{\kappa_i\} \subseteq K$ be F-independent and suppose that $\sum \lambda_i\kappa_i = 0$ for $\lambda_i \in L$. Since ϕ is injective and

$$\phi(\sum \lambda_i \otimes \kappa_i) = \sum \lambda_i \kappa_i = 0$$

we have

$$\sum \lambda_i \otimes \kappa_i = 0$$

Theorem 0.9.2 now implies that $\lambda_i = 0$ for all i.

[$2 \Rightarrow 3$] Let $\{\kappa_i\}$ and $\{\lambda_j\}$ be F-independent. If

$$\sum_{i,j} a_{i,j} \kappa_i \lambda_j = 0$$

with $a_{i,j} \in F$ then since $\{\kappa_i\}$ is also L-independent, the coefficients of κ_i must equal 0, that is,

$$\sum_j a_{i,j} \lambda_j = 0$$

for all i. Since the λ_i's are also F-independent, we get $a_{i,j} = 0$ for all i, j.

[$3 \Rightarrow 4$] This follows from the fact that if $\{\kappa_i\}$ spans K over F and $\{\lambda_j\}$ spans L over F then $\{\kappa_i \lambda_j\}$ spans $K[L]$ over F.

[$4 \Rightarrow 1$] The map ϕ sends a basis $\{\kappa_i \otimes \lambda_j\}$ for $K \otimes L$ to a basis $\{\kappa_i \lambda_j\}$ for $K[L]$ and is therefore injective.

Thus, each of 1) to 4) is equivalent, and by symmetry we may add the equivalent statement that any F-independent subset of L is also K-independent. It is clear that 2) implies 5).

[$5 \Rightarrow 1$] Let $\{\kappa_i\}$ be a basis for K over F that is L-independent. Let $\{\lambda_j\}$ be a basis for L over F. Then $\{\kappa_i \lambda_j\}$ is a basis for $K[L]$ over F, for if

$$\sum_{i,j} a_{i,j} \kappa_i \lambda_j = 0$$

with $a_{i,j} \in F$ then since $\{\kappa_i\}$ is L-independent, we have

$$\sum_j a_{i,j} \lambda_j = 0$$

for all i. Since the λ_i's are also F-independent, $a_{i,j} = 0$ for all i, j. Finally, ϕ takes the basis $\kappa_i \otimes \lambda_j$ to the basis $\kappa_i \lambda_j$ and so is injective.

As to 6), it is clear that multiplication $\phi: K \otimes L \to KL$ is injective if and only if each map $\phi_0: K_0 \otimes L_0 \to K_0 L_0$ is injective. Alternatively, if K and L are linearly disjoint, then so are K_0 and L_0, for if $\{\kappa_i\} \subseteq K_0 \subseteq K$ is F-

independent, then it is L-independent and hence also L_0-independent. Conversely, if $\{\kappa_i\} \subseteq K$ were F-independent but failed to be L-independent, then some finite subset $\{\kappa_1, \ldots, \kappa_n\}$ would be L-dependent as well, say

$$\sum_{i=1}^{n} \lambda_i \kappa_i = 0$$

for $\lambda_i \in L$, not all 0. Let $K_0 = F(\kappa_1, \ldots, \kappa_n)$ and $L_0 = F(\lambda_1, \ldots, \lambda_n)$. Since $\{\kappa_1, \ldots, \kappa_n\} \subseteq K_0$ is F-independent, it must also be L_0-independent by the linear disjointness of K_0 and L_0. Thus, $\lambda_i = 0$ for all i, a contradiction.

For 7), suppose that K and L are linearly disjoint and $\alpha \in K \cap L$. Then we have $F < F(\alpha) < K$ and $F < F(\alpha) < L$ where $F(\alpha)$ is a finite intermediate field in each case. It follows from part 6) that $F(\alpha)$ is linearly disjoint with itself. Therefore, if $\mathcal{B}$ is a basis for $F(\alpha)$ over F, it is also a basis for $F(\alpha)$ over $F(\alpha)$ and so $|\mathcal{B}| = 1$, that is, $F(\alpha) = F$ and $\alpha \in F$. Thus, $K \cap L = F$.□

Corollary 6.7.2 (Linear disjointness in the finite case) *Let $F < E$ and suppose that K and L are intermediate fields of finite degree over F.*

1) *K and L are linearly disjoint if and only if*

$$[KL : F] = [K : F][L : F]$$

2) *If one of $F < K$ or $F < L$ is Galois, then K and L are linearly disjoint if and only if*

$$K \cap L = F$$

Proof. For part 1), if K and L are linearly disjoint, then part 4) of Theorem 6.7.1 implies that the degree condition above holds. Conversely, if this degree condition holds, and if $\{\kappa_i\}$ is a basis for K over F and $\{\lambda_j\}$ is a basis for L over F, then since the set $\{\kappa_i \lambda_j\}$ spans KL and has size $[KL : F]$, it must also be a basis for KL. Hence, K and L are linearly disjoint.

Alternatively, we have remarked that the multiplication map $\phi: K \otimes L \to KL$ is surjective and so it is injective if and only if $\dim(K \otimes L) = \dim(KL)$, which by Corollary 0.9.5 is equivalent to

$$\dim_F(K) \cdot \dim_F(L) = \dim_F(KL)$$

Part 2) follows from part 1) and Corollary 6.5.3. □

Exercises

1. Find the Galois group of the polynomial $x^3 - 2$ over $\mathbb{Q}$. Find the subgroups and intermediate fields.
2. Prove that a pair of order-reversing maps $(\Pi: P \to Q, \Omega: Q \to P)$ between partially ordered sets is a Galois connection if and only if

$$q \le p^* \Leftrightarrow p \le q'$$

for all $p \in P$ and $q \in Q$, where $p^* = \Pi p$ and $q' = \Omega q$.

3. Let $F < K < E$. Prove that $G_K(E) = G_{\text{fix}(G_K(E))}(E)$.
4. If $\lambda\colon \mathcal{L} \to \mathcal{M}$ is an order-reversing bijection between two lattices, verify that $\lambda(\bigwedge a_i) = \bigvee(\lambda a_i)$ and $\lambda(\bigvee a_i) = \bigwedge(\lambda a_i)$. *Hint*: first show that λ^{-1} is also order-reversing.
5. If $F < K < E$ and $F < L < E$ where $F < E$ is algebraic and $F < K$ and $F < L$ are Galois. Show that $K \cap L < E$ is a Galois extension.
6. If $F < E$ is abelian, show that for every intermediate field $F < K < E$ we have $F \lhd K$.
7. Let $K < E$ and $L < E$ be Galois extensions. Let $G_K(E)G_L(E)$ be the join of $G_K(E)$ and $G_L(E)$ in the lattice $\mathcal{G}$ of all subgroups of $G_{K\cap L}(E)$ and let $G_K(E) \vee G_L(E)$ be the join in the lattice $\overline{\mathcal{G}}$ of all *closed* subgroups of $G_{K\cap L}(E)$. Show that $G_K(E)G_L(E)$ is finite if and only if $G_{K\cap L}(E)$ is finite, in which case $G_K(E)G_L(E) = G_K(E) \vee G_L(E)$.
8. Let $F < E$ be finite. Let $G_1 \lhd G_2 < G_F(E)$. Show that

$$G_{\text{fix}(G_2)}(\text{fix}(G_1)) \approx \frac{G_2}{G_1}$$

9. Let $F < E$ and let K and L be intermediate fields with $[K : F] = 2^m$ and $[L : F] = 2^n$. Show that $[KL : F]$ need not have degree a power of 2. *Hint*: The group S_4 has subgroups $A = \langle \sigma \in S_4 \mid \sigma t_4 = t_4 \rangle$ and $B = \langle \sigma \in S_4 \mid \sigma t_1 = t_1 \rangle$. Consider the generic polynomial

$$p(x) = (x - t_1)(x - t_2)(x - t_3)(x - t_4)$$

where $t_1, \ldots, t_4$ are independent variables over F.

10. Find an example of an infinite algebraic extension whose Galois group is finite.
11. Let $t_1, \ldots, t_n$ be independent transcendentals over F and consider the generic polynomial

$$g(x) = \prod_i (x - t_i)$$

Suppose that $g(x)$ has coefficients $s_0, \ldots, s_n$. Then t_i is algebraic over $F(s_0, \ldots, s_n)$ and so $F(s_0, \ldots, s_n) < F(t_1, \ldots, t_n)$ is algebraic. Show that the extension is Galois. Show that the degree of the extension is at most $n!$. Show that the Galois group of this extension is isomorphic to the symmetric group S_n.

12. Prove Corollary 6.5.3.
13. Let $F < E$ be finite and Galois. Let p be a prime for which $[E : F] = p^k m$, with $p \nmid m$. Show that for any $0 \le i < k$, there is an intermediate field K for which $[K : F] = p^i m$.

14. Let F be a perfect field. Define the ***p*-order** of a positive integer n to be the largest exponent e for which $p^e \mid n$. Suppose that $F < E$ is a finite extension and that p is a prime. Suppose that $[E : F]$ has p-order k. Show that for any $0 \le i < k$, F has an extension K_i whose degree has p-order i. Show also that if $[E : F]$ is not a power of p, then $[K_i : F]$ is not a power of p.
15. Let $F < E$ be a finite Galois extension and let $F < K$. Then $[EK : K]$ divides $[E : F]$. Use the following to show that the assumption that $F < E$ be Galois is essential. Let α be the real cube root of 2, let $\omega \neq 1$ be a cube root of 1. Let $F = \mathbb{Q}$, $E = \mathbb{Q}(\alpha\omega)$ and $K = \mathbb{Q}(\alpha)$.
16. Prove the following statements about abelian and cyclic extensions.
    1) If $F < E$ and $F < K$ are abelian, then $F < EK$ is abelian.
    2) If $F < E$ is abelian (cyclic) and $F < K$, then the lifting $K < EK$ is abelian (cyclic).
    3) If $F < K < E$ with $F < E$ abelian (cyclic), then $K < E$ and $F < K$ are abelian (cyclic).
17. Let $f(x) \in F[x]$ with roots $\alpha_1, \ldots, \alpha_n$ in $\overline{F}$. Let $F < E < \overline{F}$. We can consider the splitting field $S_F = F(\alpha_1, \ldots, \alpha_n)$ of $f(x)$ over F as well as the splitting field $S_E = E(\alpha_1, \ldots, \alpha_n)$ of $f(x)$ over E. Note that $S_F < S_E$. Let us examine the Galois groups $G_E(S_E)$ and $G_F(S_F)$.
    a) If $\sigma \in G_E(S_E)$, show that $\sigma|_{S_F} \in G_L(S_F)$, where
    $$L = S_F \cap \mathrm{fix}_{S_E}(G_E(S_E)) = \mathrm{fix}_{S_F}(G_E(S_E))$$
    b) Let $\phi\colon G_E(S_E) \to G_L(S_F)$ be defined by $\phi\sigma = \sigma|_{S_F}$. Show that ϕ is an isomorphism.
18. Referring to Theorem 6.5.4, show that if $\mathcal{F}$ is an arbitrary family then the map
    $$\phi\colon G_F(\bigvee E_i) \to \prod G_F(E_i)$$
    defined by
    $$\pi_i(\phi\sigma) = \sigma|_{E_i}$$
    is an isomorphism if
    $$E_j \cap \Big(\bigvee_{i \neq j} E_i \Big) = F \text{ for all } j \in I$$
19. Prove that $G_F(E)$ is a topological group under the Krull topology. Show that $G_F(E)$ is totally disconnected.
20. a) Show that in every Galois extension $F < E$, there is a largest abelian subextension F^{ab}, that is, $F < F^{\mathrm{ab}} < E$, $F < F^{\mathrm{ab}}$ is abelian and if $F < K < E$ with $F < K$ abelian then $K < F^{\mathrm{ab}}$.
    b) If G is a group, the subgroup G' generated by all **commutators** $[\alpha, \beta] = \alpha\beta(\beta\alpha)^{-1} = \alpha\beta\alpha^{-1}\beta^{-1}$, for $\alpha, \beta \in G$, is called the

commutator subgroup. Show that G' is the smallest subgroup of G for which G/G' is abelian.

c) Let $F \lhd E$. If the commutator subgroup $G_F(E)'$ of a Galois group $G_F(E)$ is closed, that is, if $G_F(E)' = G_K(E)$ for some $F < K < E$, then $K = F^{\text{ab}}$.

21. Let $F \lhd E$. Show that the separable closure F^{sc} of F in E and the purely inseparable closure F^{ic} of F in E are linearly disjoint over F. Moreover, if $F < K < E$ and if K and F^{ic} are linearly disjoint over F then $F < E$ is separable.
22. Let $F < E$ and suppose that S is a set of elements that are algebraically independent over E. Then $F(S)$ and E are linearly disjoint over F.
23. Let $F < K$ and let $F < E < L$. Assume that K and L are contained in a larger field. Then K and L are linearly disjoint over F if and only if K and E are linearly disjoint over F and KE and L are linearly disjoint over E.
24. The following concept is analogous to, but weaker than, that of linear disjointness. Let $F < K$ and $F < L$ be extensions, with K and L contained in a larger field. We say that K is **free from L over F** if whenever $S \subseteq K$ is a finite set of *algebraically* independent elements over F, then S is also algebraically independent over L.

a) The definition given above is not symmetric, but the concept is. In particular, show that if K is free from L over F, then $[KL : L]_t = [K : F]_t$. Let T be a finite F-algebraically independent set of elements of L. Show that T is algebraically independent over K.

b) Let $F < K$ and $F < L$ be field extensions, contained in a larger field. Prove that if K and L are linearly disjoint over F, then they are also free over F.

c) Find an example showing that the converse of part b) does not hold.

Chapter 7

Galois Theory III: The Galois Group of a Polynomial

In this chapter, we pass from the highly theoretical material of the previous chapter to the somewhat more concrete, where we apply the results of the previous chapter to some special Galois correspondences.

7.1 The Galois Group of a Polynomial

The **Galois group of a polynomial** $p(x) \in F[x]$, denoted by $G_F(p(x))$, is defined to be the Galois group of a splitting field S for $p(x)$ over F. If

$$p(x) = p_1^{e_1}(x) \cdots p_k^{e_k}(x)$$

is a factorization of $p(x)$ into powers of distinct irreducible polynomials over F, then S is also a splitting field for the polynomial $q(x) = p_1(x) \cdots p_k(x)$.

Moreover, the extension $F < S$ is separable (and hence Galois) if and only if each $p_i(x)$ is a separable polynomial. To see this, let S_i be the splitting field for $p_i(x)$ satisfying $F < S_i < S$. Then if $F < S$ is separable, so is the lower step $F < S_i$ and therefore so is $p_i(x)$. Conversely, if each factor $p_i(x)$ is separable over F, then S is separably generated over F and so $F < S$ is separable.

Note that each $\sigma \in G_F(S)$ is uniquely determined by its action on the roots of $p(x)$, since these roots generate S, and this action is a permutation of the roots. In fact, if α and β are roots of $p(x)$, then there is a $\sigma \in G(p(x))$ that sends α to β. Hence, the Galois group $G(p(x))$ acts *transitively* on the roots of $p(x)$. However, not all permutations of the roots of $p(x)$ need correspond to an element of $G_F(S)$. Of course, σ must send a root of an irreducible factor of $p(x)$ to another root of the same irreducible factor, but even if $p(x)$ is itself irreducible, not all permutations of the roots of $p(x)$ correspond to elements of the Galois group. Thus, the Galois group $G_F(S)$ is isomorphic to a *transitive subgroup* of the symmetric group S_n, where $n = \deg(p(x))$.

Let $p(x) = f(x)g(x)$ where $\deg(f) > 0$ and let E_p be the splitting field for $p(x)$ over F and E_f the splitting field for $f(x)$ over F. We clearly have $F \lhd E_f < E_p$ and $F \lhd E_p$ and so Theorem 6.4.2 implies that $G_{E_f}(E_p) \lhd G_F(E_p)$ and

$$G_F(E_f) \approx \frac{G_F(E_p)}{G_{E_f}(E_p)}$$

or, in another notation,

$$G_F(f(x)) \approx \frac{G_F(p(x))}{G_{E_f}(p(x))}$$

Theorem 7.1.1 *Let $p(x) = f(x)g(x)$ where $\deg(f(x)) > 0$. The Galois group of $f(x)$ is isomorphic to a quotient group of the Galois group of $p(x)$*

$$G_F(f(x)) \approx \frac{G_F(p(x))}{G_{E_f}(p(x))}$$

where E_f is a splitting field for $p(x)$.□

7.2 Symmetric Polynomials

In this section, we discuss the relationship between the roots of a polynomial and its coefficients. It is well known that the constant coefficient of a polynomial $p(x)$ is the product of its roots and the linear term of $p(x)$ is the negative of the sum of the roots. We wish to expand considerably on these statements.

The Generic Polynomial and Elementary Symmetric Functions

If F is a field and $t_1, \ldots, t_n$ are algebraically independent over F, the polynomial

$$g(x) = \prod_{i=1}^{n} (x - t_i)$$

is referred to as a **generic polynomial** over F of degree n. Since the roots $t_1, \ldots, t_n$ of the generic polynomial $g(x)$ are algebraically independent, this polynomial is, in some sense, the most general polynomial of degree n. Accordingly, it should (and does) have the most general Galois group S_n, as we will see.

It can be shown by induction that the generic polynomial can be written in the form

$$g(x) = x^n - s_1 x^{n-1} + \cdots + (-1)^n s_n$$

where the coefficients $s_k \in F(t_1, \ldots, t_n)$ are given by

$$s_1 = \sum_i t_i, \quad s_2 = \sum_{i<j} t_i t_j, \quad s_3 = \sum_{i<j<k} t_i t_j t_k \quad , \ldots, \quad s_n = \prod_{i=1}^{n} t_i$$

and are called the **elementary symmetric polynomials** in the variables t_i.

As an example of what can be gleaned from the generic polynomial, we deduce immediately the following lemma.

Lemma 7.2.1 *Let $p(x) \in F[x]$. The coefficients of $p(x)$ are, except for sign, the elementary symmetric polynomials of the roots of $p(x)$. In particular, if*

$$p(x) = x^n - s_1 x^{n-1} + \cdots + (-1)^n s_n$$

has roots $r_1, \ldots, r_n$ in a splitting field, then

$$s_k = \sum_{i_1 < \cdots < i_k} r_{i_1} \cdots r_{i_k}$$

□

Since the extension $F(s_1, \ldots, s_n) < F(t_1, \ldots, t_n)$ is algebraic, the elementary symmetric polynomials $s_1, \ldots, s_n$ are also algebraically independent over F, that is, there is no nonzero polynomial over F satisfied by $s_1, \ldots, s_n$.

Theorem 7.2.2 *The elementary symmetric polynomials $s_1, \ldots, s_n$ are algebraically independent over F.*

Proof. Since $F < F(s_1, \ldots, s_n) < F(t_1, \ldots, t_n)$, where the upper step is algebraic, Theorem 4.3.2 implies that $S = \{s_1, \ldots, s_n\}$ contains a transcendence basis for $F(t_1, \ldots, t_n)$ over F. But $\{t_1, \ldots, t_n\}$ is a transcendence basis and so $[F(t_1, \ldots, t_n):F]_t = n$. Hence, S is a transcendence basis. □

The Galois Group of the Generic Polynomial

Let us compute the Galois group G of $F(t_1, \ldots, t_n)$ over $F(s_1, \ldots, s_n)$. Since $F(t_1, \ldots, t_n)$ is a splitting field for $g(x)$ over $F(s_1, \ldots, s_n)$, and since $g(x)$ has no multiple roots, the extension

$$F(s_1, \ldots, s_n) < F(t_1, \ldots, t_n)$$

is finite and Galois and so

$$|G| = [F(t_1, \ldots, t_n) : F(s_1, \ldots, s_n)] \le n!$$

We claim that G is isomorphic to the symmetric group S_n. Let $\sigma \in S_n$. For any $f(t_1, \ldots, t_n) \in F(t_1, \ldots, t_n)$, define a map $\sigma^*: F(t_1, \ldots, t_n) \to F(t_1, \ldots, t_n)$ by

$$\sigma^*(f(t_1, \ldots, t_n)) = f(t_{\sigma(1)}, \ldots, t_{\sigma(n)})$$

Since the t_i's are algebraically independent over F, this is a well-defined automorphism of $F(t_1, \ldots, t_n)$ over F, which fixes the elementary symmetric polynomials s_k. Thus, σ^* is an automorphism of $F(t_1,\ldots,t_n)$ over $F(s_1,\ldots,s_n)$, that is, $\sigma^* \in G$, where

$$\sigma^*(g(t_1,\ldots,t_n)) = g^\sigma(t_{\sigma(1)},\ldots,t_{\sigma(n)}) = g(t_{\sigma(1)},\ldots,t_{\sigma(n)})$$

for any $g(t_1,\ldots,t_n) \in F(s_1,\ldots,s_n)(t_1,\ldots,t_n)$.

Moreover, each σ^* is distinct, since if $\sigma^* = \tau^*$, then $t_{\sigma(i)} = t_{\tau(i)}$ for all i and so $\sigma = \tau$. It follows that G is isomorphic to S_n and

$$[F(t_1,\ldots,t_n) : F(s_1,\ldots,s_n)] = n!$$

Theorem 7.2.3 *Let $t_1,\ldots,t_n$ be algebraically independent over F and let $s_1,\ldots,s_n$ be the elementary symmetric polynomials in $t_1,\ldots,t_n$.*

1) *The extension $F(s_1,\ldots,s_n) < F(t_1,\ldots,t_n)$ is Galois of degree $n!$, with Galois group G isomorphic to the symmetric group S_n.*
2) $\text{fix}(G) = F(s_1,\ldots,s_n)$, *that is, any rational function in $t_1,\ldots,t_n$ that is fixed by the maps σ^* is a rational function in $s_1,\ldots,s_n$.*
3) *The generic polynomial $g(x)$ is irreducible over $F[s_1,\ldots,s_n]$.*

Proof. To prove part 3), observe that if $g(x)$ were equal to $a(x)b(x)$ where $\deg(a(x)) = d > 0$ and $\deg(b(x)) = e > 0$, then the Galois group of $g(x)$ would have size at most $d!e! < (d+e)! = n!$. Hence $g(x)$ is irreducible. □

Symmetric Polynomials

Now we are ready to define symmetric polynomials (and rational functions).

Definition *A rational function $f(t_1,\ldots,t_n) \in F(t_1,\ldots,t_n)$ is* **symmetric** *in $t_1,\ldots,t_n$ if*

$$f(t_{\sigma(1)},\ldots,t_{\sigma(n)}) = f(t_1,\ldots,t_n)$$

for all permutations $\sigma \in S_n$, that is, if $f \in \text{fix}(G) = F(s_1,\ldots,s_n)$, where G is the Galois group of the extension $F(s_1,\ldots,s_n) < F(t_1,\ldots,t_n)$.□

A famous theorem of Isaac Newton describes the symmetric *polynomials.*

Theorem 7.2.4 (Newton's Theorem) *Let $t_1,\ldots,t_n$ be algebraically independent over F and let $s_1,\ldots,s_n$ be the elementary symmetric polynomials in $t_1,\ldots,t_n$.*

1) *A polynomial $p(t_1,\ldots,t_n) \in F[t_1,\ldots,t_n]$ is symmetric in $t_1,\ldots,t_n$ if and only if it is a polynomial in $s_1,\ldots,s_n$, that is, if and only if*

$$p(t_1,\ldots,t_n) = q(s_1,\ldots,s_n)$$

for some polynomial $q(x_1,\ldots,x_n)$ over F. Moreover, if $p(t_1,\ldots,t_n)$ has integer coefficients, then so does $q(s_1,\ldots,s_n)$.

2) *Let $p(x) \in F[x]$. Then the set of* symmetric *polynomials over F in the roots of $p(x)$ is equal to the set of polynomials over F in the coefficients of $p(x)$.*

In particular, any symmetric polynomial over F in the roots of $p(x)$ is an element of F.

3) *Let $p(x) \in \mathbb{Z}[x]$ be a polynomial with integer coefficients. Then the set of* symmetric *polynomials over $\mathbb{Z}$ in the roots of $p(x)$ is equal to the set of polynomials over $\mathbb{Z}$ in the coefficients of $p(x)$. In particular, any symmetric polynomial over $\mathbb{Z}$ in the roots of $p(x)$ is an integer.*

Proof. Statements 2) and 3) follow from statement 1) and Lemma 7.2.1. If $p(t_1, \ldots, t_n)$ has the form $q(s_1, \ldots, s_n)$, then it is clearly symmetric. For the converse, the proof consists of a procedure that can be used to construct the polynomial $q(x_1, \ldots, x_n)$. Unfortunately, while the procedure is quite straightforward, it is recursive in nature and not at all practical.

We use induction on n. The theorem is true for $n = 1$, since $s_1 = t_1$. Assume that the theorem is true for any number of variables less than n and let $p(t_1, \ldots, t_n)$ be symmetric. By collecting powers of t_n, we can write

$$p(t_1, \ldots, t_n) = p_0 + p_1 t_n + p_2 t_n^2 + \cdots + p_k t_n^k$$

where each p_i is a polynomial in $t_1, \ldots, t_{n-1}$. Since p is symmetric in $t_1, \ldots, t_{n-1}$ and $t_1, \ldots, t_n$ are independent, each of the coefficients p_i is symmetric in $t_1, \ldots, t_{n-1}$. By the inductive hypothesis, we may express each p_i as a polynomial in the elementary symmetric polynomials on $t_1, \ldots, t_{n-1}$. If these elementary symmetric polynomials are denoted by $u_1, \ldots, u_{n-1}$, then

$$p(t_1, \ldots, t_n) = q_0 + q_1 t_n + q_2 t_n^2 + \cdots + q_k t_n^k \tag{7.2.1}$$

where each q_i is a polynomial in $u_1, \ldots, u_{n-1}$, with integer coefficients if p has integer coefficients.

Note that the symmetric functions s_i can be expressed in terms of the symmetric functions u_i as follows

$$\begin{aligned} s_1 &= u_1 + t_n \\ s_2 &= u_2 + u_1 t_n \\ &\vdots \\ s_{n-1} &= u_{n-1} + u_{n-2} t_n \\ s_n &= u_{n-1} t_n \end{aligned} \tag{7.2.2}$$

These expressions can be solved for the u_i's in terms of the s_i's, giving

$$\begin{aligned} u_1 &= s_1 - t_n \\ u_2 &= s_2 - u_1 t_n = s_2 - s_1 t_n + t_n^2 \\ u_3 &= s_3 - u_2 t_n = s_3 - s_2 t_n + s_1 t_n^2 - t_n^3 \\ &\vdots \\ u_{n-1} &= s_{n-1} - u_{n-2} t_n = s_{n-1} - s_{n-2} t_n + \cdots + (-1)^{n-1} t_n^{n-1} \end{aligned}$$

and from the last equation in (7.2.2),

$$0 = s_n - u_{n-1}t_n = s_n - s_{n-1}t_n + \cdots + (-1)^n t_n^n \qquad (7.2.3)$$

Substituting these expressions for the u_i's into (7.2.1) gives

$$p(t_1, \ldots, t_n) = r_0 + r_1 t_n + r_2 t_n^2 + \cdots + r_k t_n^k$$

where each r_i is a polynomial in $s_1, \ldots, s_{n-1}$ and t_n, with integer coefficients if p has integer coefficients. Again, we may gather together powers of t_n, to get

$$p(t_1, \ldots, t_n) = g_0 + g_1 t_n + g_2 t_n^2 + \cdots + g_m t_n^m$$

where each g_i is a polynomial in $s_1, \ldots, s_{n-1}$, with integer coefficients if p has integer coefficients. If $m \geq n$, we may reduce the degree in t_n by using (7.2.3), which also introduces the term s_n. Hence,

$$p(t_1, \ldots, t_n) = h_0 + h_1 t_n + h_2 t_n^2 + \cdots + h_{n-1} t_n^{n-1} \qquad (7.2.4)$$

where each h_i is a polynomial in $s_1, \ldots, s_n$, with integer coefficients if p has integer coefficients.

Since the left side of (7.2.4) is symmetric in the t_i's, we may interchange t_n and t_i, for each $i = 1, \ldots, n-1$, to get

$$p(t_1, \ldots, t_n) = h_0 + h_1 t_i + h_2 t_i^2 + \cdots + h_{n-1} t_i^{n-1}$$

valid for all $i = 1, \ldots, n$. Hence, the polynomial

$$P(x) = h_0 + h_1 x + h_2 x^2 + \cdots + h_{n-1} x^{n-1} - p(t_1, \ldots, t_n)$$

has degree (in x) at most $n-1$ but has n distinct roots $t_1, \ldots, t_n$, whence it must be the zero polynomial. Thus, $h_i = 0$ for $i \geq 1$ and $p(t_1, \ldots, t_n) = h_0 = h_0(s_1, \ldots, s_n)$, as desired. □

Example 7.2.1 Let $p(x) = x^n - p_1 x^{n-1} + \cdots + (-1)^n p_n$ be a polynomial with roots $r_1, \ldots, r_n$ in a splitting field. For $k \geq 1$, the polynomials

$$u_k = r_1^k + r_2^k + \cdots + r_n^k$$

are symmetric in the roots of $p(x)$, and so Theorem 7.2.4 implies that the u_k's can be expressed as polynomials in the elementary symmetric polynomials $p_1, \ldots, p_n$ of the roots. One way to derive an expression relating the u_k's to the p_k's is by following the proof of Theorem 7.2.4. In the exercises, we ask the reader to take another approach to obtain the so-called **Newton identities**

$$u_k - u_{k-1}p_1 + u_{k-2}p_2 + \cdots + (-1)^{k-1} u_1 p_{k-1} + (-1)^k k p_k = 0$$

for $k \geq 1$. These identities can be used to compute recursively the u_k's in terms of the p_i's. □

7.3 The Fundamental Theorem of Algebra

The Galois correspondence can be used to provide a simple proof of the fundamental theorem of algebra.

As an aside, the history of the fundamental theorem is quite interesting. It seems that attempts to prove the fundamental theorem began with d'Alembert in 1746, based on geometric properties of the complex numbers and the concept of continuity, which was not well understood at that time.

In 1799, Gauss gave a critique of the existing "proofs" of the fundamental theorem, showing that they had serious flaws, and attempted to produce a rigorous proof. However, his proof also had gaps, since he suffered from the aforementioned lack of complete understanding of continuity. Subsequently, in 1816, Gauss gave a second proof that minimized the use of continuity, *assuming* a form of the intermediate value theorem.

It was not until Weierstrass put the basic properties of continuity on a rigorous foundation, in about 1874, that d'Alembert's proof and the second proof of Gauss could be made completely rigorous.

We will also assume a form of the intermediate value theorem, namely, that if $p(x)$ is a real polynomial, and if $p(a)$ and $p(b)$ have opposite signs, for $a < b$, then there is a $c \in (a, b)$ for which $p(c) = 0$. From this, one can deduce that any odd degree real polynomial must have a real root and is therefore reducible over $\mathbb{R}$. It follows that any nontrivial finite extension of $\mathbb{R}$ must have even degree, since it must contain an element whose minimal polynomial has even degree.

We also require some knowledge of complex numbers, namely, that every complex number has a complex square root, which can be seen from a geometric perspective: $z = re^{i\theta}$ implies $z^{1/2} = r^{1/2}e^{i\theta/2}$. Hence, no complex quadratic $p(x)$ is irreducible over $\mathbb{C}$, since the method of completing the square shows that the roots of $p(x)$ lie in $\mathbb{C}$. It follows that $\mathbb{C}$ has no extensions of degree 2.

Theorem 7.3.1 (The fundamental theorem of algebra) *Any nonconstant polynomial over $\mathbb{C}$ has a root in $\mathbb{C}$, that is, $\mathbb{C}$ is algebraically closed.*

Proof. We first show that it is sufficient to prove the theorem for real polynomials. Let $p(x) \in \mathbb{C}[x]$ be nonconstant. Consider the polynomial $r(x) = p(x)\overline{p}(x)$, where the overbar denotes complex conjugation of the coefficients. Then $r(x)$ is a real polynomial and $r(x)$ has a complex root if and only if $p(x)$ has a complex root. Hence, we may assume that $p(x) \in \mathbb{R}[x]$.

Now consider the tower $\mathbb{R} < \mathbb{C} < E$, where E is a splitting field for $q(x) = (x^2 + 1)p(x)$ over $\mathbb{R}$. Since $[\mathbb{C} : \mathbb{R}] = 2$ divides $[E : \mathbb{R}]$, we conclude that $[E:\mathbb{R}] = 2^k m$, for some $k \geq 1$ with m odd. Our goal is to show that $E = \mathbb{C}$, showing that $p(x)$ splits over $\mathbb{C}$.

Let H be a 2-Sylow subgroup of $G_{\mathbb{R}}(q(x))$. Then $|H| = 2^k$ and so

$$[\mathrm{fix}(H):\mathbb{R}] = (G_{\mathbb{R}}(q(x)):H) = m$$

Since $\mathbb{R}$ has no nontrivial extensions of odd degree, we deduce that $m = 1$ and $G = G_{\mathbb{R}}(q(x))$ is a 2-group of order $2^k \geq 2$.

Thus, we have the tower

$$\{\iota\} < G_{\mathbb{C}}(q(x)) < G$$

in which $|G_{\mathbb{C}}(q(x))| = 2^{k-1}$. Therefore, according to Theorem 0.2.19, $G_{\mathbb{C}}(q(x))$ has a subgroup of any order dividing 2^{k-1}. But $G_{\mathbb{C}}(q(x))$ cannot have a subgroup of order 2^{k-2}, that is, index 2

$$\{\iota\} < H < G_{\mathbb{C}}(q(x)) < G$$

because then

$$2 = [\mathrm{fix}(H):\mathrm{fix}(G_{\mathbb{C}}(q(x))] = [\mathrm{fix}(H):\mathbb{C}]$$

which is not possible. Hence, $|G_{\mathbb{C}}(q(x))| = 1$ and so $|G| = 2$, which implies that $[E:\mathbb{R}] = 2$, whence $E = \mathbb{C}$.$\square$

7.4 The Discriminant of a Polynomial

We have seen that the Galois group $G_F(p(x))$ of a polynomial of degree n is isomorphic to a subgroup of the symmetric group S_n and that the Galois group of a generic polynomial is isomorphic to S_n itself. A special symmetric function of the roots of $p(x)$, known as the *discriminant*, provides a tool for determining whether the Galois group is isomorphic to a subgroup of the alternating group A_n.

Let $p(x)$ be a polynomial over F, with roots $r_1, \ldots, r_n$ in a splitting field E. Let

$$\delta = \prod_{i<j} (r_i - r_j)$$

The **discriminant** of $p(x)$ is $\Delta = \delta^2$, which is clearly symmetric in the roots. Note that $\Delta \neq 0$ if and only if $p(x)$ has no multiple roots.

Let us assume that $\Delta \neq 0$. Then $p(x)$ is the product of distinct separable polynomials, implying that $F < E$ is a Galois extension. Hence,

$$\mathrm{fix}(G_F(p(x))) = F$$

Since $\sigma\Delta = \Delta$ for all $\sigma \in G_F(p(x))$, we deduce that $\Delta \in F$. (Newton's theorem also implies that $\Delta \in F$.)

Each transposition of the roots sends δ to $-\delta$, and so for any $\sigma \in G_F(p(x))$,

$$\sigma\delta = (-1)^\sigma \delta$$

where $(-1)^\sigma$ is 1 if σ is an even permutation and -1 if σ is an odd permutation. Thus, the location of δ can give us some information about the parity of the permutations in the Galois group.

If $\operatorname{char}(F) = 2$, then $\sigma\delta = \delta$ for all $\sigma \in G_F(p(x))$ and so δ is always in the base field F. This is not very helpful. But if $\operatorname{char}(F) \neq 2$, then $\sigma \in G_F(p(x))$ fixes δ if and only if σ is an even permutation. Put another way, $\delta \in F = \operatorname{fix}(G_F(p(x)))$ if and only if $G_F(p(x))$ contains only even permutations, that is, $G_F(p(x)) < A_n$.

If $\delta \notin F$ then $G_F(p(x))$ must contain an odd permutation. It is not hard to show that if a subgroup of S_n contains an odd permutation then the subgroup has even order and exactly half of its elements are even.

Hence, if $\delta \notin F$ then $G = G_F(p(x))$ has even order and $|G \cap A_n| = |G|/2$, that is,

$$(G : G \cap A_n) = 2$$

Since all groups are closed, it follows that

$$[\operatorname{fix}(G \cap A_n) : F] = (G : G \cap A_n) = 2$$

Since $[F(\delta) : F] = 2$ and $F(\delta) \subseteq \operatorname{fix}(G \cap A_n)$, we have

$$F(\delta) = \operatorname{fix}(G \cap A_n)$$

Thus, $F(\delta)$ is the fixed field of the subgroup of even permutations in $G_F(p(x))$. Let us summarize.

Theorem 7.4.1 *Let $p(x) \in F[x]$ have degree n and splitting field E. Let $\sqrt{\Delta}$ be any square root of the discriminat Δ of $p(x)$.*

1) $\Delta = 0$ if and only if $p(x)$ has multiple roots in E.

2) Assume that $\Delta \neq 0$ and $\operatorname{char}(F) \neq 2$.

a) $\sqrt{\Delta} \in F$ if and only if $G_F(p(x))$ is isomorphic to a subgroup of A_n.

b) $\sqrt{\Delta} \notin F$ if and only if $G_F(p(x))$ is isomorphic to a subgroup of S_n that contains half odd and half even permutations. In this case,

$$\operatorname{fix}(G_F(p(x)) \cap A_n) = F(\sqrt{\Delta})$$

3) If $\Delta \neq 0$ and $\operatorname{char}(F) = 2$, then $\sqrt{\Delta} \in F$ but $G_F(p(x))$ need not be isomorphic to a subgroup of A_n.

Proof. For part 3), recall that the generic polynomial $g(x) = (x - t_1)\cdots(x - t_n)$ has Galois group S_n over $F(s_1, \ldots, s_n)$.□

The usefulness of Theorem 7.4.1 comes from the fact that Δ can actually be computed without knowing the roots of $p(x)$ explicitly. This follows from the fact that δ is the *Vandermonde determinant*

$$\delta = \begin{vmatrix} 1 & 1 & 1 & 1 \\ r_1 & r_2 & \cdots & r_n \\ \vdots & \vdots & \cdots & \vdots \\ r_1^{n-1} & r_2^{n-1} & \cdots & r_n^{n-1} \end{vmatrix}$$

Multiplying this by its transpose gives

$$\Delta = \begin{vmatrix} u_0 & u_1 & \cdots & u_{n-1} \\ u_1 & u_2 & \cdots & u_n \\ \vdots & \vdots & \cdots & \vdots \\ u_{n-1} & u_n & \cdots & u_{2n-2} \end{vmatrix}$$

where $u_i = r_1^i + r_2^i + \cdots + r_n^i$. Newton's identities can then be used to determine the u_i's in terms of the coefficients of the polynomial in question (see Example 7.2.1 and the exercises). We will see some examples of this in the next section.

7.5 The Galois Groups of Some Small-Degree Polynomials

We now examine the Galois groups of some small-degree polynomials.

The Quadratic

Quadratic extensions (extensions of degree 2) hold no surprises. Let

$$p(x) = x^2 + bx + c = (x - r)(x - s)$$

be a quadratic over F, with splitting field E. To compute the discriminant, observe that $u_1 = r + s = b$ and

$$u_2 = r^2 + s^2 = (r + s)^2 - 2rs = b^2 - 2c$$

Hence

$$\Delta = \begin{vmatrix} 2 & b \\ b & b^2 - 2c \end{vmatrix} = 2(b^2 - 2c) - b^2 = b^2 - 4c$$

a familiar quantity.

Multiple Roots

If $\Delta = 0$, then $p(x)$ has a double root r and

$$p(x) = (x - r)^2 = x^2 - 2rx + r^2$$

The root r will lie in F for most well-behaved base fields F. In particular, if $\operatorname{char}(F) \neq 2$, then $-2r \in F$ implies $r \in F$. If $\operatorname{char}(F) = 2$ and F is perfect (a finite field, for example) then $p(x) = (x-r)^2$ must be reducible over F and so $r \in F$.

However, the following familiar example shows that $p(x)$ may have a multiple root not lying in F. Let $F = \mathbb{Z}_2(t^2)$ where t is transcendental over $\mathbb{Z}_2$ and let

$$p(x) = x^2 - t^2 = (x-t)^2$$

Since $t \notin \mathbb{Z}_2(t^2)$, this polynomial is irreducible over $\mathbb{Z}_2(t^2)$, but has a multiple root $t \notin F$.

No Multiple Roots

If $\Delta \neq 0$, then $p(x)$ has distinct roots and there are two possibilities:

1) The roots lie in F, $p(x)$ is reducible and $G_F(p(x))$ is trivial.
2) The roots do not lie in F, $p(x)$ is irreducible and $G_F(p(x)) \approx S_2$ is generated by the transposition (rs) of the roots.

Thus, when $\operatorname{char}(F) \neq 2$, we can tell whether the roots lie in F by looking at the discriminant: If $\sqrt{\Delta} \in F$, then $G_F(p(x)) \approx S_2 = \mathbb{Z}_2$ and possibility 2) obtains. Of course, this is also evident from the quadratic formula

$$r, s = \frac{-b \pm \sqrt{b^2 - 4c}}{2} = \frac{-b \pm \sqrt{\Delta}}{2}$$

If $\sqrt{\Delta} \neq F$, then $G_F(p(x)) = \{\iota\} = A_2$. Hence the roots lie in F if and only if $\sqrt{\Delta} \in F$. (We can now rest assured that what we tell our children about quadratic equations is actually true.)

Theorem 7.5.1 *Let $p(x) \in F[x]$ have degree 2.*
1) If $\Delta = 0$ then $p(x) = (x-r)^2$ has a double root r. If $\operatorname{char}(F) \neq 2$ or F is perfect, then $r \in F$. In any case, $G_F(p(x))$ is trivial.
2) If $\Delta \neq 0$ then $p(x)$ has distinct roots and there are two possibilities:
   a) The roots lie in F, $p(x)$ is reducible and $G_F(p(x))$ is trivial.
   b) The roots do not lie in F, $p(x)$ is irreducible and $G_F(p(x)) \approx S_2$ is generated by the transposition (rs) of the roots.

   When $\operatorname{char}(F) \neq 2$, we can distinguish the two cases as follows: Case 1) holds if $\sqrt{\Delta} \in F$ and case 2) holds if $\sqrt{\Delta} \neq F$. $\square$

Let us turn now to a more interesting case.

The Cubic

Let

$$p(x) = x^3 + bx^2 + cx + d = (x - r)(x - s)(x - t) \in F[x]$$

have splitting field E. Then $p(x)$ is irreducible if and only if none of its roots lie in F.

If $p(x)$ splits over F then $E = F$ and its Galois group is trivial. If $p(x)$ is reducible but does not split, then it can be factored over F:

$$p(x) = (x - \alpha)(x^2 + px + q)$$

where $q(x) = x^2 + px + q$ is irreducible over F. Hence, $[E : F] = 2$ and the Galois group is isomorphic to $\mathbb{Z}_2$.

Now let us assume that $p(x)$ is irreducible. A lengthy computation gives

$$\Delta = -4b^3d + b^2c^2 + 18bcd - 4c^3 - 27d^2$$

If $\Delta = 0$, then $p(x)$ has multiple roots and since each root must have the same multiplicity, we are left with $\mathrm{char}(F) = 3$ and

$$p(x) = (x - r)^3 = x^3 - r^3$$

Hence, the extension $F < F(r) = E$ is purely inseparable of degree 3 and the Galois group is trivial.

If $\Delta \neq 0$, then $p(x)$ has no multiple roots and is therefore separable. Hence, $F < E$ is Galois and

$$3 \leq |G_F(p(x))| = [E : F] \leq 3!$$

which leaves the possibilities $[E : F] = 3$ and $[E : F] = 6$.

We can now give a complete analysis for the cubic. Note that when $\mathrm{char}(F) \neq 2$, knowledge of irreducibility and the value of $\sqrt{\Delta}$ determine the Galois group and the splitting field.

Theorem 7.5.2 (The cubic) *Let $p(x) \in F[x]$ have degree 3, with splitting field E and Galois group $G = G_F(p(x))$. Then there are four mututally exclusive possibilities, each of which can be characterized in four equivalent ways:*

1) *a)* $[E : F] = 1$
   b) *$E = F$ is the splitting field for $p(x)$*
   c) $G_F(p(x)) = \{\iota\} \approx A_2$
   d) *(For* $\mathrm{char}(F) \neq 2$*) $p(x)$ is reducible and $\sqrt{\Delta} \in F$.*

2) *a)* $[E : F] = 2$

  b) *$p(x)$ is reducible and $E = F(r)$ is a splitting field for $p(x)$, where r is a root not in F.*
  c) $G_F(p(x)) \approx \mathbb{Z}_2 \approx S_2$
  d) *(For* $\operatorname{char}(F) \neq 2$) *$p(x)$ is reducible and $\sqrt{\Delta} \notin F$.*
3) a) $[E : F] = 3$
  b) *$p(x)$ is irreducible and $E = F(r)$ is the splitting field for $p(x)$, for any root r*
  c) $G_F(p(x)) \approx \mathbb{Z}_3 \approx A_3$
  d) *(For* $\operatorname{char}(F) \neq 2$) *$p(x)$ is irreducible and $\sqrt{\Delta} \in F$.*
4) a) $[E : F] = 6$
  b) *$p(x)$ is irreducible and $E = F(\sqrt{\Delta}, r)$ is the splitting field for $p(x)$, for any root r*
  c) $G_F(p(x)) \approx S_3$
  d) *(For* $\operatorname{char}(F) \neq 2$) *$p(x)$ is irreducible and $\sqrt{\Delta} \notin F$.*

Proof. We leave proof to the reader.□

We know that $\Delta \in F$. For $F = \mathbb{Q}$, we can learn more about the roots of a cubic by looking at the sign of Δ. A cubic $p(x)$ over $\mathbb{Q}$ has either one reat root r and two nonreal roots $\{a + bi, a - bi\}$ or three real roots r, s and t. In the former case,

$$\delta = [(r - a) - bi][(r - a) + bi]2bi = |(r - a) + bi|^2 2bi$$

and so $\Delta < 0$. In the latter case, $\delta = [(r - s)(r - t)(s - t)]^2 > 0$.

Theorem 7.5.3 (The cubic over $\mathbb{Q}$) Let $p(x) \in \mathbb{Q}[x]$ have degree 3. Then
1) $\Delta < 0$ *if and only if $p(x)$ has exactly one real root*
2) $\Delta > 0$ *if and only if $p(x)$ has three real roots.*□

Example 7.5.1 Let $p(x) = x^3 - 2x^2 - x + 1$ over $\mathbb{Q}$. Any rational root of $p(x)$ must be ± 1 (Theorem 1.2.3) and so $p(x)$ is irreducible. The discriminant is $\Delta = 49 > 0$, so $p(x)$ has three real roots. Since $\sqrt{49} \in \mathbb{Q}$, we have $G_{\mathbb{Q}}(p(x)) \approx \mathbb{Z}_3$ and $p(x)$ has splitting field $\mathbb{Q}(r)$, for any root r.

On the other hand, for any prime p, the polynomial $p(x) = x^3 - p$ is irreducible over $\mathbb{Q}$ and has discriminant $\Delta = -27p^2 < 0$, whose square root is not in $\mathbb{Q}$. Hence, $p(x)$ has one real root and two nonreal roots, the Galois group of $p(x)$ is isomorphic to S_3 and $p(x)$ has splitting field $\mathbb{Q}(\sqrt{-3}, \sqrt[3]{p})$.□

**The Quartic*

Since the Galois group of an irreducible quartic polynomial is isomorphic to a transitive subgroup of S_4, we should begin by determining all such subgroups. Theorem 0.3.2 implies that if G is a transitive subgroup of S_4 then

$$|G| = 4, 8, 12 \text{ or } 24$$

Here is a list.

1) **(Order 4: cyclic group)** The cyclic group $\mathbb{Z}_4$ occurs as a subgroup of S_4. The elements of S_4 of order 4 are the 4-cycles $\sigma = (1abc) = (1c)(1b)(1a)$. The three subgroups of S_4 isomorphic to $\mathbb{Z}_4$ are

$$Z_1 = \{\iota, (1234), (13)(24), (1432)\}$$
$$Z_2 = \{\iota, (1342), (14)(23), (1243)\}$$
$$Z_3 = \{\iota, (1423), (12)(34), (1324)\}$$

2) **(Order 4: Klein four-group)** The Klein four-group $\mathbb{Z}_2 \times \mathbb{Z}_2$ occurs as a subgroup of S_4. In particular, let

$$V = \{\iota, (12)(34), (13)(24), (14)(23)\}$$

which is isomorphic to $\mathbb{Z}_2 \times \mathbb{Z}_2$. We leave it to the reader to show that V is normal in S_4. Note also that $V < A_4$. This and the previous case exhaust all nonisomorphic groups of order 4. The group S_4 contains other isomorphic copies of the Klein four group, such as

$$\{\iota, (12), (34), (12)(34)\}$$

However, suppose that such a subgroup S is transitive. Every nonidentity element $\sigma \in S$ has order 2 and so is a product of disjoint 2-cycles (transpositions). Hence, σ is a transposition or a product of two disjoint transpositions. But a transposition links only two elements of $\{1, 2, 3, 4\}$ together and a product of disjoint transpositions links two pairs of elements together. Since there are $\binom{4}{2} = 6$ pairs that must be linked, we deduce that S contains no transpositions and therefore must be V.

3) **(Order 8: dihedral group)** The dihedral group of symmetries of the square, thought of as permutations of the corners of the square, occurs as a subgroup of S_4 of order 8. These subgoups are Sylow subgroups

$$D_1 = \{\iota, (12)(34), (13)(24), (14)(23), (24), (13), (1234), (1432)\}$$
$$D_2 = \{\iota, (12)(34), (13)(24), (14)(23), (14), (23), (1243), (1342)\}$$
$$D_3 = \{\iota, (12)(34), (13)(24), (14)(23), (12), (34), (1423), (1324)\}$$

Note that $V < D_i$, for each i.

4) **(Order 12: alternating group)** The alternating group A_4 is the only subgroup of S_4 of order 12.

5) **(Order 24: symmetric group)** Of course, S_4 is the only subgroup of S_4 of order 24.

Now let

$$p(x) = x^4 + ax^3 + bx^2 + cx + d$$

be an irreducible quartic over F and assume that $\operatorname{char}(F) \neq 2, 3$. This will insure that $4 \neq 0$, that $p(x)$ is separable and that all irreducible cubic polynomials that we may encounter are separable.

Replacing x by $x - a/4$ will eliminate the cubic term, resulting in a polynomial of the form

$$q(x) = x^4 + px^2 + qx + r$$

which is often referred to as the **reduced polynomial** for $p(x)$. The polynomials $p(x)$ and $q(x)$ have the same splitting field and hence the same Galois group, and their sets of roots are easily computed, one from the other. Let E be the splitting field of $q(x)$, let $r_1, \ldots, r_4$ be the roots of $q(x)$ in E and let $G = G_F(E)$ be its Galois group.

For convenience, we identify G with its isomorphic image in S_4. For example, the permutation (12) interchanges r_1 and r_2.

To analyze the quartic $q(x)$, we want to find a strategically placed intermediate field. One way to do this is to find a strategically placed subgroup of the Galois group, one that has nice intersection properties with the candidates listed above.. The alternating group A_4 immediately springs to mind, but this may be too large. In fact, if $\sqrt{\Delta} \in F$ then G is a subgroup of A_4. So let us try the Klein four group V, which gives us a subgroup $V \cap G$ of G, as shown in Figure 7.5.1.

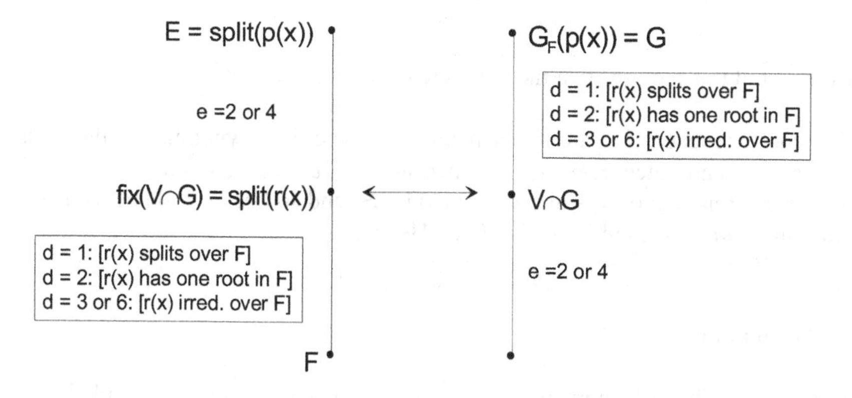

Figure 7.5.1

Comparing with the candidates for G, we have

1) $V \cap Z_1 = \{\iota, (13)(24)\} \approx \mathbb{Z}_2$
2) $V \cap Z_2 = \{\iota, (14)(23)\} \approx \mathbb{Z}_2$

3) $V \cap Z_3 = \{\iota, (12)(34)\} \approx \mathbb{Z}_2$
3) $V \cap V = V = \{\iota, (12)(34), (13)(24), (14)(23)\}$
4) $V \cap D_i = V$, for $i = 1, 2, 3$
5) $V \cap A_4 = V$
6) $V \cap S_4 = V$

Thus,

1) $|G| = 4, 8, 12$ or 24
2) $|V \cap G| = 2$ or 4
3) $(G : V \cap G) = 1, 2, 3, 4, 6, 12$

We next determine the fixed field $K = \text{fix}(V \cap G)$. Each element of V fixes the expressions

$$\begin{aligned} u &= (r_1 + r_2)(r_3 + r_4) \\ v &= (r_1 + r_3)(r_2 + r_4) \\ w &= (r_1 + r_4)(r_2 + r_3) \end{aligned}$$

and so $F(u, v, w) < K$. By checking each permutation in S_4, it is not hard to see that no permutation outside of V fixes u, v and w. Thus,

$$G_{F(u,v,w)}(E) < V \cap G$$

Taking fixed fields gives $K < F(u, v, w)$ and so

$$K = \text{fix}(V \cap G) = F(u, v, w)$$

We would like to show that K is the splitting field for the cubic polynomial

$$r(x) = (x - u)(x - v)(x - w)$$

over F, but this requires that the coefficients of $r(x)$ lie in F.

The coefficients of $r(x)$ are the elementary symmetric polynomials of the roots u, v and w and since every $\sigma \in S_4$ permutes u, v and w, it follows that *any* symmetric function of u, v and w is fixed by S_4 and so lies in F. Thus, K is the splitting field for the cubic $r(x) \in F[x]$. Hence,

$$(G : V \cap G) = [K : F] = 1, 2, 3 \text{ or } 6$$

as shown in Figure 7.5.1.

Definition The polynomial $r(x) = (x - u)(x - v)(x - w)$ is called the **resolvent cubic** of $q(x) = x^4 + px^2 + qx + r$.$\square$

The Coefficients of the Resolvent Cubic

Now let us determine the coefficients of the resolvent cubic $r(x)$. First note that since $q(x)$ has no cubic term, it follows that $r_1 + r_2 + r_3 + r_4 = 0$. Then if we write

$$r_{ij} = r_i + r_j$$

then $u = -r_{12}^2, v = -r_{13}^2$ and $w = -r_{14}^2$. Now write $q(x)$ as a product of quadratic polynomials over E, say

$$q(x) = (x^2 + ax + b)(x^2 - ax + c)$$

where the linear coefficients are negatives of each other since $q(x)$ has no cubic term, and where the roots of the first factor are r_1 and r_2. Then $a = -r_{12}$ and so

$$a^2 = r_{12}^2 = -u$$

Multiplying out the expression for $q(x)$ and equating coefficients gives the equations

$$\begin{aligned} b + c - a^2 &= p \\ ac - ab &= q \\ bc &= r \end{aligned}$$

Solving the first two for b and c and substituting into the third gives

$$a^6 + 2pa^4 + (p^2 - 4r)a^2 - q^2 = 0$$

and so $a^2 = -u$ satisfies the polynomial

$$s(x) = x^3 + 2px^2 + (p^2 - 4r)x - q^2$$

and u satisfies the polynomial

$$t(x) = x^3 - 2px^2 + (p^2 - 4r)x + q^2$$

But we can repeat this arguement, factoring $q(x)$ into a product of quadratics for which the roots of the first quadratic are r_1 and r_3, say

$$q(x) = (x^2 + a'x + b')(x^2 - a'x + c')$$

and so $a' = -r_{13}$ and $(a')^2 = -v$. The same algebra as before leads to the fact that $t(v) = 0$. Similarly, $t(w) = 0$ and so $t(x)$ is the resolvent cubic of $q(x)$.

Final Analysis of the Quartic

The first thing to note is that the discriminants of $q(x)$ and $r(x)$ are equal: $\Delta_q = \Delta_r$. We leave verification of this as an exercise. Let G_q be the Galois group of $q(x)$ and let G_r be the Galois group of $r(x)$. The following can be gleaned from Theorem 7.5.2.

1) If $r(x)$ is reducible and $\sqrt{\Delta_r} \in F$ (in which case $r(x)$ splits over F), then $(G : V \cap G) = [K : F] = 1$ and so $|V \cap G| = 4 = |V|$. Hence, $G = V$.
2) If $r(x)$ is reducible and $\sqrt{\Delta_r} \notin F$ (in which case $r(x)$ has a single root in F), then $(G : V \cap G) = [K : F] = 2$ and there are two possibilities. If

$|V \cap G| = 2$ then $|G| = 4$ and so $G = Z_i$ for $i = 1, 2$ or 3, or $G = V$. But $G = V$ is not possible, so $G = Z_i$. Note that in this case, since E is the splitting field for $p(x)$ over K and $[E : K] = 2$ the polynomial $p(x)$ must have an irreducible quadratic factor over K. If $|V \cap G| = 4$ then $|G| = 8$ and $G = D_i$, for $i = 1, 2$ or 3. In this case, $p(x)$ is irreducible over K.

3) If $r(x)$ is irreducible and $\sqrt{\Delta_r} \in F$, then $G_r \approx A_3$ and $G_p < A_4$. Hence $(G : V \cap G) = [K : F] = 3$, which implies that $|V \cap G| = 4$ and so $|G| = 12$. Thus $G = A_4$.
4) If $r(x)$ is irreducible and $\sqrt{\Delta_r} \notin F$, then $G_r \approx S_3$ and G_q is not a subgroup of A_4. Hence $(G : V \cap G) = [K : F] = 6$. and so $|G| = 12$ or 24. But $G \not\approx A_4$ and so $G = S_4$.

Theorem 7.5.4 (The quartic) *Let*

$$p(x) = x^4 + ax^3 + bx^2 + cx + d$$

be an irreducible quartic over a field F, with $\text{char}(F) \neq 2, 3$. *Let E be the splitting field for $p(x)$ over F. Let*

$$q(x) = x^4 + px^2 + qx + r$$

be obtained from $p(x)$ by substituting $x - a/4$ for x and let

$$r(x) = x^3 - 2px^2 + (p^2 - 4r)x + q^2$$

be the resolvent cubic of $q(x)$. Let G_p be the Galois group of $p(x)$ and let G_r be the Galois group of $r(x)$. Then $\Delta_p = \Delta_r$.
If $r(x)$ is reducible over F then

1) *If $\sqrt{\Delta_r} \in F$, then $G_p = V$.*
2) *If $\sqrt{\Delta_r} \notin F$, there are two possibilities. Let $K = \text{fix}(V \cap G_p)$.*
   a) *$|V \cap G_p| = 2$ and $G_p = Z_i$, for $i = 1, 2$ or 3, which occurs if and only if $p(x)$ is reducible over K, in which case $p(x)$ has an irreducible quadratic factor over K. In this case, $[K : F] = [E : K] = 2$.*
   b) *$|V \cap G_p| = 4$ and $G_p = D_i$, for $i = 1, 2$ or 3, which occurs if and only if $p(x)$ is irreducible over K. In this case, $[K : F] = 2$ and $[E : K] = 4$.*

If $r(x)$ is irreducible over F then

3) *If $\sqrt{\Delta_r} \in F$, then $G_p = A_4$.*
4) *If $\sqrt{\Delta_r} \notin F$, then $G_p = S_4$.* □

The Quartic $x^4 + bx^2 + d$

Consider the special quartic

$$p(x) = x^4 + bx^2 + d$$

and let

$$\overline{p}(x) = x^2 + bx + d$$

If we denote the roots of $p(x)$ in E by $\alpha, -\alpha, \beta, -\beta$, in this order, then $E = F(\alpha, \beta)$ and

$$b = -(\alpha^2 + \beta^2), \quad d = \alpha^2\beta^2$$

The roots α^2 and β^2 of $\overline{p}(x)$ are given by

$$\alpha^2, \beta^2 = \frac{-b \pm \sqrt{b^2 - 4d}}{2}$$

The square root of the discriminant of $p(x)$ is

$$\sqrt{\Delta_p} = (\alpha - \beta)(\alpha + \alpha)(\alpha + \beta)(\beta + \alpha)(\beta + \beta)(-\alpha + \beta) = -4\alpha\beta(\alpha^2 - \beta^2)^2$$

and since $(\alpha^2 - \beta^2)^2$ is invariant under each $\sigma \in G(p(x))$, it must lie in the base field F. Hence, $\sqrt{\Delta_p} \in F$ if and only if $\alpha\beta \in F$, or equivalently, $\sqrt{d} \in F$.

Let us also note that $\sqrt{d} = \alpha\beta$ is fixed by every possible choice of $V \cap G$. For instance, $V \cap Z_1 = \{\iota, (13)(24)\}$ sends $\alpha\beta$ to $\beta\alpha$ and $V \cap Z_2 = \{\iota, (14)(23)\}$ sends $\alpha\beta$ to $(-\beta)(-\alpha) = \alpha\beta$. It follows that $\sqrt{d} \in K = \text{fix}(V \cap G)$.

The irreducibility of $p(x)$ over F can be determined as follows. Certainly if $\overline{p}(x)$ is reducible over F, then so is $p(x)$. On the other hand, if $\overline{p}(x)$ is irreducible then its roots α^2 and β^2 do not lie in F, whence $p(x)$ cannot have a linear factor over F and, if so reducible, must have the form

$$p(x) = x^4 + bx^2 + d = (x^2 + ux + v)(x^2 - ux + w)$$

where, as seen by equating coefficients, $u(v - w) = 0$. However, if $u = 0$ then

$$p(x) = (x^2 + v)(x^2 + w)$$

which gives

$$\overline{p}(x) = (x + v)(x + w)$$

contradicting the irreducibility of $\overline{p}(x)$. Thus, $u \neq 0$ and $v = w$. We can summarize as follows:

1) If $\sqrt{b^2 - 4d} \in F$ then $\overline{p}(x)$, and therefore $p(x)$, is reducible.
2) If $\sqrt{b^2 - 4d} \notin F$ then $p(x)$ is reducible if and only if it has the form

$$q(x) = x^4 + bx^2 + d = (x^2 + ux + v)(x^2 - ux + v)$$

where $v^2 = d$ and $2v - u^2 = b$.

For example, let $p(x) = x^4 + 6x^2 + 4$ over $\mathbb{Q}$. Then $\sqrt{b^2 - 4d} = \sqrt{20} \notin \mathbb{Q}$. From 2), we have $v = \pm 2$ and

$$u^2 = 2v - 6 = \pm 4 - 6 = -2, -10$$

and since the latter has no solutions in $\mathbb{Q}$, we see that $p(x)$ is irreducible over $\mathbb{Q}$.

Let us now assume that $p(x)$ is irreducible. It follows that $\overline{p}(x)$ is also irreducible and $\sqrt{b^2 - 4d} \notin F$. Recall also that $\sqrt{\Delta_p} \in F$ if and only if $\sqrt{d} \in F$, and that $\sqrt{d} \in K = \operatorname{fix}(V \cap G)$.

The resolvent cubic for $p(x)$ (which is already in reduced form) is

$$r(x) = x[x^2 - 2bx + (b^2 - 4d)]$$

which is definitely reducible. Hence, Theorem 7.5.4 tells us the following.

1) If $\sqrt{d} \in F$, then $G_p = V$.
2) If $\sqrt{d} \notin F$, there are two possibilities. Let $K = \operatorname{fix}(V \cap G_p)$.
   a) $|V \cap G_p| = 2$ and $G_p = Z_i$, for $i = 1, 2$ or 3, which occurs if and only if $p(x)$ is reducible over K, in which case $p(x)$ has an irreducible quadratic factor over K. In this case, $[K : F] = [E : K] = 2$.
   b) $|V \cap G_p| = 4$ and $G_p = D_i$, for $i = 1, 2$ or 3, which occurs if and only if $p(x)$ is irreducible over K. In this case, $[K : F] = 2$ and $[E : K] = 4$.

Case 1) above is straightforward. Referring to case 2), we have $\sqrt{d} \notin F$ and $\sqrt{d} \in K$. But in both cases, $[K : F] = 2$ and so $K = F(\sqrt{d})$. Also, it appears that we could use some more information about when $p(x)$ is irreducible over K.

Lemma 7.5.5 *Assume that $r(x)$ is reducible and $\sqrt{d} \notin F$. Then $K = F(\sqrt{d})$ and*
1) $p(x)$ is irreducible over K if and only if $\overline{p}(x)$ is irreducible over K.
2) $p(x)$ is irreducible over K if and only if $\sqrt{d(b^2 - 4d)} \notin F$.

Proof. For part 1), if $\overline{p}(x)$ is reducible over $F(\sqrt{d})$, then clearly $p(x)$ is reducible over $F(\sqrt{d})$. Conversely, suppose that $p(x)$ is reducible over $F(\sqrt{d})$ and

$$p(x) = x^4 + bx^2 + d = (x^2 + ux + v)(x^2 - ux + w)$$

where $u(v - w) = 0$. If $v = w$, then $v^2 = vw = d$ and so $\sqrt{d} \in F$, contrary to assumption. Thus $u = 0$ and

$$p(x) = x^4 + bx^2 + d = (x^2 + v)(x^2 + w)$$

which implies that $\overline{p}(x)$ is reducible over $F(\sqrt{d})$. If $p(x)$ has a linear factor over $F(\sqrt{d})$, then we can assume that $\alpha, -\alpha \in F(\sqrt{d})$ and so

$$p(x) = (x^2 - \alpha^2)(x^2 - \beta^2)$$

which shows that $\overline{p}(x)$ is irreducible over $F(\sqrt{d})$.

Finally, it is clear that the quadratic $\overline{p}(x)$ is reducible over $F(\sqrt{d})$ if and only if $\sqrt{b^2 - 4d} \in F(\sqrt{d})$. But under the assumption that $\sqrt{b^2 - 4d} \notin F$, we have

$$\sqrt{b^2 - 4d} \in F(\sqrt{d}) \Leftrightarrow \sqrt{d(b^2 - 4d)} \in F$$

For if $\sqrt{b^2 - 4d} = a + b\sqrt{d}$ then squaring gives

$$b^2 - 4d = a^2 + b^2 d + 2ab\sqrt{d}$$

and since $\sqrt{d} \notin F$, we must have $ab = 0$. But $b \neq 0$ since $\sqrt{b^2 - 4d} \notin F$ and so $a = 0$, whence $b^2 - 4d = b^2 d$ and so $\sqrt{d(b^2 - 4d)} = bd \in F$. Conversely, if $\sqrt{d(b^2 - 4d)} = f \in F$ then $\sqrt{b^2 - 4d} = f/\sqrt{d} \in F(\sqrt{d})$.□

We can now give a complete analysis for this quartic.

Theorem 7.5.6 (The irreducible quartic $x^4 + bx^2 + d$) *Let*

$$p(x) = x^4 + bx^2 + d$$

be a quartic over a field F, with $\operatorname{char}(F) \neq 2, 3$. *Let E be the splitting field for $p(x)$ over F and let G be its Galois group.*

1) *If $\sqrt{d} \in F$ then $r(x)$ splits over F and $G \approx V$.*
2) *If $\sqrt{d} \notin F$, then there are two possibilities:*
   a) *If $\sqrt{d(b^2 - 4d)} \in F$, then $p(x)$ has an irreducible quadratic factor and $G = Z_i$, for $i = 1, 2$ or 3.*
   b) *If $\sqrt{d(b^2 - 4d)} \notin F$, then $G = D_i$, for $i = 1, 2$ or 3.*□

Exercises

1. Prove that part 4a) and part 4b) of Theorem 7.5.2 are equivalent.
2. Let p be a prime. Let $p(x) \in \mathbb{Q}[x]$ be an irreducible polynomial of degree p with exactly two nonreal roots. Prove that the Galois group of $p(x)$ is S_p. *Hint*: Recall that S_p is generated by a p-cycle and a transposition. Use Cauchy's theorem on G. What is the transposition?
3. Let $p(x) = x^n - a_1 x^{n-1} + \cdots + a_n$ where $a_1, \ldots, a_n$ are algebraically independent over F. Show that $p(x)$ is irreducible over $F(a_1, \ldots, a_n)$, separable and its Galois group is isomorphic to S_n. Thus, if the roots of $p(x)$ are $t_1, \ldots, t_n$ then $a_1, \ldots, a_n$ are algebraically independent over F if and only if $t_1, \ldots, t_n$ are.

4. If $p(x)$ is a quartic polynomial with resolvent cubic $q(x)$ then $\delta_{p(x)} = -\delta_{q(x)}$.
5. Find the Galois groups of the following polynomials over $\mathbb{Q}$:
   a) $x^4 - 10x^2 + 1$
   b) $x^4 - 4x + 2$
   c) $x^5 - 6x + 3$
6. Suppose that $p(x) \in \mathbb{Q}[x]$ is irreducible over $\mathbb{Q}$ and that $G = G(p(x))$ is isomorphic to S_3. What are the possible degrees of $p(x)$?
7. Suppose that $p(x) \in \mathbb{Q}[x]$ is irreducible of degree d and let α be a root of $p(x)$ in $\mathbb{C}$. What are the possibilities for $[\mathbb{Q}(\alpha^3) : \mathbb{Q}]$, expressed in terms of d?
8. If $p(x) \in F[x]$ has roots $r_1, \ldots, r_n$ then $\Delta = (-1)^{n(n-1)/2}\prod_i p'(r_i)$.
9. Let $p(x) = (x - r)(x - s)(x - t)$, where r, s and t are algebraically independent over $\mathbb{Z}_2$. Let s_1, s_2, s_3 be the elementary symmetric polynomials on r, s and t. Show that $\sqrt{\Delta} \in F(s_1, s_2, s_3)$ but the Galois group of $p(x)$ over $F(s_1, s_2, s_3)$ is isomorphic to S_3.
10. Let

$$p(x) = x^n - s_1 x^{n-1} + \cdots + (-1)^n s_n$$

be the generic polynomial with algebraically independent roots $r_1, \ldots, r_n$. Let $u_i = r_1^i + r_2^i + \cdots + r_n^i$. Since the u_i's are symmetric polynomials in the roots of $p(x)$, Theorem 7.2.4 implies that they can be expressed as symmetric polynomials in the elementary symmetric polynomials $s_1, \ldots, s_n$. **Newton's identities** are

$$u_k s_0 - u_{k-1} s_1 + u_{k-2} s_2 + \cdots + (-1)^{k-1} u_1 s_{k-1} + (-1)^k k s_k = 0$$

valid for $k \geq 1$, where $s_0 = 1$ and $s_i = 0$ for $i > n$. Note that for $k > n$, this reduces to

$$u_k s_0 - u_{k-1} s_1 + u_{k-2} s_2 + \cdots + (-1)^n u_{k-n} s_n = 0$$

Prove these identities as follows:

a) For $k > n$, consider the sum $\sum r_i^{k-n} p(r_i)$.

b) For $k = n$, consider the sum $\sum p(r_i)$.

c) For $1 \leq k < n$, proceed by induction on n. Let

$$u_i^{(n)} = r_1^i + r_2^i + \cdots + r_n^i$$

and write the coefficients of $p(x)$ as $s_i^{(n)}$. Then Newton's identites are

$$u_k^{(n)} s_0^{(n)} - u_{k-1}^{(n)} s_1^{(n)} + u_{k-2}^{(n)} s_2^{(n)} + \cdots + (-1)^{k-1} u_1^{(n)} s_{k-1}^{(n)} + (-1)^k k s_k^{(n)} = 0$$

Denote the left side of this by $N_k^{(n)}(r_1, \ldots, r_n)$. Show that

$$N_k^{(n)}(r_1, \ldots, r_{n-1}, 0) = N_k^{(n-1)}(r_1, \ldots, r_{n-1})$$

Hence, $r_n \mid N_k^{(n)}(r_1, \ldots, r_n)$. Show that $r_1 \cdots r_n \mid N_k^{(n)}(r_1, \ldots, r_n)$. Is this possible?

d) Let $p(x) = a + bx + x^n$. Find the values of u_i and find the discriminant of $p(x)$.

11. This exercise concerns the issue of when a value that is expressed in terms of nested radicals

$$\alpha = \sqrt{r + s\sqrt{t}}$$

where $r, s, t \in F$ ($\operatorname{char}(F) \neq 2, 3$) can be written in terms of at most two unnested radicals. For instance, we have

$$\sqrt{5 + \sqrt{21}} = \frac{1}{2}(\sqrt{6} + \sqrt{14})$$

but the number $\sqrt{7 + 2\sqrt{5}}$ cannot be so written. Note that α is a root of the quartic

$$q(x) = x^4 - 2rx^2 + (r^2 - s^2t) = [x^2 - (r + s\sqrt{t})][x^2 - (r - s\sqrt{t})]$$

Assume that $q(x)$ is irreducible over F. Show that $\alpha \in F(\sqrt{p}, \sqrt{q})$ for some p and q in F if and only if

$$\sqrt{r^2 - s^2t} \in F$$

Chapter 8

A Field Extension as a Vector Space

In this chapter, we take a closer look at a finite extension $F < E$ from the point of view that E is a vector space over F. It is clear, for instance, that any $\sigma \in G_F(E)$ is a linear operator on E over F. However, there are many linear operators that are not field automorphisms. One of the most important is multiplication by a fixed element of E, which we study next.

8.1 The Norm and the Trace

Let $F < E$ be finite and let $\alpha \in E$. The multiplication map $\widehat{\alpha}\colon E \to E$ defined by $\widehat{\alpha}\beta = \alpha\beta$ is an F-linear operator on E, since

$$\widehat{\alpha}(a\beta + b\gamma) = a\widehat{\alpha}\beta + b\widehat{\alpha}\gamma$$

for all $a, b \in F$ and $\beta, \gamma \in E$. We wish to find a basis for E over F under which the matrix of $\widehat{\alpha}$ has a nice form.

Note that if $r(x) \in F[x]$, then $r(\widehat{\alpha})\beta = r(\alpha)\beta$ for all $\beta \in E$ and so $r(\alpha) = 0$ as an element of E if and only if $r(\widehat{\alpha})$ is the zero operator on E. Hence, the set of polynomials over F satisfied by $\widehat{\alpha}$ is precisely the same as the set of polynomials satisfied by α. In particular, the minimal polynomial of α in the sense of fields is the same as the minimal polynomial of the linear operator $\widehat{\alpha}$.

The vector subspace $F(\alpha)$ of E is invariant under the linear operator $\widehat{\alpha}$, since $\widehat{\alpha}(p(\alpha)) = \alpha p(\alpha) \in F(\alpha)$. If $\mathcal{B} = (\beta_1, \ldots, \beta_d)$ is an ordered basis for $F(\alpha)$ over F and if

$$\widehat{\alpha}\beta_i = \sum_{j=1}^{d} b_{i,j}\beta_j$$

then the matrix of $\widehat{\alpha}|_{F(\alpha)}$ with respect to $\mathcal{B}$ is $M = (b_{i,j})$. If $(\gamma_1, \ldots, \gamma_e)$ is an ordered basis for E over $F(\alpha)$ where $e = [E : F(\alpha)]$, then the sequence of products

$$\mathcal{C} = (\gamma_1\beta_1, \gamma_1\beta_2, \ldots, \gamma_1\beta_d, \ldots, \ldots, \gamma_e\beta_1, \gamma_e\beta_2, \ldots, \gamma_e\beta_d)$$

is an ordered basis for E over F. To compute the matrix of $\widehat{\alpha}$ with respect to $\mathcal{C}$, note that

$$\widehat{\alpha}(\gamma_k\beta_i) = \alpha\gamma_k\beta_i = \gamma_k(\alpha\beta_i) = \sum_{j=1}^{d} b_{i,j}\gamma_k\beta_j$$

and so each of the subspaces $V_k = \langle\gamma_k\beta_1, \gamma_k\beta_2, \ldots, \gamma_k\beta_d\rangle$ is also invariant under $\widehat{\alpha}$. Hence, the matrix of $\widehat{\alpha}|_{V_k}$ is also equal to M, and the matrix of $\widehat{\alpha}$ with respect to the ordered basis $\mathcal{C}$ has the block diagonal form

$$\mathcal{M}(\widehat{\alpha}) = \begin{bmatrix} M & 0 & 0 & 0 \\ 0 & M & 0 & 0 \\ 0 & 0 & \ddots & 0 \\ 0 & 0 & 0 & M \end{bmatrix} \tag{8.1.1}$$

It follows that if the characteristic polynomial of $\widehat{\alpha}|_{F(\alpha)}$ is $p(x)$, then the characteristic polynomial of $\widehat{\alpha}$ is

$$q_\alpha(x) = p(x)^{[E:F(\alpha)]}$$

The well-known *Cayley–Hamilton theorem* implies that $q_\alpha(\widehat{\alpha}) = 0$ and therefore $p(\alpha) = 0$. But $p(x)$ is monic and has degree $[F(\alpha):F] = \deg(\min(\alpha, F))$, whence $p(x) = \min(\alpha, F)$.

Theorem 8.1.1 Let $F < E$ be finite and let $\alpha \in E$. If $\widehat{\alpha}: E \to E$ is the F-linear operator on E defined by $\widehat{\alpha}\beta = \alpha\beta$, then the characteristic polynomial of $\widehat{\alpha}$ is

$$q_\alpha(x) = [\min(\alpha, F)]^{[E:F(\alpha)]}$$ □

We recall from linear algebra that if $\tau: V \to V$ is a linear operator on a finite–dimensional vector space V over F, the *trace* of τ is the sum of the eigenvalues of τ and the *norm* (*determinant*) of τ is the product of the eigenvalues of τ, in both cases counting multiplicities. Recall also that (as with all symmetric polynomials in the roots of a polynomial) the trace and the norm lie in the base field F. We are motivated to make the following definition.

Definition Let $F < E$ be finite and let $\alpha \in E$. The **trace** of α over $F < E$, denoted by $\mathrm{Tr}_{E/F}(\alpha)$, is the trace of the F-linear operator $\widehat{\alpha}$ and the **norm** of α over $F < E$, denoted by $N_{E/F}(\alpha)$, is the norm of $\widehat{\alpha}$.□

Note that the trace and norm of α depend on the extension field E, and not just on the element α itself.

Since the trace of a linear operator is the sum of the roots of its characteristic polynomial and the norm is the product of these roots, Theorem 8.1.1 allows us to express the trace and norm in terms of the roots of the minimal polynomial of $\widehat{\alpha}$ on the subfield $F(\alpha)$. Let $F < E$ be finite, let $\alpha \in E$ and let

$$p_\alpha(x) = \min(\alpha, F) = x^d + a_{d-1}x^{d-1} + \cdots + a_0$$

have roots $t_1, \ldots, t_d$ in a splitting field. It follows from Theorem 8.1.1 that

$$\mathrm{Tr}_{E/F}(\alpha) = [E : F(\alpha)]\sum_{i=1}^{d} t_i = -[E : F(\alpha)]a_{d-1}$$

and

$$N_{E/F}(\alpha) = \prod_{i=1}^{d} t_i^{[E:F(\alpha)]} = [(-1)^d a_0]^{[E:F(\alpha)]}$$

We remark that many authors simply define the trace and norm of α directly from these formulas.

In terms of *distinct* roots of $p(x)$, if these are $r_1, \ldots, r_s$, then each of these roots has multiplicity $[F(\alpha) : F]_i = p^d$, where d is the radical exponent of $p(x)$ (Theorem 3.5.1) and so

$$\mathrm{Tr}_{E/F}(\alpha) = [E : F(\alpha)][F(\alpha) : F]_i \sum_{i=1}^{s} r_i$$

and

$$N_{E/F}(\alpha) = \prod_{i=1}^{s} r_i^{[F(\alpha):F]_i [E:F(\alpha)]}$$

We can also express the trace and norm in terms of embeddings. Let

$$\hom_F(E, \overline{E}) = \{\sigma_1, \ldots, \sigma_n\}$$

where $n = [E : F]_s$. If $\alpha \in E$ and $p(x) = \min(\alpha, F)$, then $\sigma_1\alpha, \ldots, \sigma_n\alpha$ is a list of the roots of $p(x)$ in $\overline{F}$. However, each distinct root appears $[E : F(\alpha)]_s$ times in this list, since this is the number of ways to extend an embedding of $F(\alpha)$ to an embedding of E, and each such extension has the same value at α. Hence,

$$\sum_{i=1}^{n} \sigma_i\alpha = [E : F(\alpha)]_s \sum_{i=1}^{s} r_i$$

and

$$\prod_{i=1}^{n} \sigma_i\alpha = \prod_{i=1}^{s} r_i^{[E:F(\alpha)]_s}$$

These formulas will provide another expression for the norm and the trace. Let us summarize.

Theorem 8.1.2 *Let $F < E$ be finite and let $\alpha \in E$ with $p(x) = \min(\alpha, F) = x^d + a_{d-1}x^{d-1} + \cdots + a_0$.*
1) If $p(x)$ has roots $t_1, \ldots, t_d$ and distinct roots $r_1, \ldots, r_s$ then

$$\begin{aligned}\operatorname{Tr}_{E/F}(\alpha) &= [E : F(\alpha)]\sum_{i=1}^{d} t_i \\ &= [E : F(\alpha)][F(\alpha) : F]_i \sum_{i=1}^{s} r_i \\ &= -[E : F(\alpha)]a_{d-1}\end{aligned}$$

and

$$\begin{aligned}N_{E/F}(\alpha) &= \prod_{i=1}^{d} t_i^{[E:F(\alpha)]} \\ &= \prod_{i=1}^{s} r_i^{[F(\alpha):F]_i[E:F(\alpha)]} \\ &= [(-1)^d a_0]^{[E:F(\alpha)]}\end{aligned}$$

2) If $\hom_F(E, \overline{E}) = \{\sigma_1, \ldots, \sigma_n\}$ then

$$\operatorname{Tr}_{E/F}(\alpha) = [E : F]_i \sum_{i=1}^{n} \sigma_i \alpha = \begin{cases} \sum_{i=1}^{n} \sigma_i \alpha & \text{if } F < E \text{ is separable} \\ 0 & \text{if } F < E \text{ is inseparable} \end{cases}$$

and

$$N_{E/F}(\alpha) = \prod_{i=1}^{n} (\sigma_i \alpha)^{[E:F]_i}$$

Proof. As for the first statement in part 2), if $F < E$ is inseparable, then $[E : F]_i > 1$, $\operatorname{char}(F) = p \neq 0$ and $p \mid [E : F]_i$, whence $\operatorname{Tr}_{E/F}(\alpha) = 0$.□

Theorem 8.1.2 can be used to derive some basic properties of the trace and the norm.

Theorem 8.1.3 *Let $F < E$ be finite.*
1) The trace is an F-linear functional on E, that is, for all $\alpha, \beta \in E$ and $a, b \in F$,

$$\operatorname{Tr}_{E/F}(a\alpha + b\beta) = a\operatorname{Tr}_{E/F}(\alpha) + b\operatorname{Tr}_{E/F}(\beta)$$

2) *The norm is multiplicative, that is, for all* $\alpha, \beta \in E$,

$$N_{E/F}(\alpha\beta) = N_{E/F}(\alpha)N_{E/F}(\beta)$$

Also, for all $a \in F$,

$$N_{E/F}(a\alpha) = a^{[E:F]}\, N_{E/F}(\alpha)$$

3) *If* $a \in F$ *then*

$$\mathrm{Tr}_{E/F}(a) = [E:F]a \quad \textit{and} \quad N_{E/F}(a) = a^{[E:F]}$$

4) *If* $F < E < L$ *are finite and if* $\alpha \in L$ *then*

$$\mathrm{Tr}_{L/F}(\alpha) = \mathrm{Tr}_{E/F}(\mathrm{Tr}_{L/E}(\alpha)) \quad \textit{and} \quad N_{L/F}(\alpha) = N_{E/F}(N_{L/E}(\alpha))$$

Proof. We prove part 4), leaving the rest for the reader. Let $F < E < L < \overline{F}$ and let

$$\hom_E(L, \overline{F}) = \{\sigma_1, \ldots, \sigma_n\}$$

and

$$\hom_F(E, \overline{F}) = \{\tau_1, \ldots, \tau_m\}$$

Extend each τ_j to an embedding $\overline{\tau_j}\colon \overline{F} \to \overline{F}$ and consider the products $\overline{\tau_j}\sigma_i$, each of which is an embedding of L into $\overline{F}$ over F, that is,

$$\overline{\tau_j}\sigma_i \in \hom_F(L, \overline{F})$$

Note that these embeddings are distinct, for if $\overline{\tau_j}\sigma_i = \overline{\tau_u}\sigma_v$, then $\overline{\tau_j}^{-1}\overline{\tau_u} = \sigma_i\sigma_v^{-1}$ fixes E and so $(\overline{\tau_j}^{-1}\overline{\tau_u})|_E = \iota$, that is, $\tau_j = \tau_u$, which implies that $\overline{\tau_j} = \overline{\tau_u}$. Hence, $\sigma_i = \sigma_v$.

Moreover, since

$$\begin{aligned} \left|\hom_F(L, \overline{F})\right| &= [L:F]_s \\ &= [L:E]_s[E:F]_s \\ &= \left|\hom_E(L, \overline{F})\right|\left|\hom_F(E, \overline{F})\right| \end{aligned}$$

it follows that

$$\{\overline{\tau_j}\sigma_i\} = \hom_F(L, \overline{F})$$

Now, for the norm statement, we have from Theorem 8.1.2,

$$\begin{aligned} N_{E/F}(N_{L/E}(\alpha)) &= \prod_{j=1}^{m}\tau_j\left[\prod_{i=1}^{n}(\sigma_i\alpha)^{[L:E]_i}\right]^{[E:F]_i} \\ &= \left[\prod_{j=1}^{m}\tau_j\left(\prod_{i=1}^{n}\sigma_i\alpha\right)\right]^{[L:F]_i} \\ &= \left[\prod_{j=1}^{m}\prod_{i=1}^{n}(\overline{\tau_j}\sigma_i)\alpha\right]^{[L:F]_i} \\ &= N_{L/F}(\alpha) \end{aligned}$$

Proof of the statement about the trace is similar.□

*8.2 Characterizing Bases

Let $F < E$ be finite and separable. Our goal in this section is to describe a condition that characterizes when a set $\{\alpha_1, \ldots, \alpha_n\}$ of vectors in E is a (vector space) basis for E over F.

Bilinear Forms

In order to avoid breaking the continuity of the upcoming discussion, we begin with a few remarks about bilinear forms. For more details, see Roman, *Advanced Linear Algebra.*

If V is a vector space over F, a mapping $\langle , \rangle: V \times V \to F$ is called a **bilinear form** if it is a linear function of each coordinate, that is, if for all $x, y \in V$ and $a, b \in F$,

$$\begin{aligned} \langle ax + by, z\rangle &= a\langle x, z\rangle + b\langle y, z\rangle \\ \langle z, ax + by\rangle &= a\langle z, x\rangle + b\langle z, y\rangle \end{aligned}$$

For convenience, if $S \subseteq V$, we let

$$\langle x, S\rangle = \{\langle x, s\rangle \mid s \in S\}$$

A bilinear form is **symmetric** if $\langle x, y\rangle = \langle y, x\rangle$ for all $x, y \in V$. A vector space together with a bilinear form is called a **metric vector space**.

Definition *Let V be a metric vector space.*

1) *A vector $v \in V$ is* **degenerate** *if it is orthogonal to all vectors in V (including itself), that is, if*

$$\langle v, V\rangle = \{0\}$$

2) *The space V is* **degenerate** *(or* **singular***) if it contains a nonzero degenerate vector. Otherwise, it is* **nondegenerate** *(or* **nonsingular***).*
3) *The space V is* **totally degenerate** *(or* **totally singular***) if every vector in V is degenerate, that is, if the form is the zero function*

$$\langle x, y \rangle = 0$$

for all $x, y \in V$.□

If $\mathcal{B} = (\beta_1, \ldots, \beta_n)$ is an ordered basis for V over F, the **matrix of the form** $\langle , \rangle$ with respect to $\mathcal{B}$ is

$$M_{\mathcal{B}} = (\langle \beta_i, \beta_j \rangle)$$

The proof of the following theorem is left to the reader.

Theorem 8.2.1

1) *Let $M_{\mathcal{B}}$ be the matrix of a bilinear form on V, with respect to the ordered basis $\mathcal{B}$. If $u, v \in V$ then*

$$\langle u, v \rangle = [u]_{\mathcal{B}} M_{\mathcal{B}} [v]_{\mathcal{B}}^t$$

where $[x]_B$ is the coordinate matrix for x with respect to $\mathcal{B}$.

2) *Two matrices M and N represent the same bilinear forms on V, with respect to possibly different bases, if and only if they are* ***congruent****, that is, if and only if $M = PNP^t$ for some invertible matrix P.*

3) *A metric vector space is nonsingular (nondegenerate) if and only if any, and hence all, of the matrices that represent the form are nonsingular.*□

Characterizing Bases

As mentioned earlier, for a finite separable extension $F < E$, we wish to describe a condition that characterizes when a set $\{\alpha_1, \ldots, \alpha_n\}$ of vectors in E is a basis for E over F.

Suppose that $\mathcal{B} = \{\alpha_1, \ldots, \alpha_n\}$ are vectors in E, where $n = [E : F]$ and let

$$\hom_F(E, \overline{E}) = \{\sigma_1, \ldots, \sigma_n\}$$

We will show that $\mathcal{B}$ is a basis for E over F if and only if the following matrix is nonsingular:

$$M = M(\alpha_1, \ldots, \alpha_n) = ((\sigma_j \alpha_i)_{i,j}) = \begin{bmatrix} \sigma_1\alpha_1 & \sigma_2\alpha_1 & \cdots & \sigma_n\alpha_1 \\ \sigma_1\alpha_2 & \sigma_2\alpha_2 & \cdots & \sigma_n\alpha_2 \\ \vdots & \vdots & \cdots & \vdots \\ \sigma_1\alpha_n & \sigma_2\alpha_n & \cdots & \sigma_n\alpha_n \end{bmatrix}$$

Our plan is to express this matrix in terms of the matrix of a bilinear form. To this end, observe that for any vectors α_i and β_i in E,

$$
\begin{aligned}
\left(M(\alpha_1,\ldots,\alpha_n)M(\beta_1,\ldots,\beta_n)^t\right)_{i,j} &= \sum_k (\sigma_k\alpha_i)(\sigma_k\beta_j) \\
&= \sum_k \sigma_k(\alpha_i\beta_j) \\
&= \mathrm{Tr}_{E/F}(\alpha_i\beta_j)
\end{aligned}
$$

and so

$$
M(\alpha_1,\ldots,\alpha_n)M(\beta_1,\ldots,\beta_n)^t = \left(\mathrm{Tr}_{E/F}(\alpha_i\beta_j)\right)
$$

In particular,

$$
M(\alpha_1,\ldots,\alpha_n)M(\alpha_1,\ldots,\alpha_n)^t = \left(\mathrm{Tr}_{E/F}(\alpha_i\alpha_j)\right)
$$

We can now define a symmetric bilinear form on E by

$$
\langle \alpha, \beta \rangle = \mathrm{Tr}_{E/F}(\alpha\beta) \tag{8.2.1}
$$

This form has a rather special "all or nothing" property: If E contains a nonzero degenerate vector α, then for any $\beta, \gamma \in E$, we have

$$
\langle \beta, \gamma \rangle = \langle \alpha\alpha^{-1}\beta, \gamma \rangle = \langle \alpha, \alpha^{-1}\beta\gamma \rangle = 0
$$

and so E is totally degenerate. In other words, E is either nondegenerate or else *totally* degenerate.

We have assumed that the extension $F < E$ is finite and separable. Of course, if we drop the separability condition, then the matrix $(\sigma_j\alpha_i)$ is no longer square and therefore cannot be invertible. However, the bilinear form (8.2.1) still makes sense. As it happens, this form is nonsingular precisely when $F < E$ is separable.

Theorem 8.2.2 *Let $F < E$ be finite. The following are equivalent:*
1) $F < E$ is separable
2) E is nondegenerate.
When $F < E$ is separable, the matrix

$$
M(\alpha_1,\ldots,\alpha_n) = (\sigma_j\alpha_i)
$$

is nonsingular if and only if $\mathcal{B} = \{\alpha_1,\ldots,\alpha_n\}$ is a basis for E over F.
Proof. If $F < E$ is inseparable, then part 2) of Theorem 8.1.2 shows that the trace is identically 0, whence E is totally degenerate. Thus, if E is nondegenerate, then $F < E$ is separable.

For the converse, since $F < E$ is finite and separable, it is simple, that is, $E = F(\alpha)$. If $[E : F] = n$, then $\mathcal{B} = (1, \alpha, \ldots, \alpha^{n-1})$ is an ordered basis for E over F and

$$M_{\mathcal{B}} = (\langle \alpha^i, \alpha^j \rangle) = \left(\operatorname{Tr}_{E/F}(\alpha^i \alpha^j)\right) = M(1, \alpha, \ldots, \alpha^{n-1}) M(1, \alpha, \ldots, \alpha^{n-1})^t$$

But

$$M = M(1, \alpha, \ldots, \alpha^{n-1}) = \begin{bmatrix} 1 & 1 & \cdots & 1 \\ \sigma_1 \alpha & \sigma_2 \alpha & \cdots & \sigma_n \alpha \\ \vdots & \vdots & \cdots & \vdots \\ (\sigma_1 \alpha)^{n-1} & (\sigma_2 \alpha)^{n-1} & \cdots & (\sigma_n \alpha)^{n-1} \end{bmatrix}$$

is a Vandermonde matrix, for which it is well known that

$$\det(M) = \prod_{i<j} (\sigma_i \alpha - \sigma_j \alpha)$$

Moreover, since each $\sigma_i \in \hom_F(E, \overline{E})$ is uniquely determined by its value on the primitive element α, the elements $\sigma_i \alpha$ are distinct and so $\det(M) \neq 0$. Hence $\det(M_{\mathcal{B}})$ is also nonzero and E is nondegenerate.

For the final statement, suppose first that M is nonsingular. If

$$\sum_i a_i \alpha_i = 0$$

for $a_i \in F$, then applying σ_j gives

$$0 = \sum_i a_i \sigma_j \alpha_i = A M^{(j)}$$

where $A = (a_1, \ldots, a_n)$ and $M^{(j)}$ is the jth column of M. Hence, $AM = 0$ and the nonsingularity of M implies that $A = 0$, that is, $a_i = 0$ for all i. Hence, $\mathcal{B}$ is linearly independent and therefore a basis for E over F.

For the converse, if $\mathcal{B} = (\alpha_1, \ldots, \alpha_n)$ is an ordered basis for E over F, then the matrix of the form (8.2.1) is

$$M_{\mathcal{B}} = (\langle \alpha_i, \alpha_j \rangle) = (\operatorname{Tr}_{E/F}(\alpha_i \alpha_j)) = M(\alpha_1, \ldots, \alpha_n) M(\alpha_1, \ldots, \alpha_n)^t$$

and since $M_{\mathcal{B}}$ is nonsingular because E is nondegenerate, the matrix $M(\alpha_1, \ldots, \alpha_n)$ is also nonsingular.□

The Algebraic Independence of Embeddings

Let E and L be fields. Recall that the Dedekind independence theorem says that any set $\{\sigma_1, \ldots, \sigma_n\}$ of distinct embeddings of E into L is linearly independent over L. To put this another way, let $\lambda_i \in L$ and consider the linear polynomial

$$p(x_1, \ldots, x_n) = \lambda_1 x_1 + \cdots + \lambda_n x_n$$

Then the Dedekind independence theorem says that if $p(\sigma_1, \ldots, \sigma_n)$ is the zero map, then $p(x_1, \ldots, x_n)$ is the zero polynomial. Under certain circumstances, we can strengthen this result by removing the requirement that p be linear.

Let $F < E$ be finite and separable of degree n and let

$$\hom_F(E, L) = \{\sigma_1, \ldots, \sigma_n\}$$

If $p(x_1, \ldots, x_n)$ is a polynomial with coefficients in L, then $p(\sigma_1, \ldots, \sigma_n)$ is a function from E into L, defined by

$$p(\sigma_1, \ldots, \sigma_n)\alpha = p(\sigma_1\alpha, \ldots, \sigma_n\alpha)$$

For example, if $p(x, y) = x^3 + 2xy$ then

$$p(\sigma, \tau)\alpha = (\sigma\alpha)^3 + 2(\sigma\alpha)(\tau\alpha)$$

(Note that we are not composing embeddings, but rather taking products of values of the embeddings.)

Definition *Let $F < E$. A set $\{\sigma_1, \ldots, \sigma_n\}$ of distinct F-embeddings of E into a field L is* **algebraically independent** *over L if the only polynomial $p(x_1, \ldots, x_n)$ over L for which $p(\sigma_1, \ldots, \sigma_n)$ is the zero function is the zero polynomial.* □

Theorem 8.2.3 *Let F be an infinite field, let $F < E$ be finite and separable of degree n. Then*

$$\hom_F(E, L) = \{\sigma_1, \ldots, \sigma_n\}$$

is algebraically independent over L, and therefore so is any nonempty subset of $\hom_F(E, L)$.

Proof. Suppose that $p(x_1, \ldots, x_n)$ is a polynomial over L for which $p(\sigma_1, \ldots, \sigma_n)\alpha = 0$ for all $\alpha \in E$. Let $(\beta_1, \ldots, \beta_n)$ be a basis for E over F. Then $\alpha = \sum a_i\beta_i$ and so

$$\begin{aligned} 0 &= p(\sigma_1\alpha, \ldots, \sigma_n\alpha) \\ &= p(\sum_i a_i\sigma_1\beta_i, \ldots, \sum_i a_i\sigma_n\beta_i) \\ &= p(AM^{(1)}, \ldots, AM^{(n)}) \end{aligned}$$

where $A = (a_1, \ldots, a_n) \in F^n$ and $M = M(\beta_1, \ldots, \beta_n)$. However, Theorem 8.2.2 implies that M is invertible and so any vector in F^n has the form AM, for some $A \in F^n$, which shows that $p(x_1, \ldots, x_n)$ is zero on the infinite subfield F of E. Theorem 1.3.5 then implies that $p(x_1, \ldots, x_n)$ is the zero polynomial.□

*8.3 The Normal Basis Theorem

Let $F < E$ be a finite Galois extension of degree n. Since $F < E$ is finite and separable, there exists a $\lambda \in E$ such that $E = F(\lambda)$. As we know, the set

$\{1, \lambda, \ldots, \lambda^{n-1}\}$ is a basis for E over F. This type of basis is called a **polynomial basis**. A **normal basis** for E over F is a basis for E over F consisting of the roots of an irreducible polynomial over F.

We wish to show that any finite Galois extension has a normal basis. Theorem 8.2.2 can be reworded for finite Galois extensions as follows.

Theorem 8.3.1 *If $F < E$ is finite and Galois, with $G_F(E) = \{\sigma_1, \ldots, \sigma_n\}$ then $\{\lambda_1, \ldots, \lambda_n\}$ is a basis for E over F if and only if* $\det(\sigma_i \lambda_j) \neq 0$.□

Now, if $F < E$ is finite and Galois, it is simple and so $E = F(\lambda)$. Moreover, the roots of $\min(\lambda, F)$ are

$$\{\sigma_1 \lambda, \ldots, \sigma_n \lambda \mid \sigma_j \in G_F(E)\}$$

Theorem 8.3.1 implies that this set is a (normal) basis for E over F if and only if $\det(\sigma_i \sigma_j \lambda) \neq 0$. To find such an element $\lambda \in E$, consider the matrix

$$D = \begin{bmatrix} \sigma_1\sigma_1 & \sigma_1\sigma_2 & \cdots & \sigma_1\sigma_n \\ \sigma_2\sigma_1 & \sigma_2\sigma_2 & \cdots & \sigma_2\sigma_n \\ & & \vdots & \\ \sigma_n\sigma_1 & \sigma_n\sigma_2 & \cdots & \sigma_n\sigma_n \end{bmatrix}$$

For each i, the product $\sigma_i \sigma_j$ runs through $\sigma_1, \ldots, \sigma_n$ as j runs through $1, \ldots, n$, and so each row of D is a distinct permutation of $\sigma_1, \ldots, \sigma_n$. The same applies to the columns of D. Thus, we may write

$$D = \begin{bmatrix} \sigma_{a_{1,1}} & \sigma_{a_{1,2}} & \cdots & \sigma_{a_{1,n}} \\ \sigma_{a_{2,1}} & \sigma_{a_{2,2}} & \cdots & \sigma_{a_{2,n}} \\ & & \vdots & \\ \sigma_{a_{n,1}} & \sigma_{a_{n,2}} & \cdots & \sigma_{a_{n,n}} \end{bmatrix}$$

where for each i, the row indices $(a_{i,1}, a_{i,2}, \ldots, a_{i,n})$ form a distinct permutation of $\{1, \ldots, n\}$ and for each j, the column indices $(a_{1,j}, a_{2,j}, \ldots, a_{n,j})$ form a distinct permutation of $\{1, \ldots, n\}$. Let $x_1, \ldots, x_n$ be independent variables and consider the matrix

$$N(x_1, \ldots, x_n) = \begin{bmatrix} x_{a_{1,1}} & x_{a_{1,2}} & \cdots & x_{a_{1,n}} \\ x_{a_{2,1}} & x_{a_{2,2}} & \cdots & x_{a_{2,n}} \\ & & \vdots & \\ x_{a_{n,1}} & x_{a_{n,2}} & \cdots & x_{a_{n,n}} \end{bmatrix}$$

We claim that the polynomial $p(x_1, \ldots, x_n) = \det(N(x_1, \ldots, x_n))$ is nonzero.

Each row of N is a distinct permutation of the variables $x_1, \ldots, x_n$ and similarly for each column. Thus $N(1,0,\ldots,0)$ is a *permutation matrix*, that is, each row and each column of $N(1,0,\ldots,0)$ contains one 1 and the rest 0's. Since permutation matrices are nonsingular, we have

$$p(1, 0, \ldots, 0) = \det(N(1, 0, \ldots, 0)) \neq 0$$

Hence, $p(x_1, \ldots, x_n) \neq 0$.

If F is an infinite field, Theorem 8.2.3 implies that the distinct embeddings $\sigma_1, \ldots, \sigma_n$ of E into L are algebraically independent over L and so there exists a $\lambda \in L$ for which

$$\det(\sigma_i \sigma_j \lambda) = (\det(D))(\lambda) = p(\sigma_1, \ldots, \sigma_n)\lambda \neq 0$$

Thus, we have proven the following.

Theorem 8.3.2 *If F is an infinite field, then any finite Galois extension $F < E$ has a normal basis.* □

This result holds for finite fields as well. The proof will be given in Chapter 9.

Exercises

1. Let $F < E$ be finite. Prove that for all $\alpha, \beta \in E$,

$$\mathrm{Tr}_{E/F}(\alpha + \beta) = \mathrm{Tr}_{E/F}(\alpha) + \mathrm{Tr}_{E/F}(\beta)$$

   and

$$N_{E/F}(\alpha\beta) = N_{E/F}(\alpha)N_{E/F}(\beta)$$

2. Let $F < E$ be finite. Prove that if $\alpha \in F$, then

$$\mathrm{Tr}_{E/F}(\alpha) = [E : F]\alpha$$

   and

$$N_{E/F}(\alpha) = \alpha^{[E:F]}$$

3. If $F < E < L$ are finite and if $\alpha \in L$ show that

$$\mathrm{Tr}_{L/F}(\alpha) = \mathrm{Tr}_{E/F}(\mathrm{Tr}_{L/E}(\alpha))$$

4. Let $F < E$ be finite and let $\sigma \in \hom_F(E, L)$. If $\alpha \in E$ prove that

$$N_{\sigma E/\sigma F}(\sigma\alpha) = \sigma(N_{E/F}(\alpha))$$

   State and prove a similar statement for the trace.
5. Find a normal basis for the splitting field of $p(x) = x^4 - 5x^2 + 6$ over $\mathbb{Q}$.
6. If $F < E$ is finite and Galois, with $G_F(E) = \{\sigma_1, \ldots, \sigma_n\}$, prove without appeal to Theorem 8.2.2, but rather using the Dedekind independence theorem, that if $\{\lambda_1, \ldots, \lambda_n\}$ is a basis for E over F then $\det(\sigma_i \lambda_j) \neq 0$.

7. Let $F < E$ be a finite separable extension, with $E = F(\alpha)$. Let $p(x) = \min(\alpha, F)$ have degree n. Show that

$$|M(1, \alpha, \ldots, \alpha^{n-1})|^2 = (-1)^{n(n-1)/2} N_{E/F}(p'(\alpha))$$

8. Let $F < E$ be finite and separable with form (8.2.1) and let $\{\alpha_i\}$ be a basis for E over F. The **dual basis** $\{\beta_i\}$ to $\{\alpha_i\}$ is a basis with the property that

$$\mathrm{Tr}_{E/F}(\alpha_i \beta_j) = \langle \alpha_i, \beta_j \rangle = \delta_{i,j}$$

where $\delta_{i,j} = 1$ if $i = j$ and 0 otherwise. In matrix terms, $\{\alpha_i\}$ and $\{\beta_i\}$ are dual bases if

$$M(\alpha_1, \ldots, \alpha_n) M(\beta_1, \ldots, \beta_n)^t = I$$

A basis for E over F is called a **polynomial basis** if it has the form $\{1, \alpha, \ldots, \alpha^{n-1}\}$ for some $\alpha \in E$. Any simple algebraic extension $E = F(\alpha)$ has a polynomial basis. Let $F < E$ be finite and separable, with polynomial basis $\{1, \alpha, \ldots, \alpha^{n-1}\}$. Let

$$p(x) = \min(\alpha, F) = (x - \alpha)(a_0 + a_1 x + \cdots + a_{n-1} x^{n-1})$$

Prove that the dual basis for $\{1, \alpha, \ldots, \alpha^{n-1}\}$ is

$$\left\{ \frac{a_0}{p'(\alpha)}, \frac{a_1}{p'(\alpha)}, \ldots, \frac{a_{n-1}}{p'(\alpha)} \right\}$$

9. If V is a vector space, let V^* denote the algebraic dual space of all linear functionals on V. Note that if $\dim(V)$ is finite then $\dim(V) = \dim(V^*)$.
   a) Prove the Riesz Representation Theorem for nonsingular metric vector spaces: Let V be a finite-dimensional nonsingular metric vector space over F and let $f \in V^*$ be a linear functional on V. Then there exists a unique vector $x \in V$ such that $fx = \langle y, x \rangle$ for all $y \in V$. Hint: Let $\phi_x : V \to F$ be defined by $\phi_x(y) = \langle y, x \rangle$. Define a map $\tau : V \to V^*$ by $\tau x = \phi_x$. Show that τ is an isomorphism.
   b) Let $F < E$ be finite and separable, with form (8.2.1). Prove that for any linear functional $\tau : E \to F$ there exists a unique $\alpha \in E$ for which $\tau\beta = \mathrm{Tr}_{E/F}(\alpha\beta)$ for all $\beta \in E$.

Chapter 9
Finite Fields I: Basic Properties

In this chapter and the next, we study finite fields, which play an important role in the applications of field theory, especially to coding theory, cryptology and combinatorics. For a thorough treatment of finite fields, the reader should consult the book *Introduction to Finite Fields and Their Applications*, by Lidl and Niederreiter, Cambridge University Press, 1986.

9.1 Finite Fields Redux

If F is a field, then F^* will denote the multiplicative group of all nonzero elements of F. Let us recall some facts about finite fields that have already been established.

Theorem 9.1.1 *Let F be a finite field.*
1) *F has prime characteristic p. (Theorem 0.4.4)*
2) *F^* is cyclic. (Corollary 1.3.4)*
3) *Any finite extension of F is simple. (Theorem 2.4.3)*
4) *F is perfect, and so every algebraic extension of F is separable and the Frobenius map $\sigma_{p^k}: \alpha \mapsto \alpha^{p^k}$ is an automorphism of F, for all $k \geq 1$. (Theorem 3.4.3)* □

Lemma 9.1.2 *If F is a finite field and $[E:F] = d$ then $|E| = |F|^d$.*
Proof. If $\{\alpha_1, \ldots, \alpha_d\}$ is a basis for E over F, then each element of E has a *unique* representation of the form $a_1\alpha_1 + \cdots + a_d\alpha_d$, where $a_i \in F$. Since there are $|F|$ possibilities for each coefficient a_i, we deduce that $|E| = |F|^d$.□

Since a finite field F has prime characteristic p, we have $\mathbb{Z}_p < F$ and so Lemma 9.1.2 gives

Corollary 9.1.3 *If F is a finite field with $\operatorname{char}(F) = p$, then F has p^n elements for some positive integer n.* □

From now on, unless otherwise stated, p will represent a prime number, and q will represent a power of p.

9.2 Finite Fields as Splitting Fields

Let F be a finite field of size q. Then F^* has order $q-1$ and so every element $\alpha \in F^*$ has exponent $q-1$, that is, $\alpha^{q-1}=1$. It follows that every element of F is a root of the polynomial

$$f_q(x) = x^q - x$$

Since $f_q'(x) = -1$, this polynomial has no multiple roots and so F is precisely the set of roots of $f_q(x)$ in some splitting field. In fact, since F is a field, it is a splitting field for $f_q(x)$ over the prime subfield $\mathbb{Z}_p$. In symbols,

$$F = \text{Roots}(f_q(x)) = \text{Split}_{\mathbb{Z}_p}(f_q(x))$$

This has profound consequences for the behavior of finite fields.

Existence

We have seen that every finite field of characteristic p has $q = p^n$ elements for some $n > 0$. Conversely, let $q = p^n$. If R is the set of roots of $f_q(x)$, then R is actually a field. For if $\alpha, \beta \in R$, then $\alpha^q = \alpha$ and $\beta^q = \beta$, whence

$$(\alpha \pm \beta)^q = \alpha^q \pm \beta^q = \alpha \pm \beta$$

and

$$(\alpha\beta^{-1})^q = \alpha^q(\beta^q)^{-1} = \alpha\beta^{-1}$$

Thus $\alpha \pm \beta, \alpha\beta^{-1} \in R$. It follows that R is a field and hence a splitting field for $f_q(x)$. Furthermore, since $f_q(x)$ has no multiple roots, R has size q. Thus, for every prime power $q = p^n$, there is a field of size q.

Of course, since each finite field of size q is a splitting field for $f_q(x)$ over $\mathbb{Z}_p$, we know that all such fields are isomorphic.

It is customary to denote a finite field of size q by F_q, or $GF(q)$. (The symbol GF stands for *Galois Field,* in honor of Evariste Galois.)

Theorem 9.2.1

1) *Every finite field has size $q = p^n$, for some prime p and integer $n > 0$.*
2) *For every $q = p^n$ there is, up to isomorphism, a unique finite field $GF(q)$ of size q, which is both the set of roots of $f_q(x) = x^q - x$ and the splitting field for $f_q(x)$ over $\mathbb{Z}_p$.* □

Let us refer to the polynomial $x^q - x$ as the **defining polynomial** of the finite field $GF(q)$. In view of this theorem, we will often refer to *the* finite field $GF(q)$.

An immediate consequence of the splitting field characterization of finite fields is that any extension of finite fields is normal.

Corollary 9.2.2 *The extension $GF(q) < GF(q^n)$ is a finite Galois extension. Hence, in the Galois correspondence for $GF(q) < GF(q^n)$, all intermediate fields and all subgroups are closed.* □

9.3 The Subfields of a Finite Field

We wish to examine the subfields of a finite field $GF(q^n)$. Note that if k and n are positive integers and $n = mk + r$ for $0 \le r < k$, then

$$x^{mk+r} - 1 = (x^k - 1)x^{(m-1)k+r} + (x^{(m-1)k+r} - 1)$$

Hence, $x^k - 1$ divides $x^{mk+r} - 1$ if and only if $x^k - 1$ divides $x^{(m-1)k+r} - 1$. Repeating this shows that $x^k - 1$ divides $x^{mk+r} - 1$ if and only if $x^k - 1$ divides $x^r - 1$, that is, if and only if $r = 0$. In other words,

$$k \mid n \Leftrightarrow x^k - 1 \mid x^n - 1 \tag{9.3.1}$$

over the prime subfield $\mathbb{Z}_p$.

Theorem 9.3.1 (Subfields of $GF(q^n)$) *The following are equivalent:*

1) $d \mid n$

2) The defining polynomial of $GF(q^d)$ divides the defining polynomial of $GF(q^n)$, that is,

$$f_{q^d}(x) \mid f_{q^d}(x)$$

over the prime subfield $\mathbb{Z}_p$.

3) $GF(q^d) < GF(q^n)$

Put another way, the following lattices are isomorphic (under the obvious maps):

a) $\mathcal{L}_n = \{d \mid 1 \le d \le n, d \text{ divides } n\}$*, under division*

b) $\{f_{q^d}(x) \mid f_{q^d}(x) \text{ divides } f_{q^n}(x)\}$*, under division*

c) Subfields of $GF(q^n)$, under set inclusion.

Moreover, $GF(q^n)$ has exactly one subfield of size q^d, for each $d \mid n$.

Proof. Two applications of (9.3.1) show that

$$d \mid n \Leftrightarrow q^d - 1 \mid q^n - 1 \Leftrightarrow x^{q^d-1} - 1 \mid x^{q^n-1} - 1 \Leftrightarrow f_{q^d}(x) \mid f_{q^n}(x)$$

and so 1) and 2) are equivalent. Moreover,

$$GF(q^d) < GF(q^n) \Leftrightarrow \text{Roots}(f_{q^d}(x)) \subseteq \text{Roots}(f_{q^n}(x)) \Leftrightarrow f_{q^d}(x) \mid f_{q^n}(x)$$

and so 2) and 3) are equivalent. For the last statement, if $GF(q^n)$ has two distinct subfields of size q^d, then the polynomial $f_{q^d}(x)$ would have more than q^d roots in $GF(q^n)$. □

9.4 The Multiplicative Structure of a Finite Field

Since $GF(q)^*$ is cyclic, Theorem 0.2.11 implies the following theorem.

Theorem 9.4.1 *There are exactly $\phi(d)$ elements of $GF(q)^*$ of order d for each $d \mid q-1$ and this accounts for all of the elements of $GF(q)^*$.* □

It is customary to refer to any element of $GF(q)$ that generates the cyclic group $GF(q)^*$ as a *primitive element* of $GF(q)$. However, this brings us into conflict with the term *primitive* as used earlier to denote any element of a field that generates the field using *both* field operations (addition and multiplication). Accordingly, we adopt the following definition.

Definition *Any element of $GF(q)$ that generates the cyclic group $GF(q)^*$ is called a* **group primitive element** *of $GF(q)$. In contrast, if $F < E$, then any element $\alpha \in E$ for which $E = F(\alpha)$ is called a* **field primitive element** *of E over F.*□

Roots in a Finite Field

If $\beta \in GF(q)$, we may wish to know when β has a kth root in $GF(q)$, that is, when the equation

$$x^k = \beta \tag{9.4.1}$$

has a solution in $GF(q)$. This question has a simple answer in view of the fact that $GF(q)^*$ is cyclic. If α is a group primitive element of $GF(q)$ then $\beta = \alpha^i$ for some i and so (9.4.1) has a solution $x = \alpha^j$ if and only if

$$\alpha^{kj} = \alpha^i$$

for some integer j, that is, $\alpha^{kj-i} = 1$, which holds if and only if $q-1 \mid kj-i$, that is,

$$i = kj + n(q-1)$$

for some integer n. But this holds if and only if

$$\gcd(k, q-1) \mid i$$

Thus, equation (9.4.1) has a solution for all $\beta \in GF(q)$ if and only if $\gcd(k, q-1) = 1$, that is, if and only if k and $q-1$ are relatively prime.

Theorem 9.4.2

1) *Let α be a group primitive element of $GF(q)$. Then α^i has a kth root in $GF(q)$ if and only if*
$$\gcd(k, q-1) \mid i$$
2) *Every element of $GF(q)$ has a kth root if and only if k* and $q-1$ are relatively prime, *in which case every element has a* unique *kth root.*
3) *The function*
$$\sigma_k \colon \alpha \mapsto \alpha^k$$
is a permutation of $GF(q)$ if and only if k and $q-1$ are relatively prime. In this case, $p(x) = x^k$ is called a **permutation polynomial.** □

9.5 The Galois Group of a Finite Field

Since the extension $GF(q) < GF(q^n)$ is Galois, if G is the Galois group of $GF(q^n)$ over $GF(q)$ then
$$|G| = [GF(q^n) : GF(q)] = n$$
The structure of G could not be simpler, as we now show.

Theorem 9.5.1 *The Galois group G of $GF(q^n)$ over $GF(q)$ is cyclic of order n, generated by the* **Frobenius automorphism** $\sigma_q \colon \alpha \mapsto \alpha^q$.
Proof. We have seen that the Frobenius map σ_q is an automorphism of $GF(q^n)$. If $\alpha \in GF(q)$, then $\sigma_q \alpha = \alpha^q = \alpha$ and so σ_q fixes $GF(q)$ and is therefore in the Galois group G. Moreover, the n automorphisms
$$\iota, \sigma_q, \sigma_q^2, \ldots, \sigma_q^{n-1}$$
are distinct elements of G, for if $\sigma_q^k = \iota$ then $\alpha^{q^k} = \alpha$ for all $\alpha \in GF(q^n)$ and so $GF(q^n) < GF(q^k)$, which implies that $k \geq n$. Finally, since $|G| = n$, we see that $G = \langle \sigma_q \rangle$. □

9.6 Irreducible Polynomials over Finite Fields

Some of the most remarkable properties of finite fields stem from the fact that every finite field $GF(q)$ is not only the splitting field for the polynomial $f_q(x) = x^q - x$, but is also the *set* of roots of $f(x)$. This applies to the properties of irreducible polynomials over a finite field.

Existence of Irreducible Polynomials

As to existence, if $GF(q)$ is a finite field and d is a positive integer, then there is an irreducible polynomial of degree d over $GF(q)$. This follows from the fact that the extension $GF(q) < GF(q^d)$ is simple and so $GF(q^d) = GF(q)(\alpha)$ for some $\alpha \in GF(q^d)$. Then the minimal polynomial $p(x) = \min(\alpha, GF(q))$ is irreducible of degree d.

The Splitting Field and Roots of an Irreducible Polynomial

Let $p(x)$ be irreducible over $GF(q)$ of degree d. Let α be a root of $p(x)$. Since $GF(q) < GF(q)(\alpha)$ is normal, it follows that $p(x)$ splits in $GF(q)(\alpha)$ and so $GF(q)(\alpha) = GF(q^d)$ is a splitting field for $p(x)$. Thus, $p(x) \mid x^{q^d} - x$. Moreover,

$$p(x) \mid x^{q^n} - x \Leftrightarrow GF(q^d) < GF(q^n) \Leftrightarrow d \mid n$$

and so the degree d can be characterized as the smallest positive integer for which $p(x) \mid x^{q^d} - x$.

Since the Galois group is the cyclic group $\langle \sigma_q \rangle$, the roots of $p(x)$ are

$$\alpha, \alpha^q, \alpha^{q^2}, \ldots, \alpha^{q^{d-1}}$$

Note that d can also be characterized as the smallest positive integer for which $\alpha^{q^d} = \alpha$.

The Order of an Irreducible Polynomial

Since none of the roots of $p(x)$ is zero, the roots belong to the multiplicative group $GF(q^d)^*$. Moreover, since each root is obtained by applying an automorphism to a single root α, all roots of $p(x)$ have the same multiplicative order. Let us denote this order by $\nu = o(\alpha)$. Thus, $\alpha^m = 1$ if and only if $\nu \mid m$.

The common order ν of the roots is referred to as the **order** of the irreducible polynomial $p(x)$ and is denoted by $o(p)$. Note that this definition makes sense only for irreducible polynomials.

As an aside, if the order of $p(x)$ is $q^d - 1$, then each root of $p(x)$ is group primitive, and we say that $p(x)$ is **primitive**. Primitive polynomials play an important role in finite field arithmetic, as we will see in the next chapter.

The Relationship Between Degree and Order

The relationship between the degree d and the order ν of $p(x)$ can be gleaned as follows. First, note that

$$\alpha^{q^n} = \alpha \Leftrightarrow \alpha^{q^n - 1} = 1 \Leftrightarrow \nu \mid q^n - 1 \Leftrightarrow q^n \equiv 1 \bmod \nu$$

and since d is the smallest positive integer for which the former holds, it is also the smallest positive integer for which the latter holds, that is, the *order* of q modulo $\nu = o(p)$.

It happens that this relationship between order and degree actually characterizes irreducibility. That is, if $p(x)$ is a polynomial with root α of order ν and if $d = \deg(p)$ is equal to the order of q modulo ν, then $p(x)$ must be irreducible (in which case all roots have order ν). For if $p(x)$ is reducible, then α is a root of an irreducible factor of $p(x)$, with degree $e < d$. Hence, $e < d$ is the order of q modulo ν.

Summary

Let us summarize.

Theorem 9.6.1 *For every finite field $GF(q)$, and every positive integer d, there exists an irreducible polynomial $p(x)$ of degree d over $GF(q)$. Let $p(x)$ be irreducible of order d and let α be a root of $p(x)$ in some extension field. Let $o_n(q)$ denote the order of q in $\mathbb{Z}_n$.*

1) **(Splitting Field)** *The splitting field of $p(x)$ is $GF(q)(\alpha) = GF(q^d)$.*

2) $p(x) \mid x^{q^n} - x$ *if and only if $d \mid n$.*

3) **(Roots)** *The roots of $p(x)$ in a splitting field are*

$$\alpha, \alpha^q, \alpha^{q^2}, \ldots, \alpha^{q^{d-1}}$$

and so d is the smallest positive integer for which $\alpha^{q^d} = \alpha$.

4) **(Order of Roots)** *All roots of $p(x)$ have order the same order ν, called the* **order** *of $p(x)$.*

5) **(Degree)** *The degree d of $p(x)$ is the smallest positive integer n for which $\alpha^{q^n} = \alpha$, or equivalently, $p(x) \mid x^{q^n} - x$.*

6) **(Relationship between degree and order characterizes irreducibility)** *Let $f(x)$ be a polynomial over $GF(q)$ with order ν and degree d. Then $f(x)$ is irreducible if and only if*

$$d = o_\nu(q)$$

□

Computing the Order of a Polynomial

To compute the order ν of an irreducible polynomial $p(x)$ of degree d, we can use the fact that

$$\nu \mid q^d - 1$$

and

$$p(x) \mid x^n - 1 \Leftrightarrow \nu \mid n$$

Let

$$q^d - 1 = p_1^{e_1} \cdots p_m^{e_m}$$

where the p_i's are distinct primes. Then

$$\nu = p_1^{f_1} \cdots p_m^{f_m}$$

where $f_i \leq e_i$ and, for each i,

$$p(x) \mid x^{p_1^{e_1} \cdots p_i^{a_i} \cdots p_m^{e_m}} - 1$$

if and only if $p_1^{e_1} \cdots p_i^{f_i} \cdots p_m^{e_m} \mid p_1^{e_1} \cdots p_i^{a_i} \cdots p_m^{e_m}$, that is, if and only if $a_i \geq f_i$. Thus, f_i is the smallest nonnegative integer for which

$$p(x) \mid x^{p_1^{e_1} \cdots p_i^{f_i} \cdots p_m^{e_m}} - 1$$

Example 9.6.1 Consider the irreducible polynomial $p(x) = x^6 + x + 1$ over $GF(2)$. Since $q = 2$, we have

$$q^d - 1 = q^6 - 1 = 63 = 3^2 \cdot 7$$

Let $\nu = 3^a 7^b$. Then a is the smallest nonnegative integer for which

$$p(x) \mid x^{3^a 7} - 1$$

Division shows that

$$p(x) \nmid x^{3^0 \cdot 7} - 1, \quad p(x) \nmid x^{3^1 \cdot 7} - 1, \quad p(x) \mid x^{3^2 \cdot 7} - 1$$

and so $a = 2$. For b, division gives

$$p(x) \nmid x^{9 \cdot 7^0} - 1, \quad p(x) \mid x^{9 \cdot 7^1} - 1$$

and so $b = 1$. Thus, Thus $\nu = 3^2 \cdot 7 = 63$, showing that $p(x)$ is primitive over $GF(2)$.

As another example, the polynomial $g(x) = x^6 + x^4 + x^2 + x + 1$ is also irreducible over $GF(2)$. If $\nu = 3^a 7^b$, then

$$g(x) \nmid x^{3^0 \cdot 7} - 1, \quad f(x) \mid x^{3^1 \cdot 7} - 1$$

and so $a = 1$. Also,

$$f(x) \nmid x^{9 \cdot 7^0} - 1, \quad f(x) \mid x^{9 \cdot 7^1} - 1$$

and so $b = 1$. Thus, $\nu = 3 \cdot 7 = 21$. Note that both of these polynomials have degree 6 but they have different orders. This shows that the degree of an irreducible polynomial does not determine its order. □

*9.7 Normal Bases

Since any extension $GF(q) < GF(q^d)$ is simple, there is an $\alpha \in GF(q^d)$ for which $GF(q^d) = GF(q)(\alpha)$. Moreover, the set $\{1, \alpha, \ldots, \alpha^{d-1}\}$ is a basis for E over F. This type of basis is called a **polynomial basis**.

Since the d roots of an irreducible polynomial $p(x)$ of degree d over $GF(q)$ are distinct, it is natural to wonder whether there is an irreducible polynomial $p(x)$ whose roots form a basis for $GF(q^d)$ over $GF(q)$. Such a basis is referred to as a **normal basis**. In short, a normal basis is a basis of roots of an irreducible polynomial.

We saw in Chapter 8 that if $F < E$ is a finite Galois extension and F is an infinite field, then E has a normal basis over F. This is also true for finite fields and stems from the fact that the members of the Galois group are linearly independent.

Let $p(x)$ be irreducible of degree d over $GF(q)$. Then $GF(q^d)$ is the splitting field of $p(x)$ and the Galois group of $p(x)$ is

$$G = \{\iota, \sigma_q, \sigma_q^2, \ldots, \sigma_q^{d-1}\}$$

where σ_q is the Frobenius automorphism. But since these automorphisms are distinct, the Dedekind independence theorem tells us that they are linearly independent.

This implies that as a linear operator on $GF(q^d)$, the automorphism σ_q has minimal polynomial $x^d - 1$, for no polynomial of smaller degree can be satisfied by σ_q. But the characteristic polynomial of σ_q is monic, has degree d and is divisible by the minimal polynomial (this is the Cayley–Hamilton theorem), and so it is also equal to $x^d - 1$.

The following result from linear algebra, which we will not prove here, is just what we need.

Theorem 9.7.1 *Let $T: V \to V$ be a linear operator on a finite-dimensional vector space V over a field F. Then V contains a vector $v \in V$ for which*

$$\{v, Tv, T^2v, \ldots, T^{n-1}v\}$$

is a basis for V if and only if the minimal polynomial and characteristic polynomial of T are equal. □

This theorem implies that there is an $\alpha \in GF(q^d)$ for which

$$\text{Roots}(p(x)) = \{\alpha, \sigma_q\alpha, \sigma_q^2\alpha, \ldots, \sigma_q^{d-1}\alpha\}$$

is a (normal) basis for $GF(q^d)$ over $GF(q)$.

Theorem 9.7.2 *There exists a normal basis $\{\alpha, \alpha^q, \ldots, \alpha^{q^{n-1}}\}$ for $GF(q^n)$ over $GF(q)$.* □

*9.8 The Algebraic Closure of a Finite Field

In this section, we determine the algebraic closure of a finite field $GF(q)$. Since $GF(q) < GF(q^n)$ is algebraic for all positive integers n, an algebraic closure of $GF(q)$ must contain all of the fields $GF(q^n)$.

Since $n! \mid (n+1)!$, it follows that

$$GF(q^{n!}) < GF(q^{(n+1)!})$$

and so the union

$$\Gamma(q) = \bigcup_{n=0}^{\infty} GF(q^{n!})$$

is an extension field of $GF(q)$ that contains $GF(q^n)$, for all $n \geq 1$. Moreover, if E is a field for which $GF(q^n) < E$ for all n, then $\Gamma(q) < E$, that is, $\Gamma(q)$ is the smallest field containing each $GF(q^n)$.

Theorem 9.8.1 *The field* $\Gamma(q)$ *is the algebraic closure of* $GF(q)$.
Proof. Every element of $\Gamma(q)$ lies in some $GF(q^{n!})$, whence it is algebraic over $GF(q)$. Thus $\Gamma(q)$ is algebraic over $GF(q)$. Now let $p(x)$ be an irreducible polynomial over $\Gamma(q)$ of degree d. Then the coefficients of $p(x)$ lie in some $GF(q^{n!})$ and so $p(x)$ is irreducible as a polynomial over $GF(q^{n!})$. Hence, the splitting field for $p(x)$ is $GF(q^{n! \cdot d}) < \Gamma(q)$ and so $p(x)$ splits over $\Gamma(q)$.$\square$

Steinitz Numbers

We wish now to describe the subfields of the algebraic closure $\Gamma(q)$. Recall that a field K is a subfield of $GF(q^n)$ if and only if $K = GF(q^d)$ where $d \mid n$. The set $\mathbb{N}^+$ of positive integers is a complete lattice where $m \wedge n = \gcd(m, n)$ and $m \vee n = \operatorname{lcm}(m, n)$. If we denote by $\mathcal{F}_q$ the set of all finite fields (or more properly the set of all isomorphism classes of finite fields) that contain $GF(q)$, then $\mathcal{F}_q$ is also a complete lattice where $E \wedge F = E \cap F$ and $E \vee F = EF$.

Theorem 9.8.2 *The map* $\phi: \mathbb{N}^+ \to \mathcal{F}_q$ *defined by* $\phi(n) = GF(q^n)$ *is an order-preserving bijection. Hence, it is an isomorphism of lattices, that is,*
1) $n \mid m$ *if and only if* $GF(q^n) < GF(q^m)$
2) $GF(q^n) \cap GF(q^m) = GF(q^{n \wedge m})$
3) $GF(q^n)GF(q^m) = GF(q^{n \vee m})$
Proof. Left to the reader. $\square$

It is clear that the lattice of intermediate fields between $GF(q)$ and $GF(q^n)$ is isomorphic to the sublattice of $\mathbb{N}^+$ consisting of all positive integers dividing n. In order to describe the lattice of intermediate fields between $GF(q)$ and $\Gamma(q)$, we make the following definition.

Definition *A* **Steinitz number** *is an expression of the form*

$$S = \prod_{i=1}^{\infty} p_i^{e_i}$$

where p_i is the ith prime and $e_i \in \{0,1,2,\dots\} \cup \{\infty\}$. We denote the set of all Steinitz numbers by $\mathbb{S}$. Two Steinitz numbers are equal if and only if the exponents of corresponding prime numbers p_i are equal. □

We will denote arbitrary Steinitz numbers using uppercase letters and reserve lowercase letters strictly for ordinary positive integers. We will take certain obvious liberties when writing Steinitz numbers, such as omitting factors with exponent equal to 0. Thus, any positive integer is a Steinitz number. We next define the arithmetic of Steinitz numbers.

Definition *Let $S = \prod p_i^{e_i}$ and $T = \prod p_i^{f_i}$ be Steinitz numbers.*

1) The **product** *and* **quotient** *of S and T are defined by*

$$ST = \prod_{i=1}^{\infty} p_i^{e_i+f_i} \text{ and } S/T = \prod_{i=1}^{\infty} p_i^{e_i-f_i}$$

where $\infty - \infty = 0$.

2) We say that S **divides** *T and write $S \mid T$ if $e_i \le f_i$ for all i.* □

It is clear that $S \mid T$ if and only if $n \mid S \Rightarrow n \mid T$ for all positive natural numbers n. Also, $S = T$ if and only if $S \mid T$ and $T \mid S$.

Theorem 9.8.3 *Under the relation of "divides" given in the previous definition, the set $\mathbb{S}$ is a complete distributive lattice, with meet and join given by*

$$S \wedge T = \prod_{i=1}^{\infty} p_i^{\min(e_i,f_i)} \text{ and } S \vee T = \prod_{i=1}^{\infty} p_i^{\max(e_i,f_i)}$$

Moreover, the set of positive integers is a sublattice of $\mathbb{S}$. □

Subfields of the Algebraic Closure

We can now describe the subfields of $\Gamma(q)$. Let $\mathcal{S}(\Gamma(q))$ denote the lattice of all subfields of $\Gamma(q)$ that contain $GF(q)$.

Definition *If S is a Steinitz number, let*

$$GF(q^S) = \bigcup_{d \mid S} GF(q^d)$$

where, as indicated by the lowercase notation, d is a positive integer. □

If $\alpha, \beta \in GF(q^S)$ then $\alpha \in GF(q^k)$ for some $k \mid S$ and $\beta \in GF(q^n)$ for some $n \mid S$. Thus $\alpha, \beta \in GF(q^m)$ where $m = \operatorname{lcm}(k,n)$. It follows that $GF(q^S)$ is a subfield of $\Gamma(q)$ containing $GF(q)$.

Theorem 9.8.4 *The map $\phi: \mathbb{S} \to \mathcal{S}(\Gamma(q))$ defined by $\phi(S) = GF(q^S)$ is an order-preserving bijection. Hence, it is an isomorphism of lattices, that is,*
1) $S \mid T$ if and only if $GF(q^S) < GF(q^T)$,
2) $GF(q^S) \cap GF(q^T) = GF(q^{S \wedge T})$,
3) $GF(q^S)GF(q^T) = GF(q^{S \vee T})$.
In addition, $GF(q^S)$ is finite if and only if S is a positive integer.
Proof. We begin by showing that $n \mid S$ if and only if $GF(q^n) < GF(q^S)$. One direction follows immediately from the definition: if $n \mid S$ then $GF(q^n) < GF(q^S)$. Suppose that $GF(q^n) < GF(q^S)$. Let α be a field primitive element of $GF(q^n)$ over $GF(q)$. Then $\alpha \in GF(q^S)$ and so $\alpha \in GF(q^d)$ for some $d \mid S$. Hence $GF(q^n) = GF(q)(\alpha) < GF(q^d)$, which implies that $n \mid d$, whence $n \mid S$.

Since $T \mid S$ if and only if $n \mid T \Rightarrow n \mid S$, it follows that $T \mid S$ if and only if

$$GF(q^n) < GF(q^T) \Rightarrow GF(q^n) < GF(q^S)$$

that is, if and only if $GF(q^T) < GF(q^S)$.

To see that ϕ is injective, if $GF(q^S) = GF(q^T)$, then each field is contained in the other and so $n \mid S$ if and only if $n \mid T$, which implies that $S = T$.

To see that ϕ is surjective, let $GF(q) < F < \Gamma(q)$. We must find an S for which $GF(q^S) = F$. For each prime p_i, let e_i be the largest power of p_i for which

$$GF(q^{p_i^{e_i}}) < F \tag{9.8.1}$$

where $e_i = \infty$ if (9.8.1) holds for all positive integers e_i. Let

$$S = \prod_{i=1}^{\infty} p_i^{e_i}$$

We claim that

$$\bigvee GF(q^{p_i^{e_i}}) = \bigvee_{n \mid S} GF(q^n) = GF(q^S) \tag{9.8.2}$$

The second equality is by definition and the first field is clearly contained in the second. Also, if $n \mid S$, then

$$n = \prod_{i=1}^{r} p_i^{f_i}$$

where $f_i \leq e_i$ and so

$$GF(q^n) = \bigvee GF(q^{p_i^{f_i}}) < \bigvee GF(q^{p_i^{e_i}})$$

It follows that (9.8.2) holds. This implies that

$$GF(q^S) = \bigvee GF(q^{p_i^{e_i}}) < F$$

For the reverse inclusion, if $\alpha \in F$ then $\alpha \in GF(q^n) < F$ for some n. If

$$n = \prod_{i=1}^{r} p_i^{f_i}$$

then

$$GF(q^{p_i^{f_i}}) < GF(q^n) < F$$

and so $f_i \le e_i$ for all i, by the maximality of e_i. Hence $n \mid S$ and so $\alpha \in GF(q^n) < GF(q^S)$. This shows that $F < GF(q^S)$. Hence $F = GF(q^S)$ and so ϕ is surjective. We leave the rest of the proof to the reader. □

Since the largest Steinitz number is

$$P = \prod_{i=1}^{\infty} p_i^{\infty}$$

this corresponds to the largest subfield of $\Gamma(q)$, that is,

$$GF(q^P) = \Gamma(q)$$

Exercises

1. Determine the number of subfields of $GF(1024)$ and $GF(729)$.
2. Group primitive elements of $GF(p)$, p prime, can often be found by experimentation and the fact that if $o(\alpha) = m$ and $o(\beta) = n$ and $(m, n) = 1$ then $o(\alpha, \beta) = mn$. For instance, if $p = 31$, then by checking some small primes, we see that $o(-2) = 10$ and $o(5) = 3$, whence $o(-10) = 30$ and so $-10 = 21$ is group primitive for $GF(31)$.
   a) For $p = 41$, show that $o(2) = 20$ and $o(3) = 8$. Find an element of order 5 to pair with 3.
   b) If β is group primitive for $GF(p)$, p an odd prime, then what is $\beta^{(p-1)/2}$?
   c) Prove **Wilson's theorem**. If p is an odd prime then

   $$(p-1)! \equiv -1 \bmod p$$

   Hint: The left side is the product of all nonzero elements in $\mathbb{Z}_p$. Conisder this product from the point of view of a group primitive element β.
3. Show that except for the case of $GF(2)$, the sum of all the elements in a finite field is equal to 0.
4. Find all group primitive elements of $GF(7)$.

5. Show that the polynomial $x^4 + x^3 + x^2 + x + 1$ is irreducible over $GF(2)$. Is it primitive?
6. Let F be an arbitrary field. Prove that if F^* is cyclic then F must be a finite field.

Find the order of the following irreducible polynomials.

7. $x^4 + x^3 + x^2 + x + 1$ over $GF(2)$.
8. $x^4 + x + 1$ over $GF(2)$.
9. $x^8 + x^4 + x^3 + x^2 + 1$ over $GF(2)$.
10. $x^8 + x^5 + x^4 + x^3 + 1$ over $GF(2)$.
11. $x^8 + x^7 + x^5 + x + 1$ over $GF(2)$.
12. $x^4 + x + 2$ over $GF(3)$.
13. $x^4 + x^3 + x^2 + 1$ over $GF(3)$.
14. $x^5 - x + 1$ over $GF(3)$.

15. Show that every element in $GF(q^n)$ has a unique q^ith root, for $i = 1, \ldots, n-1$.
16. If $2 \not| \, q$, show that exactly one-half of the nonzero elements of $GF(q)$ have square roots.
17. Show that if $\alpha \in GF(q)$ and n is a positive integer, then $x^q - x + \alpha$ divides $x^{q^n} - x + n\alpha$.
18. Find a normal basis for $GF(8)$ over $GF(2)$. *Hint*: Let α be a root of the irreducible polynomial $p(x) = x^3 + x^2 + 1$.
19. Show that $\Gamma(q) = \bigcup_{n=0}^{\infty} GF(q^n)$.
20. Let a_n be any strictly increasing infinite sequence of positive integers. Prove that $\Gamma(q) = \bigcup_{n=0}^{\infty} GF(q^{a_n})$.
21. Show that $\Gamma(q^n) = \Gamma(q^m)$.
22. Let F be a field F satisfying $GF(q) < F < \Gamma(q)$. Show that all the proper subfields of F are finite if and only if F is finite or $F = GF(q^S)$ where $S = r^{\infty}$ for some prime r.
23. Show that $\Gamma(q)$ has no maximal subfields.
24. Show that $[\Gamma(q) : F]$ is not finite for any proper subfield $F < \Gamma(q)$.
25. Show that $\Gamma(q)$ has an uncountable number of nonisomorphic subfields.
26. Let $S \mid T$. Show that $[GF(q^T) : GF(q^S)]$ is finite if and only if T/S is finite, in which case the two numbers are equal.

Chapter 10
Finite Fields II: Additional Properties

10.1 Finite Field Arithmetic

There are various ways in which to represent the elements of a finite field. Since every finite field F is simple, it has the form $F = GF(p)(\alpha)$ for some $\alpha \in F$ and so the elements of F are polynomials in α of degree less than $\deg(\alpha)$. Another way to represent the elements of a finite field is to use the fact that $GF(q)^*$ is cyclic, and so its elements are all powers of a group primitive element.

It is clear that addition is more easily performed when field elements are written as polynomials and multiplication is more easily performed when all elements are written as a power of a single group primitive element. Fortunately, the two methods can be combined to provide an effective means for doing finite field arithmetic.

Example 10.1.1 Consider the finite field $GF(16)$ as an extension of $GF(2)$. The polynomial

$$p(x) = x^4 + x + 1$$

is irreducible over $GF(2)$. To see this, note that if $p(x)$ is reducible, it must have either a linear or a quadratic factor. But since $p(0) \neq 0$ and $p(1) \neq 0$, it has no linear factors. To see that $p(x)$ has no quadratic factors, note that there are precisely four quadratic polynomials over $GF(2)$, namely,

$$x^2, x^2 + 1, x^2 + 1, x^2 + x + 1$$

and it is easy to check that no product of any two of these polynomials equals $p(x)$.

Thus, letting α be a root of $p(x)$, we can represent the elements of $GF(16)$ as the 16 binary polynomials of degree 3 or less in α, as follows:

Constant:	$0, 1$
Linear:	$\alpha, \alpha+1$
Quadratic:	$\alpha^2,\ \alpha^2+1, \alpha^2+\alpha, \alpha^2+\alpha+1$
Cubic:	$\alpha^3, \alpha^3+1, \alpha^3+\alpha, \alpha^3+\alpha^2, \alpha^3+\alpha+1,$
	$\alpha^3+\alpha^2+1, \alpha^3+\alpha^2+\alpha, \alpha^3+\alpha^2+\alpha+1$

Addition of elements of $GF(16)$ is quite simple, since it is just addition of polynomials, but multiplication requires reduction modulo $p(\alpha)$, using the relation $\alpha^4 = \alpha + 1$. On the other hand, observe that

$$\begin{aligned} \alpha^{15} &= (\alpha^5)^3 \\ &= (\alpha \cdot \alpha^4)^3 \\ &= (\alpha \cdot (\alpha{+}1))^3 \\ &= \alpha^3(\alpha{+}1)^3 \\ &= \alpha^3 \cdot (\alpha^3+\alpha^2+\alpha+1) \\ &= \alpha^6+\alpha^5+\alpha^4+\alpha^3 \\ &= (\alpha^3+\alpha^2\ )+(\alpha^2+\alpha)+(\alpha+1)+\alpha^3 \\ &= (\alpha^3+\alpha^2)+(\alpha^2+\alpha)+(\alpha+1)+\alpha^3 \\ &= 1 \end{aligned}$$

and so $o(\alpha) \mid 15$. Since $\alpha^3 \neq 1$ and $\alpha^5 \neq 1$, we conclude that α is group primitive and

$$GF(16) = \{0, 1, \alpha, \ldots, \alpha^{14}\}$$

With this representation, multiplication is all but trivial, but addition is cumbersome.

We can link the two representations of $GF(16)$ by computing a table showing how each element α^k can be represented as a polynomial in α of degree at most 3. Using the fact that $\alpha^4 = 1 + \alpha$, we have

$$\begin{aligned} \alpha^4 &= \alpha+1 \\ \alpha^5 &= \alpha \cdot \alpha^4 = \alpha(\alpha+1) = \alpha^2+\alpha \\ \alpha^6 &= \alpha \cdot \alpha^5 = \alpha^3+\alpha^2 \\ \alpha^7 &= \alpha \cdot \alpha^6 = \alpha^4+\alpha^3 = \alpha^3+\alpha+1 \end{aligned}$$

and so on. The complete list, given in Table 10.1.1, is known as a **field table** for $GF(16)$. As is customary, we write only the exponent k for α^k, and $a_3a_2a_1a_0$ for the polynomial $a_3\alpha^3 + a_2\alpha^2 + a_1\alpha + a_0$.

Table 10.1.1

k	$a_3a_2a_1a_0$
0	0001
1	0010
2	0100
3	1000
4	0011
5	0110
6	1100
7	1011
8	0101
9	1010
10	0111
11	1110
12	1111
13	1101
14	1001

Computations using this table are quite straightforward; for example,

$$\begin{aligned}(\alpha^8 + \alpha^4 + 1)(\alpha^3 + \alpha) &= (0101 + 0011 + 0001)(1000 + 0010)\\ &= (0111)(1010)\\ &= \alpha^{10} \cdot \alpha^9 = \alpha^{19} = \alpha^4 = \alpha + 1\end{aligned}$$

Thus, the key to doing arithmetic in a finite field is having a group primitive element, along with its minimal (primitive) polynomial. In general, the task of finding primitive polynomials is not easy. There are various methods that achieve some measure of success in certain cases, and we mention one such method at the end of Section 11.2. Fortunately, extensive tables of primitive polynomials and field tables have been constructed.

Let us use the primitive polynomial $p(x)$ and the field table for $GF(16)$ to compute the minimal polynomial over $GF(2)$ for each element of $GF(16)$. We begin by computing sets of conjugates, using Theorem 9.6.1 and the fact that $\alpha^{16} = \alpha$,

Conjugates of α: $\alpha, \alpha^2, \alpha^4, \alpha^8$
Conjugates of α^3: $\alpha^3, \alpha^6, \alpha^{12}, \alpha^{24} = \alpha^9$
Conjugates of α^5: α^5, α^{10}
Conjugates of α^7: $\alpha^7, \alpha^{14}, \alpha^{28} = \alpha^{13}, \alpha^{56} = \alpha^{11}$

Letting $m_k(x)$ be the minimal polynomial for α^k, we have, for example

$$m_5(x) = m_{10}(x) = (x - \alpha^5)(x - \alpha^{10}) = x^2 - (\alpha^5 + \alpha^{10})x + \alpha^{15}$$

The field table for $GF(16)$ gives

$$\alpha^5 + \alpha^{10} = (0110) + (0111) = (0001) = \alpha^0 = 1$$

and since $\alpha^{15} = 1$, we have

$$m_5(x) = m_{10}(x) = x^2 + x + 1$$

The other minimal polynomials are computed similarly. The complete list is

$$\begin{aligned}
&m_0(x) = x + 1 \\
&m_1(x) = m_2(x) = m_4(x) = m_8(x) = x^4 + x + 1 \\
&m_3(x) = m_6(x) = m_9(x) = m_{12}(x) = x^4 + x^3 + x^2 + x + 1 \\
&m_5(x) = m_{10}(x) = x^2 + x + 1 \\
&m_7(x) = m_{11}(x) = m_{13}(x) = m_{14}(x) = x^4 + x^3 + 1
\end{aligned}$$

Being able to factor polynomials of the form $x^n - 1$ is important for a variety of applications of finite field theory, especially to coding theory. Since the roots of $x^{15} - 1$ over $GF(2)$ are precisely the elements of $GF(16)^*$, we have

$$x^{15} - 1 = m_0(x)m_1(x)m_3(x)m_5(x)m_7(x)$$

Of course, in order to obtain this factorization, we worked in the splitting field $GF(16)$. Let us turn to a method for factoring polynomials over $\mathbb{Z}_p = GF(p)$ that does not require working in any extension of $\mathbb{Z}_p$.

Factoring over $\mathbb{Z}_p$: Berlekamp's Algorithm

Berlekamp's algorithm is an algorithm for factoring polynomials over $\mathbb{Z}_p$. Suppose that $f(x)$ is a polynomial over $\mathbb{Z}_p$ of degree d. Let us first show that we can reduce the problem of factoring $f(x)$ to one of factoring a polynomial with no repeated factors.

We know that $f(x)$ has a repeated factor if and only if $f(x)$ and $f'(x)$ have a common factor. Write

$$f(x) = \gcd(f(x), f'(x)) \cdot \frac{f(x)}{\gcd(f(x), f'(x))}$$

Let $d(x) = \gcd(f(x), f'(x))$. If $d(x) = 1$ then $f(x)$ has no repeated factors. If $d(x) = f(x)$ then $f'(x) = 0$ and so

$$f(x) = h(x^p) = (h(x))^p$$

and we can factor $h(x)$ (or repeat the process). Otherwise, $d(x)$ is a nonconstant polynomial with degree less than that of $f(x)$ and $f(x)/d(x)$ has no repeated factors. Thus, we can consider the polynomials $d(x)$ and $f(x)/d(x)$ separately. For the former polynomial, we can repeat the above argument until the factoring problem reduces to one of factoring polynomials with no repeated factors.

So let us suppose that $f(x)$ is the product of distinct irreducible factors. (Actually, the factoring algorithm that we are about to describe does not require this restriction on $f(x)$, but the formula for the number of irreducible factors that we will present does.)

Suppose that we can find a nonconstant polynomial $g(x)$ of degree less than d for which

$$f(x) \mid g(x)^p - g(x)$$

Since $\mathbb{Z}_p = \{0, 1, \ldots, p-1\}$ is the set of roots of $x^p - x$, we have

$$x^p - x = x(x-1)\cdots(x-p+1)$$

and so

$$g(x)^p - g(x) = g(x)(g(x)-1)\cdots(g(x)-p+1)$$

Also, if in general, $a \mid b_1 \cdots b_k$, where $b_1, \ldots, b_k$ are pairwise relatively prime, then

$$a = \gcd(a, b_1)\cdots\gcd(a, b_k)$$

Hence, since the polynomials $g(x) - k$ are pairwise relatively prime for $k = 0, \ldots, p-1$, we have

$$f = \gcd(f, g)\gcd(f, g-1)\cdots\gcd(f, g-p+1)$$

Note that the degree of each of these factors is at most $\deg(g) < d$ and so this factorization of $f(x)$ is nontrivial. Note also that the Euclidean algorithm can be used to find the gcd of the pairs of polynomials in the previous factorization and so if we can find such a polynomial $g(x)$, then we will have an algorithm for finding a nontrivial factorization of $f(x)$.

A polynomial $g(x)$ for which $f(x) \mid g(x)^p - g(x)$ is called an ***f*-reducing polynomial**. We are interested in nonconstant f-reducing polynomials with degree less than the degree of $f(x)$, since these polynomials provide factorizations of $f(x)$.

To find such an f-reducing polynomial $g(x)$, write

$$g(x) = g_0 + g_1 x + \cdots + g_{d-1} x^{d-1}$$

Then since we are working over a field of characteristic p, and since $g_k^p = g_k$ modulo p, it follows that $g(x)^p = g(x^p)$ and so

$$g(x)^p - g(x) = g(x^p) - g(x) = \sum_{i=0}^{d-1} g_i(x^{ip} - x^i)$$

Now suppose that $x^{ip} = a_i(x)f(x) + r_i(x)$, where $\deg(r_i) < d$. Then $g(x)^p - g(x)$ is divisible by $f(x)$ if and only if

$$f(x) \Big| \sum_{i=0}^{d-1} g_i(r_i(x) - x^i)$$

but since the right hand sum has degree less that that of $f(x)$, this is equivalent to

$$\sum_{i=0}^{d-1} g_i(r_i(x) - x^i) = 0$$

and this is equivalent to a system of linear equations. To express this system in matrix form, suppose that $r_i(x) = r_{i,0} + r_{i,1}x + \cdots + r_{i,d-1}x^{d-1}$. Then the previous equation is equivalent to the system

$$\sum_{i=0}^{d-1} g_i(r_{i,j} - \delta_{i,j}) = 0$$

for $j = 0, \ldots, d-1$. In matrix terms, if $M = (r_{i,j})$, $I_d = (\delta_{i,j})$ and

$$G = \begin{bmatrix} g_0 & \cdots & g_{d-1} \end{bmatrix}$$

is the row matrix of coefficients of $g(x)$, then this system is

$$G(M - I_d) = 0$$

Example 10.1.1 Consider the polynomial

$$f(x) = 1 + x + x^2 + x^3 + x^4 + x^5 + x^6$$

over $\mathbb{Z}_2$. First, we find the polynomials $r_i(x)$ by dividing x^{2i} by $f(x)$, to get

$$\begin{aligned}
r_0(x) &= 1 \\
r_1(x) &= x^2 \\
r_2(x) &= x^4 \\
r_3(x) &= 1 + x + x^2 + x^3 + x^4 + x^5 \\
r_4(x) &= x \\
r_5(x) &= x^3
\end{aligned}$$

Hence,

$$M - I_6 = \begin{bmatrix} 0 & 0 & 0 & 0 & 0 & 0 \\ 0 & 1 & 1 & 0 & 0 & 0 \\ 0 & 0 & 1 & 0 & 1 & 0 \\ 1 & 1 & 1 & 0 & 1 & 1 \\ 0 & 1 & 0 & 0 & 1 & 0 \\ 0 & 0 & 0 & 1 & 0 & 1 \end{bmatrix}$$

and our system is

$$\begin{aligned} g_3 &= 0 \\ g_1 + g_3 + g_4 &= 0 \\ g_1 + g_2 + g_3 &= 0 \\ g_5 &= 0 \\ g_2 + g_3 + g_4 &= 0 \\ g_3 + g_5 &= 0 \end{aligned}$$

whose solution is

$$g_0 \text{ arbitrary; } g_1 = g_2 = g_4;\ g_3 = g_5 = 0$$

The only nonconstant solution is

$$g(x) = g_0 + x + x^2 + x^4$$

where $g_0 = 0, 1$. It follows that

$$g(x)^2 - g(x) = x^8 + x = (x^2 + x)f(x)$$

and so, using Euclid's algorithm for the gcd, we get the factorization

$$\begin{aligned} f(x) &= \gcd(f(x), g(x))\gcd(f(x), g(x) - 1) \\ &= \gcd(f(x), x + x^2 + x^4)\gcd(f(x), 1 + x + x^2 + x^4) \\ &= (1 + x + x^3)(1 + x^2 + x^3) \end{aligned}$$

□

The Number of Irreducible Factors

Knowledge of the number of irreducible factors of $f(x)$ would help us determine when the factorization algorithm has produced a complete factorization of $f(x)$ into irreducible factors. Suppose that

$$f(x) = p_1(x) \cdots p_k(x)$$

where the $p_i(x)$ are distinct monic, irreducible polynomials over $\mathbb{Z}_p$.

Let $\mathcal{F}$ be the set of f-reducing polynomials with degree less than that of $f(x)$, including the constant polynomials. Note that $\mathcal{F}$ is isomorphic to the null space $\text{null}(M - I)$ of the matrix $M - I$ of the Berlekamp algorithm.

If $g \in \mathcal{F}$, then

$$p_1(x)\cdots p_k(x) \mid g(x)(g(x)-1)\cdots(g(x)-p+1)$$

and since the polynomials on the right are relatively prime, each $p_i(x)$ divides precisely one of these polynomials, say $p_i(x) \mid g(x) - u_i$. This is a system of congruences

$$\begin{aligned} g(x) &\equiv u_1 \operatorname{mod} p_1(x) \\ &\vdots \\ g(x) &\equiv u_k \operatorname{mod} p_k(x) \end{aligned}$$

and since the $p_i(x)$'s are relatively prime (this is where we use the fact that the $p_i(x)$ are distinct), the Chinese remainder theorem tells us that there is a unique solution $g(x)$ modulo $f(x)$, that is, a unique solution of degree less than that of $f(x)$. In other words, there is at most one f-reducing polynomial $g(x)$ for each k-tuple $(u_1, \ldots, u_k)$. But if $g(x)$ is a solution to this system, then

$$p_i(x) \mid g(x) - u_i$$

for all $i = 1, \ldots, k$ and so $f(x) \mid \prod_{i=1}^{n}(g(x) - i)$, whence $g \in \mathcal{F}$. It follows that there is precisely one f-reducing polynomial for each k-tuple $(u_1, \ldots, u_k)$ in $\mathbb{Z}_p^k$. Hence,

$$p^k = |F_p^k| = |\mathcal{F}| = p^{|\dim(\text{null}(M-I))|} = p^{d-\text{rk}(M-I)}$$

that is, the number of distinct irreducible factors of $f(x)$ is

$$k = d - \text{rk}(M - I)$$

Example 10.1.2 The matrix $M - I$ from Example 10.1.1 has rank 4, which can be determined by applying elementary row operations to reduce the matrix to echelon form. Hence, the nullity is $6 - 4 = 2$ and so the factorization in that example is complete.□

*10.2 The Number of Irreducible Polynomials

Of course, if F is a finite field, then there is only a finite number of polynomials of a given degree d over F. It is possible to obtain an explicit formula for the number of irreducible polynomials of degree d over $GF(q)$ by using Möbius inversion. (See the appendix for a discussion of Möbius inversion.) First, we need the following result.

Theorem 10.2.1 Let $GF(q)$ be a finite field, and let n be a positive integer. Then the product of all monic irreducible polynomials over $GF(q)$, whose degree divides n is

$$f_{q^n}(x) = x^{q^n} - x$$

Proof. According to Theorem 9.6.1, an irreducible polynomial $p(x)$ divides

$f_{q^n}(x)$ if and only if $\deg(p(x)) \mid n$. Hence, $f_{q^n}(x)$ is a product of irreducible polynomials whose degrees divide n and every irreducible polynomial whose degree divides n divides $f_{q^n}(x)$. Since no two such irreducible polynomials have any roots in common and since $f_{q^n}(x)$ has no multiple roots, the result follows.□

Let us denote the number of monic irreducible polynomials of degree d over $GF(q)$ by $N_q(d)$. By counting degrees, Theorem 10.2.1 gives the following.

Corollary 10.2.2 For all positive integers d and n, we have

$$q^n = \sum_{d|n} dN_q(d)$$

□

Now we can apply Möbius inversion to get an explicit formula for $N_q(d)$. Classical Möbius inversion is

$$g(n) = \sum_{d|n} f(d) \Rightarrow f(n) = \sum_{d|n} g(d)\mu\left(\frac{n}{d}\right) \qquad (10.2.1)$$

where the Möbius function μ is defined by

$$\mu(m) = \begin{cases} 1 & \text{if } m = 1 \\ (-1)^k & \text{if } m = p_1p_2\cdots p_k \text{ for distinct primes } p_i \\ 0 & \text{otherwise} \end{cases}$$

Corollary 10.2.3 The number $N_q(n)$ of monic irreducible polynomials of degree n over $GF(q)$ is

$$N_q(n) = \frac{1}{n}\sum_{d|n} \mu\left(\frac{n}{d}\right)q^d = \frac{1}{n}\sum_{d|n} \mu(d)q^{n/d}$$

Proof. Letting $g(n) = q^n$ and $f(d) = dN_q(d)$ in (10.2.1) gives the result.□

Example 10.2.1 The number of monic irreducible polynomials of degree 12 over $GF(q)$ is

$$\begin{aligned} N_q(12) &= \frac{1}{12}\left(\mu(1)q^{12} + \mu(2)q^6 + \mu(3)q^4 + \mu(4)q^3 + \mu(6)q^2 + \mu(12)q\right) \\ &= \frac{1}{12}\left(q^{12} - q^6 - q^4 + q^2\right) \end{aligned}$$

The number of monic irreducible polynomials of degree 4 over $GF(2)$ is

$$N_2(4) = \frac{1}{4}\left(\mu(1)2^4 + \mu(2)2^2 + \mu(4)2^1\right) = 3$$

as we would expect from the results of Example 10.1.1.□

Möbius inversion can also be used to find the *product* of all monic irreducible polynomials of degree d over $GF(q)$. Let us denote this product by $I(q,d;x)$. Then Theorem 10.2.1 is equivalent to

$$x^{q^n} - x = \prod_{d|n} I(q,d;x)$$

Applying the multiplicative version of Möbius inversion gives the following.

Corollary 10.2.4 *The product $I(q,n;x)$ of all monic irreducible polynomials of degree n over $GF(q)$ is*

$$I(q,n;x) = \prod_{d|n}\left(x^{q^d} - x\right)^{\mu(n/d)} = \prod_{d|n}\left(x^{q^{n/d}} - x\right)^{\mu(d)} \qquad \square$$

Example 10.2.2 For $q = 2$ and $n = 4$, we get

$$\begin{aligned} I(2,4;x) &= (x^{16} - x)^{\mu(1)}(x^4 - x)^{\mu(2)}(x^2 - x)^{\mu(4)} \\ &= \frac{x^{16} - x}{x^4 - x} = \frac{x^{15} - 1}{x^3 - 1} = x^{12} + x^9 + x^6 + x^3 + 1 \end{aligned} \qquad \square$$

*10.3 Polynomial Functions

Finite fields have the special property that *any* function from a finite field F to itself can be represented by a polynomial. As a matter of fact, this property actually *characterizes* finite fields from among all commutative rings (finite and infinite)!

Since $GF(q)$ has size q, there are precisely q^q functions from $GF(q)$ to itself. Among these functions are the *polynomial functions* $\alpha \mapsto p(\alpha)$ where $p(x) \in GF(q)[x]$. We will denote this polynomial function by $p(x)$ as well. If $p(x)$ and $q(x)$ are polynomial functions on $GF(q)$ then $p(x) = q(x)$ as functions if and only if $p(\alpha) = q(\alpha)$ for all $\alpha \in GF(q)$, which holds if and only if

$$x^q - x \mid p(x) - q(x)$$

Thus, two polynomials represent the same function if and only if they are congruent modulo $x^q - x$. Since every polynomial is congruent modulo $x^q - x$ to precisely one polynomial of degree less than q (namely, its remainder after dividing by $x^q - x$), and since there are q^q polynomials of degree less than q, we have the following theorem. (Proof of the last statement in part 2 of the theorem is left to the reader.)

Theorem 10.3.1

1) Two polynomials over $GF(q)$ represent the same polynomial function on $GF(q)$ if and only if they are congruent modulo $x^q - x$.

2) *Every function $f: GF(q) \to GF(q)$ is a polynomial function, for a unique polynomial of degree less than q. In fact, the unique polynomial of degree less than q that represents f is*

$$p_f(x) = \sum_{\alpha \in GF(q)} f(\alpha)(1 - (x - \alpha)^{q-1}) \qquad \square$$

(The representation of f given in part 2) above is the *Lagrange interpolation formula* as applied to finite fields.) Part 2) has a very interesting converse as well.

Theorem 10.3.2 *If R is a commutative ring and if every function $f: R \to R$ is a polynomial function, that is, $f(x) = p(x)$ for some $p(x) \in R[x]$, then R is a finite field.*

Proof. First, we show that R is finite. Suppose that $|R| = \lambda$. The number of functions from R to itself is λ^λ and the number of polynomials over R is the same as the number of finite sequences with elements from R, which is $\aleph_0\lambda$. Since distinct functions are represented by distinct polynomials, we must have $\lambda^\lambda \le \aleph_0\lambda$, which happens only when λ is finite. Thus, R is a finite set.

Now let $r, a \in R$ with $r \neq 0$. Define a function $f_{r,a}: R \to R$ by

$$f_{r,a}(x) = \begin{cases} a & \text{if } x = r \\ 0 & \text{if } x \neq r \end{cases}$$

By hypothesis, there exists a polynomial $a_0 + a_1x + \cdots + a_nx^n$ for which

$$a_0 + a_1r + \cdots + a_nr^n = a$$

and

$$a_0 + a_1x + \cdots + a_nx^n = 0, \text{ for } x \neq r$$

Setting $x = 0$ gives $a_0 = 0$ and so

$$r(a_1 + a_2r + \cdots + a_nr^{n-1}) = a$$

Thus, we conclude that for any $r \neq 0$ and any $a \in R$, there is a $u \in R$ for which $ru = a$. In other words, the map $\phi_r: R \to R$ defined by $\phi_r s = rs$ is surjective. Since R is a finite set, ϕ_r must also be injective. Hence, $rs = 0$, $r \neq 0$ implies that $s = 0$ and so R has no zero divisors. In addition, since ϕ_r is surjective, there exists a $u \in R$ for which $\phi_r u = r$, that is, $ru = r$. If $a \in R$ then $aru = ar$ and since R is commutative and has no zero divisors, we may cancel r to get $au = a$. Thus $u \in R$ is the multiplicative identity of R. Hence R is a finite integral domain, that is, a finite field. $\square$

*10.4 Linearized Polynomials

We now turn to a discussion of linear operators on $GF(q^n)$ over $GF(q)$. We will see that all such linear operators can be expressed as polynomial functions of a very special type.

Definition *A polynomial of the form*

$$L(x) = \sum_{i=0}^{m} \alpha_i x^{q^i}$$

with coefficients $\alpha_i \in GF(q^n)$ *is called a* **linearized polynomial**, *or a* **q-polynomial**, *over* $GF(q^n)$. □

The term *linearized polynomial* comes from the following theorem, whose proof is left to the reader.

Theorem 10.4.1 *Let* $L(x)$ *be a linearized polynomial over* $GF(q^n)$. *If* $\alpha, \beta \in GF(q^n)$ *and* $a, b \in GF(q)$, *then*

$$L(a\alpha + b\beta) = aL(\alpha) + bL(\beta)$$

Thus, the polynomial function $L(x)\colon GF(q^n) \to GF(q^n)$ *is a linear operator on* $GF(q^n)$ *over* $GF(q)$. □

The roots of a q-polynomial in a splitting field have some rather special properties, which we give in the next two theorems.

Theorem 10.4.2 *Let* $L(x)$ *be a nonzero q-polynomial over* $GF(q^n)$, *with splitting field* $GF(q^s)$. *Then each root of* $L(x)$ *in* $GF(q^s)$ *has the same multiplicity, which must be either* 1 *or else a power of* q. *Furthermore, the roots of* $L(x)$ *form a vector subspace of* $GF(q^s)$ *over* $GF(q)$.

Proof. Since $L'(x) = \alpha_0$, if $\alpha_0 \neq 0$ then all roots of $L(x)$ are simple. On the other hand, suppose that $\alpha_0 = \alpha_1 = \cdots = \alpha_{k-1} = 0$ but $\alpha_k \neq 0$. Then since $\alpha_i \in GF(q^n)$, we have

$$\alpha_i^{q^{nk}} = \alpha_i$$

and so

$$L(x) = \sum_{i=k}^{m} \alpha_i x^{q^i} = \sum_{i=k}^{m} \alpha_i^{q^{nk}} x^{q^i} = \left(\sum_{i=k}^{m} \alpha_i^{q^{(n-1)k}} x^{q^{i-k}} \right)^{q^k}$$

which is the q^kth power of a linearized polynomial with nonzero constant term, and therefore has only simple roots. Hence, each root of $L(x)$ has multiplicity q^k. We leave proof of the fact that the roots form a vector subspace of $GF(q^s)$ to the reader. □

The following theorem, whose proof we omit, is a sort of converse to Theorem 10.4.1. (For a proof of this theorem, and more on q-polynomials, see the book by Lidl and Niederreiter (1986).)

Theorem 10.4.3 *Let U be a vector subspace of $GF(q^n)$ over $GF(q)$. Then for any nonnegative integer k, the polynomial*

$$L(x) = \prod_{\alpha \in U} (x - \alpha)^{q^k}$$

is a q-polynomial over $GF(q^n)$. □

If $L(x)$ is a q-polynomial, then as a function, we have

$$L: \alpha \mapsto L(\alpha) = \sum_{i=0}^{m} \alpha_i \alpha^{q^i} = \sum_{i=0}^{m} \alpha_i \sigma_q^i \alpha$$

where σ_q is the Frobenius automorphism. Thus, as an operator

$$L = \sum_{i=0}^{m} \alpha_i \sigma_q^i$$

Since $\sigma_q^n = \iota$ we may reduce the expression for L to a polynomial in σ_q of degree at most $n-1$. In fact, adding 0 coefficients if necessary, we can say that every q-polynomial function on $GF(q^n)$ has the **standard form**

$$L = \sum_{i=0}^{n-1} \alpha_i \sigma_q^i$$

for $\alpha_i \in GF(q^n)$.

There are q^{n^2} such q-polynomial functions on $GF(q^n)$, and this happens also to be the number of linear operators on $GF(q^n)$ over $GF(q)$. Moreover, since the maps σ_q^i are linearly independent over $GF(q^n)$, we deduce that each q-polynomial in standard form represents a unique linear operator. Thus, we have characterized the linear operators on $GF(q^n)$ over $GF(q)$.

Theorem 10.4.4 *Every linear operator on $GF(q^n)$ over $GF(q)$ can be represented by a unique q-polynomial in standard form*

$$L(x) = \sum_{i=0}^{n-1} \alpha_i x^{q^i}$$

for some $\alpha_i \in GF(q^n)$. □

Exercises

1. Construct two distinct finite field tables for $GF(8)$ over $GF(2)$.
2. Factor the polynomial
$$f(x) = x^8 + x^6 + x^4 + x^3 + 1$$
over $\mathbb{Z}_2$.
3. Factor
$$f(x) = x^5 + x^4 + x^3 + x^2 + 1$$
over $\mathbb{Z}_2$.
4. Factor $x^5 + x^2 + 1$ over $\mathbb{Z}_2$.
5. Calculate $N_q(20)$.
6. Show that
$$N_q(n) \leq \frac{1}{n}(q^n - q)$$
and
$$q^n = \sum_{d|n} d\, N_q(d) \leq n\, N_q(n) + \sum_{k=0}^{\lfloor n/2 \rfloor} q^k \leq n\, N_q(n) + q^{1+n/2}$$
Hence, $N_q(n) \geq \frac{1}{n}(q^n - q^{1+n/2})$. Finally, show that $N_q(n) \approx q^n/n$.
7. Show that the unique polynomial of degree less than q that represents the function $f: GF(q) \to GF(q)$ is
$$p_f(x) = \sum_{\alpha \in GF(q)} f(\alpha)(1 - (x - \alpha)^{q-1})$$
8. Prove that a linearized polynomial over $GF(q^n)$ is a linear operator on $GF(q^n)$ over $GF(q)$.
9. Prove that the roots of a q-polynomial over $GF(q^n)$ form a vector subspace of the splitting field $GF(q^s)$ over $GF(q)$.
10. Prove that the greatest common divisor of two q-polynomials over $GF(q^n)$ is a q-polynomial, but the least common multiple need not be a q-polynomial.

Chapter 11
The Roots of Unity

Polynomials of the form $x^n - u$, where $0 \neq u \in F$, are known as **binomials**. Even though binomials have a simple form, their study is quite involved, as is evidenced by the fact that the Galois group of a binomial is often nonabelian. As we will see, an understanding of the binomial $x^n - 1$ is key to an understanding of all binomials.

We will have use for the following definition.

Definition *The* **exponent characteristic** $\text{expchar}(F)$ *of a field* F *is defined to be* 1 *if* $\text{char}(F) = 0$ *and* $\text{char}(F)$ *otherwise.*□

11.1 Roots of Unity

The roots of the binomial $x^n - 1$ over a field F are referred to as the n**th roots of unity over** F. Throughout this section, we will let F be a field with $p = \text{expchar}(F)$, S a splitting field for $x^n - 1$ over F and U_n the set of nth roots of unity over F, located in S. Note that if $n = kp$ then

$$x^n - 1 = x^{kp} - 1 = (x^k - 1)^p$$

and so the nth roots of unity are the same as the kth roots of unity, taken with a higher multiplicity. *Thus, from now on, we assume that* $(n, p) = 1$.

Theorem 11.1.1 *The set* U_n *of* n*th roots of unity over* F *is a cyclic group of order* n *under multiplication. Moreover, if* $(m, n) = 1$ *then*

$$U_{mn} = U_m U_n$$

where the product $U_n U_m$ *of groups is direct.*

Proof. Clearly $\alpha, \beta \in U_n$ implies $\alpha\beta, \alpha^{-1} \in U_n$. Hence, U_n is a finite subgroup of the multiplicative group S^* of nonzero elements of the field S. By Corollary 1.3.4, U_n is cyclic. Since $(n, p) = 1$, we have

$$D(x^n - 1) = nx^{n-1} \neq 0$$

showing that $x^n - 1$ is separable, and so $|U_n| = n$.

For the second part, if $\alpha \in U_m \cap U_n$ then $\alpha^m = 1 = \alpha^n$ and since $(m, n) = 1$ there exist $a, b \in \mathbb{Z}$ such that $am + bn = 1$, whence

$$\alpha = \alpha^{am+bn} = \alpha^{am}\alpha^{bn} = 1$$

which shows that $U_m \cap U_n = \{1\}$. Hence, the mn products in the group $U_m U_n$ are distinct and since $U_m U_n < U_{mn}$, it follows that $U_{mn} = U_m U_n$. □

Definition *An element $\omega \in U_n$ of order n, that is, a generator of U_n, is called a* **primitive nth root of unity** *over F. We shall denote the set of all primitive nth roots of unity over F by Ω_n and reserve the notation ω_n for a primitive nth root of unity.* □

Note that a primitive nth root of unity ω, being a group primitive element, is also a field primitive element of S, that is

$$F(\omega) = F(U_n) = S$$

However, in general, S has field primitive elements that are not primitive nth roots of unity.

Theorem 11.1.2

1) *If $\omega \in \Omega_n$ then*

$$\Omega_n = \{\omega^k \mid 1 \le k < n, (n, k) = 1\}$$

and $|\Omega_n| = \phi(n)$. Hence, there is a bijection from Ω_n onto the abelian (but not necessarily cyclic) group $\mathbb{Z}_n^*$ of all elements of $\mathbb{Z}_n$ that are relatively prime to n, that is, to the group of units of $\mathbb{Z}_n$.

2) *If $d \mid n$ then $\Omega_d = \Omega_n^{n/d}$.*
3) *If $(n, m) = 1$ then $\Omega_{mn} = \Omega_m \Omega_n$.*

Proof. Part 1) follows from the fact that if $o(\omega) = n$ then $o(\omega^k) = n$ if and only if $(n, k) = 1$. For part 2), if $\omega \in \Omega_n$ then

$$o(\omega^{n/d}) = \frac{n}{(n, n/d)} = d$$

and so $\omega^{n/d} \in \Omega_d$. Thus $\Omega_n^{n/d} \subseteq \Omega_d$. For the reverse inclusion, since $\omega^{n/d}$ has order d, the set

$$A = \{(\omega^{n/d})^k \mid (k, d) = 1, 1 \le k < d\}$$

consists of $\phi(d)$ distinct roots of unity of order d and so $A = \Omega_d$. But each element of A belongs to $\Omega_n^{n/d}$, since

$$o(\omega^{nk/d}) = \frac{n}{(n, nk/d)} = \frac{n}{(nd/d, nk/d)} = \frac{n}{(n/d)(d,k)} = \frac{n}{n/d} = d$$

and so $\Omega_d = A \subseteq \Omega_n^{n/d}$.

For part 3), since

$$o(\omega_m\omega_n) = \operatorname{lcm}(o(\omega_n), o(\omega_m)) = nm$$

we have $\omega_m\omega_n \in \Omega_{mn}$ and so $\Omega_m\Omega_n \subseteq \Omega_{mn}$. Now, since the products in U_mU_n are distinct, so are the products in $\Omega_m\Omega_n$. Hence

$$|\Omega_m\Omega_n| = \phi(m)\phi(n) = \phi(mn) = |\Omega_{mn}|$$

which shows that $\Omega_m\Omega_n = \Omega_{mn}$.□

11.2 Cyclotomic Extensions

The term *cyclotomy* is the process of dividing a circle into equal parts, which is precisely the effect obtained by plotting the nth roots of unity over $\mathbb{Q}$ in the complex plane.

Definition *Let F be a field. A splitting field S of $x^n - 1$ over F is called a* **cyclotomic extension of order** n *of F.* □

Since

$$S = F(U_n) = F(\Omega_n) = F(\omega)$$

for $\omega \in \Omega_n$ is the splitting field of a separable polynomial, it follows that $F < S$ is a finite Galois extension and

$$[S : F] = \deg(\min(\omega, F)) = |G_F(S)|$$

Now, any $\sigma \in G_F(S)$ is uniquely determined by its value on a fixed $\omega \in \Omega_n$, and since σ preserves order, $\sigma\omega$ must be one of the $\phi(n)$ primitive roots of unity in S, that is,

$$\sigma\omega = \omega^{k(\sigma)}$$

where $k(\sigma) \in \mathbb{Z}_n^*$. Since

$$\omega^{k(\sigma\tau)} = (\sigma\tau)\omega = \sigma(\omega^{k(\tau)}) = (\sigma\omega)^{k(\tau)} = \omega^{k(\sigma)k(\tau)}$$

it follows that

$$k(\sigma\tau) = k(\sigma)k(\tau) \bmod n$$

and so the map $k\colon G_F(S) \to \mathbb{Z}_n^*$ is a homomorphism. Since $k(\sigma) = 1$ implies that $\sigma = \iota$, the map k is a monomorphism and thus $G_F(S)$ is isomorphic to a subgroup of $\mathbb{Z}_n^*$.

Theorem 11.2.1 *If $F < S$ is a cyclotomic extension of order n, then $G_F(S)$ is isomorphic to a subgroup of $\mathbb{Z}_n^*$, the group of units of $\mathbb{Z}_n$. Hence, $G_F(S)$ is abelian and $[S : F] \mid \phi(n)$.* $\square$

Since the structure of $\mathbb{Z}_n^*$ is clearly important, we record the following theorem, whose proof is left as an exercise.

Theorem 11.2.2 *Let $n = \prod p_i^{e_i}$, where the p_i's are distinct primes. Then*

$$\mathbb{Z}_n^* \approx \prod \mathbb{Z}_{p_i^{e_i}}^*$$

Moreover, $\mathbb{Z}_n^*$ is cyclic if and only if $n = p^e, 2p^e$ or 4, where p is an odd prime. $\square$

Corollary 11.2.3 *A cyclotomic extension $F < S$ is abelian and if $n = p^e$, $2p^e$ or 4, where p is an odd prime, then $F < S$ is cyclic.* $\square$

Cyclotomic Polynomials

To investigate the properties of cyclotomic extensions further, we factor the polynomial $x^n - 1$. Since each root of this polynomial is a primitive dth root of unity for some $d \mid n$, we define the dth **cyclotomic polynomial** $Q_d(x)$ to be the polynomial whose roots are precisely the primitive dth roots of unity. Thus, if ω_d is a primitive dth root of unity, then

$$Q_d(x) = \prod_{\substack{1 \le k < d \\ (k,d)=1}} (x - \omega_d^k)$$

It follows that $\deg(Q_d(x)) = \phi(d)$ and

$$x^n - 1 = \prod_{d \mid n} Q_d(x)$$

since each side is the product of the linear factors $x - \alpha$, as α varies over all nth roots of unity. Note that cyclotomic polynomials are not necesssarily irreducible, and we will explore this issue as soon as we have recorded the basic properties of these polynomials.

Note also that the cyclotomic polynomial $Q_n(x)$ is defined only for $(n, p) = 1$ where $p = \text{expchar}(F)$.

Theorem 11.2.4 *Let $Q_n(x)$ be the nth cyclotomic polynomial over F.*

1) $\deg(Q_n(x)) = \phi(n)$.

2) *The following product formula holds:*

$$x^n - 1 = \prod_{d|n} Q_d(x) \tag{11.2.1}$$

3) $Q_n(x)$ *is monic and has coefficients in the prime subfield of* F.
4) *If* $F = \mathbb{Q}$ *then the coefficients of* $Q_n(x)$ *are integers.*
5) *The cyclotomic polynomials are given by*

$$Q_n(x) = \prod_{d|n} (x^d - 1)^{\mu(n/d)} = \prod_{d|n} (x^{n/d} - 1)^{\mu(d)}$$

where μ *is the Möbius function, defined by*

$$\mu(d) = \begin{cases} 1 & \text{if } d = 1 \\ (-1)^k & \text{if } d = p_1 p_2 \cdots p_k \text{ for distinct primes } p_i \\ 0 & \text{otherwise} \end{cases}$$

Note that some of the exponents $\mu(d)$ *may be equal to* -1, *and so a little additional algebraic manipulation may be required to obtain* $Q_n(x)$ *as a product of polynomials.*

Proof. Parts 3) and 4) can be proved by induction, using formula (11.2.1). In particular, let F' be the prime subfield of F. It is clear from the definition that $Q_n(x)$ is monic. Since $Q_1(x) = x - 1$, the result is true for $n = 1$. If p is a prime then

$$Q_p(x) = \frac{x^p - 1}{x - 1} = x^{p-1} + x^{p-2} + \cdots + x + 1$$

and the result holds for $n = p$. Assume that 3) and 4) hold for all proper divisors of n. Then

$$x^n - 1 = Q_n(x) \prod_{\substack{d|n \\ d<n}} Q_d(x) = Q_n(x) R(x)$$

By the induction hypothesis, $R(x)$ has coefficients in F', and therefore so does $Q_n(x)$. Moreover, if $F = \mathbb{Q}$, then $R(x)$ has integer coefficients and since $R(x)$ is monic (and therefore primitive), Theorem 1.2.2 implies that $Q_n(x)$ has integer coefficients. Part 5) follows by Möbius inversion. (See the appendix for a discussion of Möbius inversion.)□

Example 11.2.1 Formula (11.2.1) can be used to compute cyclotomic polynomials rather readily, starting from the fact that

$$Q_1(x) = x - 1$$

and

$$Q_p(x) = x^{p-1} + x^{p-2} + \cdots + x + 1$$

for p prime. Thus, for example,

$$Q_4(x) = \frac{x^4 - 1}{Q_1(x)Q_2(x)} = \frac{x^4 - 1}{(x-1)(x+1)} = x^2 + 1$$

$$Q_6(x) = \frac{x^6 - 1}{Q_1(x)Q_2(x)Q_3(x)} = \frac{x^6 - 1}{(x-1)(x+1)(x^2+x+1)} = x^2 - x + 1$$

and

$$Q_{15}(x) = \frac{x^{15} - 1}{Q_1(x)Q_2(x)Q_3(x)Q_5(x)} = x^8 - x^7 + x^5 - x^4 + x^3 - x + 1$$

This gives us, for instance, the following factorization of $x^{15} - 1$ into cyclotomic polynomials:

$$\begin{aligned} x^{15} - 1 &= (x-1)(x^2+x+1)(x^4+x^3+x^2+x+1) \\ &\quad \cdot (x^8 - x^7 + x^5 - x^4 + x^3 - x + 1) \end{aligned}$$

The Möbius inversion formula gives

$$\begin{aligned} Q_6(x) &= \prod_{d \mid 6} (x^d - 1)^{\mu(n/d)} \\ &= (x^6-1)^{\mu(1)}(x^3-1)^{\mu(2)}(x^2-1)^{\mu(3)}(x-1)^{\mu(6)} \\ &= (x^6-1)(x^3-1)^{-1}(x^2-1)^{-1}(x-1) \\ &= \frac{(x^6-1)(x-1)}{(x^3-1)(x^2-1)} \\ &= x^2 - x + 1 \end{aligned}$$

□

Part 4) of Theorem 11.2.4 describes a factorization of $x^n - 1$ within the prime subfield of F. In general, however, this is not a prime factorization since $Q_n(x)$ is not irreducible. For instance, comparing Examples 11.2.1 and 10.1.1 shows that $Q_{15}(x)$ is reducible over $GF(2)$.

When Is the Galois Group as Large as Possible?

We have seen that if $F < S$ is a cyclotomic extension of order n, then $G_F(S)$ is isomorphic to a subgroup of $\mathbb{Z}_n^*$, which has order $\phi(n)$. Thus, $G_F(S)$ is isomorphic to the full group $\mathbb{Z}_n^*$ if and only if

$$[F(\omega_n) : F] = |G_F(S)| = \phi(n)$$

that is, if and only if the cyclotomic polynomial $Q_n(x)$ is irreducible, in which case $Q_n(x) = \min(\omega_n, F)$.

Theorem 11.2.5 *Let S be the splitting field for $x^n - 1$ over F. Then $G_F(S)$ is isomorphic to $\mathbb{Z}_n^*$ if and only if the nth cyclotomic polynomial $Q_n(x)$ is irreducible over F,* in which case $Q_n(x) = \min(\omega_n, F)$.□

The Irreducibility of Cyclotomic Polynomials

With regard to the irreducibility of cyclotomic polynomials, we have the following important results. In particular, if $Q_n(x)$ is irreducible, then so is $Q_d(x)$ for $d \mid n$. Also, over the rational numbers, all cyclotomic polynomials are irreducible.

Note that since the Galois group of $F < F(\omega_n)$ is isomorphic to a subgroup of $\mathbb{Z}_n^*$, which has order $\phi(n)$, it follows that the degree of $F < F(\omega_n)$ divides $\phi(n)$.

Theorem 11.2.6 *Let $Q_n(x)$ be irreducible over F and let $n = km$, where $m > 1$. Then $Q_m(x)$ is also irreducible over F. As usual, we assume that $(n, \text{char}(F)) = 1$.*
Proof. Let $k = p$ be a prime and consider the tower

$$F < F(\omega_m) < F(\omega_{pm})$$

The first step has degree $a \mid \phi(m)$ and the second step has degree $b \le p$, since $\omega_{pm}^p \in \Omega_m \subseteq F(\omega_m)$. But $Q_n(x)$ irreducible implies that the degree of the full extension is $\phi(pm)$ and so

$$ab = \phi(pm)$$

with $a \mid \phi(m)$ and $b \le p$.

If $(p, m) = 1$ then $\phi(pm) = (p-1)\phi(m)$ and we have

$$ab = (p-1)\phi(m)$$

with $a \mid \phi(m)$ and $b \le p$. It follows that $b \ge p - 1$. If $b = p - 1$ then $a = \phi(m)$ and so $Q_m(x)$ is irreducible. If $b = p$ then $a = (p-1)\phi(m)/p$, which does not divide $\phi(m)$ for $p \ne 2$. If $b = p = 2$, then $n = 2m$, where m is odd. It follows that $\omega_n = \omega_{2m} = \omega_2\omega_m = -\omega_m$ and so

$$\phi(m) = \phi(2)\phi(m) = \phi(2m) = \phi(n) = [F(\omega_n) : F] = [F(\omega_m) : F]$$

and so $Q_m(x)$ is irreducible.

Finally, if $p \mid m$ then $\phi(pm) = p\phi(m)$ and so

$$ab = p\phi(m)$$

with $a \mid \phi(m)$ and $b \le p$. Hence, $a = \phi(m)$ and again $Q_m(x)$ is irreducible.

Thus, we have shown that if $n = pm$ with p prime, then $Q_m(x)$ is irreducible. Suppose that $k = p_1 \cdots p_s$, that is, $n = p_1 \cdots p_s m$. Then repeatedly applying the argument above shows that $Q_m(x)$ is irreducible.$\square$

Theorem 11.2.7 *All cyclotomic polynomials $Q_n(x)$ over the rational field $\mathbb{Q}$ are irreducible over $\mathbb{Q}$. Therefore, $G_{\mathbb{Q}}(S) \approx \mathbb{Z}_n^*$ and $[S : \mathbb{Q}] = \phi(n)$.*
Proof. Suppose that $Q_n(x) = f(x)g(x)$ is a nontrivial factorization, where we may assume that both factors are monic and have integer coefficients. Assume that $f(x)$ is irreducible and that $\omega \in \Omega_n$ is a root of $f(x)$. We show that ω^p is also a root of $f(x)$, for any prime $p \nmid n$. For if not, then $\omega^p \in \Omega_n$ must be a root of $g(x)$. Hence, ω is a root of $g(x^p)$, which implies that $f(x) \mid g(x^p)$ and

$$g(x^p) = h(x)f(x)$$

where $h(x)$ is monic and has integer coefficients. Since $a^p \equiv a \bmod p$, for any integer a, we have $g(x^p) \equiv g(x)^p \bmod p$ and so taking residues gives

$$g(x)^p \equiv h(x)f(x) \bmod p$$

or, in a different notation

$$\overline{g}(x)^p = \overline{h}(x)\overline{f}(x)$$

in $\mathbb{Z}_p[x]$. It follows that $\overline{f}(x)$ and $\overline{g}(x)$ have a common root in some extension of $\mathbb{Z}_p$. However, $\overline{f}(x)\overline{g}(x) \mid x^n - \overline{1}$, which has no multiple roots in any extension. This contradiction implies that ω^p is a root of $f(x)$.

Thus, if ω is a root of $f(x)$, then so is ω^p, where $(p, n) = 1$. If q is a prime and $(q, n) = 1$, the same argument applied to ω^p shows that ω^{pq} is also a root of $f(x)$. In fact, for any $(k, n) = 1$, it follows that ω^k is a root of $f(x)$, that is, all roots of $Q_n(x)$ are roots of $f(x)$, and so $f(x) = Q_n(x)$, whence $Q_n(x)$ is irreducible over $\mathbb{Q}$. □

Finite Fields

If the base field $F = GF(q)$ is a finite field, then we know that the cyclotomic extension $S = GF(q^d)$ is also a finite field and the Galois group $G_F(S)$ is cyclic with generator $\sigma_q\colon \alpha \to \alpha^q$:

$$G_F(S) = \{\iota, \sigma_q, \sigma_q^2, \ldots, \sigma_q^{d-1}\}$$

Since the order of $\omega \in \Omega_n$ is n, Theorem 9.6.1 implies that

$$d = [S : F] = |G_F(S)| = o_n(q)$$

From this, we also get a simple criterion to determine when a cyclotomic polynomial is irreducible.

Theorem 11.2.8 *Let S be the splitting field for $x^n - 1$ over $GF(q)$, where $(q, n) = 1$. Then*
1) $[S : F] = |G_F(S)| = o_n(q)$
2) $S = GF(q^{o_n(q)})$
3) $G_F(S) = \langle \sigma_q \rangle$ is isomorphic to the cyclic subgroup $\langle q \rangle$ of $\mathbb{Z}_n^*$.

4) *The following are equivalent*
   a) $G_F(S) \approx \mathbb{Z}_n^*$
   b) $o_n(q) = \phi(n)$
   c) *The cyclotomic polynomial* $Q_n(x)$ *is irreducible over* $GF(q)$.□

Let us consider an example.

Example 11.2.2 Since

$$o_5(2) = 4 = \phi(5)$$

the polynomial $Q_5(x)$ is irreducible over $GF(2)$ and has degree 4. Since

$$o_{15}(2) = 4 = \phi(5) < \phi(15) = 8$$

the polynomial $Q_{15}(x)$ of degree 8 is not irreducible over $GF(2)$.□

Types of Primitivity

There are three types of elements in the splitting field S of $x^n - 1$ over a finite field $GF(q)$ that are referred to as *primitive*: field primitive elements, group primitive elements and primitive roots of unity. Since each type of primitive element is field primitive, that is, $S = F(\alpha)$, each type of primitive element has degree $o_n(q)$. However, the orders of each type of primitive element differ.

If α is field primitive, that is, $S = F(\alpha)$, then
1) $\deg(\alpha) = o_n(q)$
2) $o(\alpha) \mid q^{o_n(q)} - 1$

If β is group primitive, that is, $S^* = \langle\beta\rangle$, then
1) $\deg(\beta) = o_n(q)$
2) $o(\beta) = q^{o_n(q)} - 1$

If ω is a primitive nth root of unity, that is, $U_n = \langle\omega\rangle$, then
1) $\deg(\omega) = o_n(q)$
2) $o(\omega) = n$

Given a group primitive element β of S, we can identify from among its powers

$$S^* = \{\iota, \beta, \beta^2, \ldots, \beta^{q^{o_n(q)}-2}\}$$

which are the primitive nth roots of unity. In fact, β^k is a primitive nth root of unity if and only if

$$n = o(\beta^k) = \frac{q^{o_n(q)} - 1}{(k, q^{o_n(q)} - 1)}$$

In general, the equation

$$n = \frac{m}{(k,m)}$$

is equivalent to $(k,m) = m/n$, or

$$\left(k, \frac{m}{n}n\right) = \frac{m}{n}$$

which holds if and only if $k = u(m/n)$ where $(u,n) = 1$. In this case, $m = q^{o_n(q)} - 1$ and we have the following.

Theorem 11.2.9 *Let β be a group primitive element of the cyclotomic extension $GF(q) < S$ of order n. Then β^k is a primitive nth root of unity if and only if*

$$k = u\frac{q^{o_n(q)} - 1}{n}$$

where $1 \le u < n$ and $(u,n) = 1$. □

More on Cyclotomic Polynomials

If $p(x)$ is monic and irreducible over $GF(q)$ and has order ν, then each root of $p(x)$ has order ν and thus $p(x) \mid Q_\nu(x)$. Since every monic irreducible factor of $Q_\nu(x)$ has order ν, and since these factors have no common roots, we conclude that $Q_\nu(x)$ is the product of all monic irreducible polynomials of order ν. According to Theorem 9.6.1, the degree of any such factor $p(x)$ is $o_\nu(q)$. Hence, the number of monic irreducible polynomials of order ν is $\phi(\nu)/o_\nu(q)$.

Theorem 11.2.10 *Let ν be a positive integer.*

1) *The cyclotomic polynomial $Q_\nu(x)$ over $GF(q)$ is the product of all monic irreducible polynomials of order ν over $GF(q)$.*
2) *The number of monic irreducible polynomials over $GF(q)$ of order ν is $\phi(\nu)/o_\nu(q)$, where $o_\nu(q)$ is the order of q mod ν.* □

Let us mention that the roots of the $(q^n - 1)$-st cyclotomic polynomial $Q_{q^n-1}(x)$ have order $q^n - 1 = |GF(q^n)^*|$ and so are group primitive elements of $GF(q^n)$. In other words, the monic irreducible factors of $Q_{q^n-1}(x)$ are precisely the primitive polynomials of $GF(q^n)$ over $GF(q)$. Thus, one way to find primitive polynomials is to factor this cyclotomic polynomial.

Example 11.2.3 We have at our disposal a number of tools for factoring polynomials of the form $x^n - 1$ over $GF(q)$, for $(n,q) = 1$:

1) $x^n - 1 = \prod_{d|n} Q_d(x)$
2) $Q_1(x) = x - 1$

3) For p prime,

$$Q_p(x) = x^{p-1} + x^{p-2} + \cdots + x + 1$$

4)

$$Q_n(x) = \prod_{d|n} (x^d - 1)^{\mu(n/d)} = \prod_{d|n} (x^{n/d} - 1)^{\mu(d)}$$

5) $Q_n(x)$ is irreducible over $GF(q)$ if and only if $o_n(q) = \phi(n)$.
6) Over $GF(q)$, the polynomial $Q_\nu(x)$ is the product of all monic irreducible polynomials of order ν over $GF(q)$.
7) A polynomial over $GF(q)$ is irreducible if and only if its order is $o_\nu(q)$.
8) Let $u \in GF(q)$. Then $p(x)$ is irreducible if and only if $p(x+u)$ is irreducible. Moreover, if α and β are roots of an irreducible polynomial $p(x)$, then $f(x) = p(x-u)$ is also irreducible and

$$f(\alpha + u) = p(\alpha) = 0 = p(\beta) = f(\beta + u)$$

Hence, translation by an element of the base field $\gamma \mapsto \gamma + u$ preserves the property of being conjugate (that is, being roots of the same irreducible polynomial).

To illustrate, consider the polynomial $x^{15} - 1$. Over $GF(2)$, we have

$$\begin{aligned} x^{15} - 1 &= Q_1(x)Q_3(x)Q_5(x)Q_{15}(x) \\ &= (x+1)(x^2+x+1) \\ &\quad \cdot (x^4+x^3+x^2+x+1)(x^8+x^7+x^5+x^4+x^3+x+1) \end{aligned}$$

A small table of order/degrees is useful:

$o_3(2) = 2$	$\phi(3) = 2$
$o_5(2) = 4$	$\phi(5) = 4$
$o_{15}(2) = 4$	$\phi(15) = 8$

This table shows that $Q_3(x)$ and $Q_5(x)$ are irreducible, but that $Q_{15}(x)$ is not.

However, since the roots of $Q_{15}(x)$ have order 15, the degree d of any irreducible factor of $Q_{15}(x)$ must satisfy $d = o_{15}(2) = 4$. Thus, $Q_{15}(x)$ factors into a product of two irreducible quartics, which are primitive polynomials for $GF(16)$.

To find the quartic factors of $Q_{15}(x)$, we can proceed by brute force. The quartic factors must have the form

$$g(x) = x^4 + ax^3 + bx^2 + cx + 1$$

where $a, b, c \in \{0, 1\}$. Since $Q_{15}(1) \neq 0$, we must have $a + b + c = 1$, which implies that either $a = b = c = 1$ or exactly one of a, b or c is 1.

If $a = b = c = 1$, then $g(x) = Q_5(x)$, which is not a factor of $Q_{15}(x)$, because the orders are not equal. Hence, exactly one of a, b or c is equal to 1. If $b = 1$, then

$$g(x) = x^4 + x^2 + 1 = (x^2 + x + 1)^2$$

which is not irreducible. Hence, we are left with only two possibilities, and

$$Q_{15}(x) = (x^4 + x^3 + 1)(x^4 + x + 1)$$

where the factors are irreducible over $GF(2)$.

Another approach is to observe that $Q_5(x)$ is irreducible and so therefore is

$$Q_5(x + 1) = x^4 + x^3 + 1$$

and since $Q_5(x + 1)$ does not divide $Q_5(x)$ or $Q_3(x)$, its roots have order 15 and so $Q_5(x + 1) \mid Q_{15}(x)$.

Once we have factored $x^{15} - 1$, we can find a group primitive element of its splitting field, which is $S = GF(2^{o_{15}(2)}) = GF(16)$. In particular, a group primitive element β has order 15, and so is a root of $Q_{15}(x)$. So let β be a root of the irreducible polynomial $x^4 + x + 1$ over $GF(2)$. Then

$$S = \{0, \beta, \beta^2, \ldots, \beta^{15}\} = \{a + b\beta + c\beta^2 + d\beta^3 \mid a, b, c, d \in GF(2)\}$$

where $\beta^4 = \beta + 1$. Note that β is also a primitive 15th root of unity, the other primitive 15th roots being β^u, where $u < 15$, $(u, 15) = 1$.□

*11.3 Normal Bases and Roots of Unity

Recall that a normal basis for $F < E$ is a basis for E over F that consists of the roots of an irreducible polynomial $p(x)$ over F. We have seen that in some important cases (especially $F = \mathbb{Q}$), the cyclotomic polynomials $Q_n(x)$ are irreducible over F, which leaves open the possibility that the primitive nth roots of unity Ω_n might form a normal basis for S over F. Indeed, if $Q_n(x)$ is irreducible then $Q_n(x) = \min(\omega_n, F)$ and so

$$\deg(Q_n(x)) = [S : F]$$

and since the roots of $Q_n(x)$ are distinct, there is the right number of primitive nth roots of unity and they will form a basis for S over F if and only if they span S over F.

Theorem 11.3.1 *Let F be a field with the property that $Q_m(x)$ is irreducible over F for all m. Then Ω_n is a normal basis for the cyclotomic extension $F < S$ if and only if n is the product of distinct primes.*
Proof. First, let p be prime and $(p, m) = 1$. Consider the extension

$$F(\omega_m) < F(\omega_m, \omega_p) = F(\omega_{pm})$$

Since $Q_n(x)$ is irreducible, it follows that $Q_p(x)$ is irreducible over $F(\omega_m)$ and so the powers

$$\{1, \omega_p, \omega_p^2, \ldots, \omega_p^{p-2}\}$$

form a basis for $F(\omega_{pm})$ over $F(\omega_m)$. But

$$1 + \omega_p + \cdots + \omega_p^{p-1} = \frac{\omega_p^p - 1}{\omega_p - 1} = 0$$

and so the set $\Omega_p = \{\omega_p, \omega_p^2, \ldots, \omega_p^{p-1}\}$ is a normal basis for $F(\omega_{pm})$ over $F(\omega_m)$.

Now we can proceed by induction on n. We have just seen that the result is true for n prime. Suppose that $n = pm$, where $(p, m) = 1$. Then by the inductive hypothesis, we may assume that Ω_m is a normal basis for $F(\omega_m)$ over F. Then the product $\Omega_m\Omega_p$ is a basis for $F(\omega_n)$ over F. But $\Omega_m\Omega_p = \Omega_n$ and so Ω_n is a normal basis for $F(\omega_n)$ over F.

For the converse, let $n = mp^k$ for $k \geq 2$. Since

$$Q_n(x) = Q_{mp^k}(x) = Q_{mp}(x^{p^{k-1}})$$

(an exercise) the coefficient of $x^{\phi(n)-1}$ in $Q_n(x)$ is 0, whence the sum of the roots of $Q_n(x)$, that is, the sum of the primitive nth roots of unity, is 0, showing that these roots are linearly dependent. Hence, they cannot form a basis for S over F. □

*11.4 Wedderburn's Theorem

In this section, we present an important result whose proof uses the properties of cyclotomic polynomials.

Theorem 11.4.1 (Wedderburn's Theorem) *If D is a finite division ring, then D is a field.*
Proof. Let the multiplicative group D^* act on itself by conjugation. The stabilizer of $\beta \in D^*$ is the centralizer

$$C^*(\beta) = \{\alpha \in D^* \mid \alpha\beta = \beta\alpha\}$$

and the class equation is

$$|D^*| = |Z(D^*)| + \sum \frac{|D^*|}{|C^*(\beta)|}$$

where the sum is taken over one representative β from each conjugacy class $o(\beta) = \{\alpha\beta\alpha^{-1} \mid \alpha \in G\}$ of size greater than 1. If we assume for the purposes of contradiction that $Z(D^*) \neq D^*$, then the sum on the far right is not an empty sum and $|C^*(\beta)| < |D^*|$ for some $\beta \in D^*$.

The sets

$$Z(D) = \{\beta \in D \mid \beta\alpha = \alpha\beta \text{ for all } \alpha \in D\}$$

and

$$C(\beta) = \{\alpha \in D \mid \alpha\beta = \beta\alpha\}$$

are subrings of D and, in fact, $Z(D)$ is a commutative division ring; that is, a field. Let $|Z(D)| = z \geq 2$. Since $Z(D) \subseteq C(\beta)$, we may view $C(\beta)$ and D as vector spaces over $Z(D)$ and so

$$|C(\beta)| = z^b \quad \text{and} \quad |D| = z^n$$

for integers $1 \leq b < n$. The class equation now gives

$$z^n - 1 = z - 1 + \sum_b \frac{z^n - 1}{z^b - 1}$$

and since $z^b - 1 \mid z^n - 1$, it follows that $b \mid n$.

If $Q_n(x)$ is the nth cyclotomic polynomial over $\mathbb{Q}$, then $Q_n(z)$ divides $z^n - 1$. But $Q_n(z)$ also divides each summand on the far right above, since for $b \mid n$, $b < n$, we have

$$\frac{x^n - 1}{x^b - 1} = \prod_{k \mid n} Q_k(x) \bigg/ \prod_{j \mid b} Q_j(x)$$

and $Q_n(x)$ divides the right-hand side. It follows that $Q_n(z) \mid z - 1$. On the other hand,

$$Q_n(z) = \prod_{\omega \in \Omega_n} (z - \omega)$$

and since $\omega \in \Omega_n$ implies that $|z - \omega| > z - 1$, we have a contradiction. Hence $Z(D^*) = D^*$ and D is commutative, that is, D is a field. $\square$

*11.5 Realizing Groups as Galois Groups

A group G is said to be **realizable** over a field F if there is an extension $F < E$ whose Galois group is G. Since any finite group of order n is isomorphic to a subgroup of a symmetric group S_n, we have the following.

Theorem 11.5.1 *Let F be a field. Every finite group is realizable over some extension of F.*

Proof. Let G be a group of order n. Let $t_1, \ldots, t_n$ be algebraically independent over F and let $s_1, \ldots, s_n$ be the elementary symmetric polynomials in the t_i's. Then $K = F(t_1, \ldots, t_n) > F(s_1, \ldots, s_n) = E$ is a Galois extension whose Galois group $G_E(K)$ is isomorphic to S_n. (See Theorem 7.2.3.) We may assume that G is a subgroup of $G_E(K)$ and since G is closed in the Galois correspondence, it is the Galois group of $\text{fix}(G) < K$.□

It is a major unsolved problem to determine which finite groups are realizable over the rational numbers $\mathbb{Q}$. We shall prove that any finite abelian group is realizable over $\mathbb{Q}$. It is also true that for any n, the symmetric group S_n is realizable over $\mathbb{Q}$, but we shall prove this only when $n = p$ is a prime.

Realizing Finite Abelian Groups over Q

We wish to show that any finite abelian group is realizable over the rational field $\mathbb{Q}$. Since all cyclotomic polynomials are irreducible over the rationals, the extension

$$\mathbb{Q} < \mathbb{Q}(\omega_n)$$

has Galois group $\mathbb{Z}_n^*$, which is finite and abelian. For any subgroup

$$\{\iota\} < H < \mathbb{Z}_n^*$$

we have the corresponding tower of fields

$$\mathbb{Q} < \text{fix}(H) < \mathbb{Q}(\omega_n)$$

and since the extension $\mathbb{Q} < \mathbb{Q}(\omega_n)$ is Galois and all subgroups are normal, the quotient $\mathbb{Z}_n^*/H$ is the Galois group of the extension $\mathbb{Q} < \text{fix}(H)$.

Hence, it is sufficient to show that any finite abelian group G is isomorphic to a quotient $\mathbb{Z}_n^*/H$, for some n. Since G is finite and abelian, we have

$$G \approx C(n_1) \times \cdots \times C(n_s)$$

where $C(n_i)$ is cyclic of degree n_i. If we show that $C(n_i)$ is isomorphic to a quotient of the form $\mathbb{Z}_p^*/H_i$, where the p_i's are distinct odd primes, then if $H = H_1 \times \cdots \times H_s$ and $n = p_1 \cdots p_s$, it follows that

$$G \approx \frac{\mathbb{Z}^*_{p_1}}{H_1} \times \cdots \times \frac{\mathbb{Z}^*_{p_s}}{H_s} \approx \frac{\mathbb{Z}^*_{p_1} \times \cdots \times \mathbb{Z}^*_{p_s}}{H_1 \times \cdots \times H_s} \approx \frac{\mathbb{Z}^*_{p_1 \cdots p_s}}{H} \approx \frac{\mathbb{Z}^*_n}{H}$$

as desired.

Now, if p_i is an odd prime, then $\mathbb{Z}^*_{p_i}$ is cyclic of order $p_i - 1$ and so all we need to do is find distinct odd primes p_i for which $n_i \mid p_i - 1$, because a cyclic group of order m has quotient groups of all orders dividing m.

Put another way, we seek a set of distinct primes of the form $kn_i + 1$, for $i = 1, \ldots, s$. It is a famous theorem of Dirichlet that there are infinitely many primes of the form $kn + m$ provided that $(m, n) = 1$ and so the case $m = 1$ is what we require.

First a lemma on cyclotomic polynomials.

Lemma 11.5.2 *Let p be a prime and let $(n, p) = 1$. Let $\overline{Q}_n(x)$ be the polynomial obtained from $Q_n(x)$ by taking the residue of each coefficient modulo p. Then $\overline{Q}_n(x)$ is the nth cyclotomic polynomial over $\mathbb{Z}_p$.*

Proof. Let $P_n(x)$ be the nth cyclotomic polynomial over $\mathbb{Z}_p$. If n is a prime then $Q_n(x), P_n(x)$ and $\overline{Q}_n(x)$ are all equal to

$$x^{n-1} + x^{n-2} + \cdots + 1$$

and so the result holds for n prime. Let $n > 2$ and suppose the result holds for all proper divisors of n. Since

$$x^n - 1 = \prod_{d|n} Q_d(x)$$

taking residues modulo p gives

$$x^n - \overline{1} = \prod_{d|n} \overline{Q}_d(x)$$

over $\mathbb{Z}_p$. But

$$x^n - \overline{1} = \prod_{d|n} P_d(x)$$

over $\mathbb{Z}_p$ and since $P_d(x) = \overline{Q}_d(x)$ for all $d \mid n$, $d < n$, it follows that $P_n(x) = \overline{Q}_n(x)$.□

Theorem 11.5.3 *Let n be a positive integer. Then there are infinitely many prime numbers of the form $nk + 1$, where k is a positive integer.*

Proof. Suppose to the contrary that $p_1, \ldots, p_s$ is a complete list of all primes of the form $nk + 1$. Let $m = p_1 \cdots p_s n$. Let $Q_m(x)$ be the mth cyclotomic polynomial over $\mathbb{Q}$ and consider the polynomial $Q_m(mx)$. Since $Q_n(x)$ has integer coefficients, $Q_m(mk)$ is an integer for all $k \in \mathbb{Z}^+$. Since $Q_m(mk)$ can equal 0, 1 or -1 for only a finite number of positive integers k, there exists a positive integer k for which $|Q_m(mk)| > 1$. Let p be a prime dividing $Q_m(mk)$. Since $Q_m(x) \mid x^m - 1$, we have

$$p \mid (mk)^m - 1$$

which implies that $p \not| m$, hence $p \neq p_i$ for $i = 1, \ldots, s$. To arrive at a contradiction, we show that p has the form $kn + 1$.

If $P_m(x)$ is the mth cyclotomic polynomial over $\mathbb{Z}_p$, then $p \mid Q_m(mk)$ and the previous lemma imply that

$$P_m(\overline{mk}) = \overline{Q}_m(\overline{mk}) = \overline{Q_m(mk)} = 0$$

in $\mathbb{Z}_p$, where the overbar denotes residue modulo p. Thus, $\overline{mk}$ is a primitive mth root of unity over Z_p, that is, $\overline{mk}$ has order m in $\mathbb{Z}_p^*$. Hence,

$$m \mid o(\mathbb{Z}_p^*) = p - 1$$

and so $n \mid p - 1$, that is, p has the form $nk + 1$.□

We can now put the pieces together.

Theorem 11.5.4 *Let G be a finite abelian group. Then there exists an integer n and a field E such that $\mathbb{Q} < E < \mathbb{Q}(\omega)$, where $\omega \in \Omega_n$ and such that $G_{\mathbb{Q}}(E) \approx G$.*□

Realizing S_p over Q

We begin by discussing a sometimes useful tool for showing that the Galois group of a polynomial is a symmetric group.

Let G be the Galois group of an irreducible polynomial $f(x)$ over F, thought of as a group of permutations on the set R of roots of $f(x)$. Then G acts transitively on R. Let us define an equivalence relation on R by saying that $r \sim s$ if and only if either $r = s$ or the *transposition* $(r\,s)$ is an element of G. It is easy to see that this is an equivalence relation on R. Let $[r]$ be the equivalence class containing r.

Suppose that G contains a transposition $(r\,s)$. Then for any $\sigma \in G$, we have

$$\sigma(r\,s)\sigma^{-1} = (\sigma r\,\sigma s)$$

In other words, if $s \sim r$ then $\sigma s \sim \sigma r$ and so $\sigma[r] = [\sigma s]$. It follows that $[r]$ and $[\sigma s]$ have the same cardinality and since G acts transitively on R, all equivalence classes have the same cardinality.

Hence, if $f(x)$ has a prime number of roots, then there can be only one equivalence class, which implies that $(r\,s)$ is in G for all $r, s \in R$. Since G contains every transposition, it must be the symmetric group on R. We have proved the following.

Theorem 11.5.5 *If $f(x) \in F[x]$ is a separable polynomial of prime degree p and if the Galois group G of $f(x)$ contains a transposition, then G is isomorphic to the symmetric group S_p.* □

Corollary 11.5.6 *If $f(x) \in \mathbb{Q}[x]$ is irreducible of prime degree p and if $f(x)$ has precisely two nonreal roots, then the Galois group of $f(x)$ is isomorphic to the symmetric group S_p.*

Proof. Let S be a splitting field for $f(x)$ over $\mathbb{Q}$. Complex conjugation $\sigma\colon \mathbb{C} \to \mathbb{C}$ is an automorphism of $\mathbb{C}$ leaving $\mathbb{Q}$ fixed. Moreover, since $\mathbb{Q} < S$ is normal, $\sigma \in G_{\mathbb{Q}}(S)$. Since σ leaves the $p-2$ real roots of $f(x)$ fixed, σ is a transposition on the roots of $f(x)$. Thus, the theorem applies. □

Example 11.5.1 Consider the polynomial $f(x) = x^5 - 5x + 2$, which is irreducible over $\mathbb{Q}$ by Eisenstein's criterion. A quick sketch of the graph reveals that $f(x)$ has precisely 3 real roots and so its Galois group is isomorphic to S_5.□

Corollary 11.5.6 is just what we need to establish that S_p is realizable over $\mathbb{Q}$.

Theorem 11.5.7 *Let p be a prime. There exists an irreducible polynomial $p(x)$ over $\mathbb{Q}$ of degree p such that $p(x)$ has precisely two nonreal roots. Hence, the symmetric group S_p is realizable over $\mathbb{Q}$.*

Proof. The result is easy for $p = 2$ and 3, so let us assume that $p \geq 5$. Let n be a positive integer and $m \geq 5$ be an odd integer. Let $k_1, \ldots, k_{m-2}$ be distinct even integers and let

$$q(x) = (x^2 + n)(x - k_1)\cdots(x - k_{m-2})$$

It is easy to see from the graph that $q(x)$ has $(m-3)/2$ relative maxima. Moreover, if k is an odd integer, then

$$|q(k)| \geq 2|k^2 + n| > 2$$

Let $p(x) = q(x) - 2$. Since the relative maxima of $q(x)$ are all greater than 2 and since $q(-\infty) = -\infty$ and $q(\infty) = \infty$, we deduce that $p(x)$ has at least $m-2$ real roots.

We wish to choose a value of n for which $p(x)$ has at least one nonreal root z, for then the complex conjugate $\overline{z}$ is also a root, implying that $p(x)$ has two nonreal roots and $m-2$ real roots. Let the roots of $p(x)$ in a splitting field be

$\alpha_1, \ldots, \alpha_m$. Then

$$p(x) = \prod_{i=1}^{m}(x - \alpha_i) = (x^2 + n)(x - k_1)\cdots(x - k_{m-2}) - 2$$

Equating coefficients of x^{m-1} and x^{m-2} gives

$$\sum_{i=1}^{m} \alpha_i = \sum_{i=1}^{m-2} k_i \quad \text{and} \quad \sum_{i<j} \alpha_i\alpha_j = \sum_{i<j} k_ik_j + n$$

and so

$$\begin{aligned}\sum_{i=1}^{m} \alpha_i^2 &= \Big(\sum_{i=1}^{m} \alpha_i\Big)^2 - 2\sum_{i<j} \alpha_i\alpha_j \\ &= \Big(\sum_{i=1}^{m-2} k_i\Big)^2 - 2\Big(\sum_{i<j} k_ik_j + n\Big) \\ &= \sum_{i=1}^{m-2} k_i^2 - 2n\end{aligned}$$

If n is sufficiently large, then $\sum \alpha_i^2$ is negative, whence at least one of the roots α_i must be nonreal, as desired.

It is left to show that $p(x)$ is irreducible, which we do using Eisenstein's criterion. Let us write

$$q(x) = (x^2 + n)(x - k_1)\cdots(x - k_{m-2}) = x^m + a_{m-1}x^{m-1} + \cdots + a_0$$

In the product $(x - k_1)\cdots(x - k_{m-2})$, each coefficient except the leading one is divisible by 2. Hence, we may write

$$(x - k_1)\cdots(x - k_{m-2}) = x^{m-2} + 2f(x)$$

Multiplying by $x^2 + n$ gives

$$q(x) = x^m + 2x^2f(x) + nx^{m-2} + 2nf(x)$$

Taking n to be even, we deduce that all nonleading coefficients of $q(x)$ are even. In addition, the constant term of $q(x)$ is divisible by 4 since $m \geq 5$. It follows that $p(x) = q(x) - 2$ is monic, all nonleading coefficients are divisible by 2, but the constant term is not divisible by $2^2 = 4$. Therefore $p(x)$ is irreducible and the proof is complete. □

Exercises

All cyclotomic polynomials are assumed to be over fields for which they are defined.

1. Prove that if $x^n - 1 = Q_n(x)p(x)$ where $p(x) \in \mathbb{Z}[x]$ then $Q_n(x) \in \mathbb{Z}[x]$.

2. When is a group primitive element of the cyclotomic extension S_n also a primitive nth root of unity over $GF(q)$?
3. If $(n,q) \neq 1$, how many nth roots of unity are there over $GF(q)$?
4. What is the splitting field for $x^4 - 1$ over $GF(3)$? Find the primitive 4th roots of unity in this splitting field. Do the same for the 8th roots of unity over $GF(3)$.
5. If $\alpha_1, \ldots, \alpha_n$ are the nth roots of unity over $GF(q)$ show that $\alpha_1^k + \alpha_2^k + \cdots + \alpha_n^k = 0$ for $1 \leq k < n$. What about when $k = n$?
6. If $(n,q) = 1$, prove that $f(x) = x^{n-1} + x^{n-2} + \cdots + x + 1$ is irreducible over $GF(q)$ if and only if n is prime and $Q_n(x)$ is irreducible.
7. Show that if r is a prime, then $Q_{r^n}(x) = (x^{r^n} - 1)/(x^{r^{n-1}} - 1)$.
8. Show that $\mathbb{Q}(\omega_n) \cap \mathbb{Q}(\omega_m) = \mathbb{Q}$ if $(m,n) = 1$.

Verify the following properties of the cyclotomic polynomials. As usual, p is a prime number.

9. $Q_{np}(x) = Q_n(x^p)/Q_n(x)$ for $p \not| n$.
10. $Q_{np}(x) = Q_n(x^p)$ for all $p \mid n$.
11. $Q_{np^k}(x) = Q_{np}(x^{p^{k-1}})$
12. If $n = p_1^{e_1} \cdots p_k^{e_k}$ is the decomposition of n into a product of powers of distinct primes, then

$$Q_n(x) = Q_{p_1 \cdots p_k}(x^{p_1^{e_1 - 1} \cdots p_k^{e_k - 1}})$$

13. $Q_n(0) = 1$ for $n \geq 2$.
14. $x^{\phi(n)} Q_n(x^{-1}) = Q_n(x)$ for $n \geq 2$.
15. Evaluate $Q_n(1)$.

On the structure of $\mathbb{Z}_n^*$.

16. If $n = \prod r_i$ where $r_i = p_i^{e_i}$ are distinct prime powers then

$$\mathbb{Z}_n^* \approx \prod \mathbb{Z}_{r_i}^*$$

17. Let $p \neq 2$ be prime and let $n = p^e$.
    a) Show that $|\mathbb{Z}_n^*| = p^{e-1}(p-1)$.
    b) Show that $\mathbb{Z}_n^*$ has an element of order $p - 1$. *Hint*: consider an element $a \in \mathbb{Z}_p^*$ of order $p - 1$ modulo p, which exists since $\mathbb{Z}_p$ is a field. Compute the order of $a^{p^{e-1}}$ as an element of $\mathbb{Z}_{p^e}^*$.
    c) Show that $1 + p \in \mathbb{Z}_n^*$ has order p^{e-1}. *Hint*: Show that if $p \not| a$ then

$$(1 + ap^u)^p \equiv 1 + a_1 p^{u+1}$$

    where $p \not| a_1$. Then consider the powers $(1+p)^p, (1+p)^{p^2}$, etc.
    d) Show that $\mathbb{Z}_n^*$ is cyclic.
    e) Show that $\mathbb{Z}_{2^e}^*$ is cyclic if and only if $e = 1$ or 2.
    f) Show that $\mathbb{Z}_m^*$ is cyclic if and only if $m = p^e, 2p^e$ or 4.

18. Prove that if $n > 1$ then there exists an irreducible polynomial of degree n over $\mathbb{Q}$ whose Galois group is isomorphic to C_n, the cyclic group of order n.
19. Find an integer n and a field E such that $\mathbb{Q} < E < \mathbb{Q}(\omega_n)$ with $G_{\mathbb{Q}}(E) = C_8$, the cyclic group of order 8. Here ω_n is a primitive nth root of unity over $\mathbb{Q}$.
20. Calculate the Galois group of the polynomial $f(x) = x^5 - 4x + 2$.
21. Let t be transcendental over $\mathbb{Z}_p$, p prime. Show that the Galois group of $f(x) = x^p - x - t$ is isomorphic to $\mathbb{Z}_p$.

More on Constructions

The following exercises show that not all regular n-gons can be constructed in the plane using only a straight-edge and compass. The reader may refer to the exercises of Chapter 2 for the relevant definitions.

Definition *A complex number z is* **constructible** *if its real and imaginary parts are both constructible.* □

22. Prove that the set of all constructible complex numbers forms a subfield of the complex numbers $\mathbb{C}$.
23. Prove that a complex number $z = re^{i\theta}$ is constructible if and only if the real number r and the angle θ (that is, the real number $\cos\theta$) are constructible.
24. Prove that if z is constructible, then both square roots of z are constructible. Hint: use the previous exercise.
25. Prove that a complex number z is constructible if and only if there exists a tower of fields $\mathbb{Q} < F_1 < \cdots < F_n$, each one a quadratic extension of the previous one, such that $z \in F_n$.
26. Prove that if z is constructible, then $[\mathbb{Q}(z) : \mathbb{Q}]$ must be a power of 2.
27. Show that the constructibility of a regular n-gon is equivalent to the constructibility of a primitive nth root of unity ω_n. Since the cyclotomic polynomial $Q_n(x)$ is irreducible over the rationals, we have $[\mathbb{Q}(\omega_n) : \mathbb{Q}] = \deg(Q_n(x)) = \phi(n)$.
28. Prove that $\phi(n)$ is a power of 2 if and only if n has the form

$$n = 2^k p_1 \cdots p_m$$

where p_m are distinct Fermat primes, that is, primes of the form

$$2^{2^s} + 1$$

for some nonnegative integer s. Hint: if $2^j + 1$ is prime then j must be a power of 2. Conclude that if n does not have this form, then a regular n-gon is not constructible. For instance, we cannot construct a regular n-gon for $n = 7$, 11 or 90. [Gauss proved that if n has the above form, then a regular n-gon can be constructed. See Hadlock (1978).]

Chapter 12
Cyclic Extensions

Continuing our discussion of binomials begun in the previous chapter, we will show that if S is a splitting field for the binomial $x^n - u$, then $S = F(\omega, \alpha)$ where ω is a primitive nth root of unity. In the tower

$$F < F(\omega) < F(\omega, \alpha)$$

the first step is a cyclotomic extension, which, as we have seen, is abelian and may be cyclic. In this chapter, we will see that the second step is cyclic of degree $d \mid n$ and α can be chosen so that $\min(\alpha, F(\omega)) = x^d - \alpha^d$. Nevertheless, as we will see in the next chapter, the Galois group $G_F(S)$ need not even be abelian.

In this chapter, we will also characterize cyclic extensions of degree relatively prime to $p = \text{expchar}(F)$, as well as extensions of degree p, but we will not discuss extensions of degree p^e for $e > 1$, since this case is not needed and is considerably more complex.

12.1 Cyclic Extensions

Let F be a field with $\text{expchar}(F) = p$, let $u \in F$ and let $S(n, u)$ be a splitting field for the binomial $x^n - u$ over F, where $(n, p) = 1$. Note that $x^n - u$ has n distinct roots in $S(n, u)$.

If α is a root of $x^n - u$ in $S(n, u)$ and ω is a primitive nth root of unity over F then the roots of $x^n - u$ are

$$\alpha, \omega\alpha, \ldots, \omega^{n-1}\alpha \tag{12.1.1}$$

and so $S(n, u) = F(\omega, \alpha)$. In words, all nth roots of u can be obtained by first adjoining the nth roots of unity and then adjoining any single nth root of u.

The extension $F < S(n,u)$ can thus be decomposed into a tower

$$F < F(\omega) < F(\omega,\alpha) = S(n,u)$$

where the first step is cyclotomic.

For the second step, it will simplify the notation to simply assume that $\omega \in F$. Hence,

$$F < F(\alpha) = S(n,u)$$

is finite, Galois and the base field F contains all the nth roots of unity.

As to the Galois group G of $S(n,u)$, each $\sigma \in G$ is uniquely determined by its value on α and

$$\sigma\alpha = \omega^{k(\sigma)}\alpha$$

for some $k(\sigma) \in \mathbb{Z}_n$. In fact, the map $\sigma \mapsto \omega^{k(\sigma)}$ is an embedding of G into U_n, and so G is isomorphic to a subgroup of U_n and is therefore cyclic of degree $d \mid n$. This follows easily from the assumption that F contains the nth roots of unity, for if $\sigma, \tau \in G$, then

$$(\sigma\tau)\alpha = \sigma(\omega^{k(\tau)}\alpha) = \omega^{k(\tau)}\sigma\alpha = \omega^{k(\tau)}\omega^{k(\sigma)}\alpha$$

and so $\sigma\tau \mapsto \omega^{k(\tau)}\omega^{k(\sigma)} \in U_n$.

Definition *Let* $(n, \operatorname{expchar}(F)) = 1$. *An extension* $F < F(\alpha)$ *is* **pure** *of* **type** n *if* α *is a root of a binomial* $x^n - u$ *over* F, *that is, if* $\alpha^n \in F$.□

Note that if $F < E$ is pure of type d and if $d \mid n$, then $F < E$ is also pure of type n.

We can now provide a characterization of cyclic extensions when the base field contains the nth roots of unity.

Theorem 12.1.1 *Let* $(n, \operatorname{expchar}(F)) = 1$. *Suppose that* F *contains the* n*th roots of unity and let* $F < E$. *Then the following are equivalent:*

1) $F < E$ *is pure of type* n.
2) $F < E$ *is cyclic of degree* $d \mid n$.

In this case, $\alpha \in E$ *is a root of* $x^n - u$ *for some* $u \in F$ *if and only if*

$$\min(\alpha, F) = x^d - v$$

for some $v \in F$.

Proof. We have seen that a pure extension of type n is cyclic of type $d \mid n$. For the converse, assume that $F < E$ is cyclic of degree $d \mid n$, with Galois group

$$G = \langle\sigma\rangle = \{\iota, \sigma, \ldots, \sigma^{d-1}\}$$

We are looking for a field primitive element $\alpha \in E$ that is a root of a binomial of the form $x^d - u$, for $u \in F$. The roots of any polynomial $p(x)$ have the form

$$\alpha, \sigma\alpha, \ldots, \sigma^{d-1}\alpha$$

and the roots of the binomial $x^d - u$ have the form

$$\alpha, \omega_d\alpha, \ldots, \omega_d^{d-1}\alpha$$

where ω_d is a primitive dth root of unity. Hence, if we can find an $\alpha \in E$ for which $\sigma\alpha = \omega_d\alpha$, then

$$\min(\alpha, F) = \prod_{k=0}^{d-1}(x - \alpha\omega_d^k)$$

Since the product of these roots

$$\beta = \alpha^d\omega_d^{d(d-1)/2} = \pm\alpha^d$$

is in F, we have $\alpha^d \in F$ and so $\min(\alpha, F) = x^d - \alpha^d$. Hence, $F < E$ is pure of type d, and therefore also of type n.

Thus, we are left with finding an $\alpha \in E$ for which $\sigma\alpha/\alpha$ is a primitive dth root of unity. This is the content of Hilbert's Theorem 90, which we prove next. We leave proof of the final statement of this theorem as an exercise.$\square$

Theorem 12.1.2 (Hilbert's Theorem 90) *Let $F < E$ be a finite cyclic extension of degree d, with Galois group $G = \langle\sigma\rangle$. An element $\beta \in E$ has the form*

$$\beta = \frac{\alpha}{\sigma\alpha}$$

for some nonzero $\alpha \in E$ if and only if

$$N_{E/F}(\beta) = \beta(\sigma\beta)(\sigma^2\beta)\cdots(\sigma^{d-1}\beta) = 1$$

In particular, if the base field F contains a primitive dth root of unity ω_d, then $N_{E/F}(\omega_d) = 1$ and the previous statement applies.

Proof. If $\beta = \alpha/\sigma\alpha$, then $\sigma^k\beta = \sigma^k\alpha/\sigma^{k+1}\alpha$ and so

$$\begin{aligned} N_{E/F}(\beta) &= \beta(\sigma\beta)(\sigma^2\beta)\cdots(\sigma^{d-1}\beta) \\ &= \frac{\alpha}{\sigma\alpha}\frac{\sigma\alpha}{\sigma^2\alpha}\cdots\frac{\sigma^{d-1}\alpha}{\sigma^d\alpha} \\ &= 1 \end{aligned}$$

For the converse, suppose that $N_{E/F}(\beta) = 1$. We seek an element $\alpha \in E$ for which $\beta(\sigma\alpha) = \alpha$, that is, an element α fixed by the operator $\widehat{\beta}\sigma$, where $\widehat{\beta}$ is multiplication by β. This suggests looking at the elements

$$\delta_k = \beta(\sigma\beta)(\sigma^2\beta)\cdots(\sigma^{k-1}\beta)(\sigma^k\beta)$$

for $k = 0, \ldots, d-1$, which have the property that

$$\widehat{\beta}\sigma(\delta_k) = \delta_{k+1}$$

for $k = 0, \ldots, d-2$. Hence, the sum

$$\delta = \sum_{k=0}^{d-1} \delta_k$$

is a promising candidate, since applying $\widehat{\beta}\sigma$ shifts each term to the next, except for the last term. But since $N_{E/F}(\beta) = 1$, we have

$$\widehat{\beta}\sigma(\delta_{d-1}) = \beta(\sigma\beta)(\sigma^2\beta)\cdots(\sigma^{d-1}\beta)(\sigma^d\beta) = N_{E/F}(\beta)\beta = \beta = \delta_0$$

and so applying $\widehat{\beta}\sigma$ wraps the last term to the first. Hence, $\beta(\sigma\delta) = \delta$, as desired.

However, there is a problem. We do not know that δ is nonzero. Accordingly, a change in the definition of δ_k is in order. Let

$$\alpha_k = \beta(\sigma\beta)(\sigma^2\beta)\cdots(\sigma^{k-1}\beta)(\sigma^k\gamma)$$

with $\alpha_0 = \gamma$, where γ is an as yet undetermined element of E. Then the previous analysis still applies. In particular, $\widehat{\beta}\sigma(\alpha_k) = \alpha_{k+1}$ and if

$$\alpha = \sum_{k=0}^{d-1} \alpha_k$$

then since

$$\widehat{\beta}\sigma(\alpha_{d-1}) = \beta(\sigma\beta)(\sigma^2\beta)\cdots(\sigma^{d-1}\beta)(\sigma^d\gamma) = N_{E/F}(\beta)\gamma = \gamma = \alpha_0$$

we again have $\beta(\sigma\alpha) = \alpha$. But now, since the automorphisms $\iota, \sigma, \ldots, \sigma^{d-1}$ are distinct, the Dedekind independence theorem implies that the linear combination

$$\mu = \sum_{k=0}^{d-1} \beta(\sigma\beta)(\sigma^2\beta)\cdots(\sigma^{k-1}\beta)\sigma^k$$

is nonzero, and so there must be a nonzero $\gamma \in E$ for which $\alpha = \mu\gamma$ is nonzero. This is our α.

For the last statement, if $\omega_d \in E$ is a primitive dth root of unity, then since $\omega_d \in F$, we have

$$N_{E/F}(\omega_d) = \omega_d(\sigma\omega_d)(\sigma^2\omega_d)\cdots(\sigma^{d-1}\omega_d) = (\omega_d)^d = 1$$

and the previous statement applies.□

12.2 Extensions of Degree Char(F)

There is an "additive" version of Theorem 12.1.1 that deals with cyclic extensions of degree equal to $p = \text{char}(F) > 0$, where the role of the binomial $x^n - u$ is played by the polynomial $x^p - x - u$.

Suppose that F is a field of characteristic $p \neq 0$. Let $F < E$ and suppose that $\alpha \in E$ is a root of the polynomial

$$f(x) = x^p - x - u$$

for $u \in F$. Since $k^p = k$ for all $k \in \mathbb{Z}_p$, we have

$$\begin{aligned} f(x+k) &= (x+k)^p - (x+k) - u \\ &= x^p + k^p - x - k - u \\ &= f(x) \end{aligned}$$

and so the p distinct roots of $f(x)$ are

$$\alpha, \alpha+1, \ldots, \alpha+p-1$$

Hence, $F(\alpha)$ is a splitting field of $f(x)$. (In contrast to the previous case, we need no special conditions on F, such as containing roots of unity, to ensure that if an extension of F contains one root of $f(x)$, it contains all the roots of $f(x)$.)

If $\alpha \in F$ then $f(x)$ splits in F. If $\alpha \notin F$, then $p(x) = \min(\alpha, F)$ has degree $d > 1$, with roots

$$\alpha, \alpha + e_1, \ldots, \alpha + e_{d-1}$$

where $0 \leq e_i \leq p-1$. The sum of these roots is $d\alpha + k$, for some integer k, and since this number lies in F, but since $\alpha \notin F$, it follows that $d = p$, whence $f(x) = \min(\alpha, F)$ is irreducible. In short, $f(x)$ either splits in F or is irreducible over F with splitting field $F(\alpha)$, for any root α of $f(x)$.

Since $F(\alpha)$ is a splitting field for the separable polynomial $f(x) = x^p - x - u$, it follows that $F < F(\alpha)$ is Galois. If $f(x)$ is irreducible over F and $G = G_F(F(\alpha))$, there exists a $\sigma \in G$ for which $\sigma\alpha = \alpha + 1$. Since $\sigma^i\alpha = \alpha + i$, it follows that $G = \langle\sigma\rangle = \{\iota, \sigma, \ldots, \sigma^{p-1}\}$ is the cyclic group generated by σ.

Definition *An extension $F < F(\alpha)$ of degree $p =$ char$(F) \neq 0$ is* **pure** *of* **type** *p if α is a root of an* irreducible *binomial $x^p - x - u$ over F.*□

Theorem 12.2.1 (Artin–Schreier) *Let* $\operatorname{char}(F) = p \neq 0$. *The polynomial*

$$f(x) = x^p - x - u$$

either splits in F or is irreducible over F. An extension $F < E$ is cyclic of degree p if and only if it is pure of type p.

Proof. We have seen that an extension that is pure of type p is cyclic of type p. Suppose that $F < E$ is cyclic of degree p, with Galois group $G = \langle\sigma\rangle = \{\iota, \sigma, \ldots, \sigma^{p-1}\}$. If $\alpha \in E$ has the property that $\sigma\alpha = \alpha + 1$, then the roots of $\min(\alpha, F)$ are

$$\alpha, \alpha + 1, \ldots, \alpha + p - 1$$

Moreover, since

$$\sigma(\alpha^p - \alpha) = (\alpha + 1)^p - \alpha - 1 = \alpha^p - \alpha$$

it follows that $u = \alpha^p - \alpha \in F$ and so $\min(\alpha, F) = x^p - x - u$. To find such an element, we need the additive version of Hilbert's Theorem 90.□

Theorem 12.2.2 (Hilbert's Theorem 90, Additive Version) *Let $F < E$ be a finite cyclic extension with Galois group $G = \langle\sigma\rangle$. An element $\beta \in E$ has the form $\beta = \alpha - \sigma\alpha$ for some $\alpha \in E$ if and only if* $\operatorname{Tr}_{E/F}(\beta) = 0$.

Proof. Assume that $\operatorname{Tr}_{E/F}(\beta) = 0$. Let $[E : F] = n$ and consider the map

$$\tau = \beta\sigma + [\beta + (\sigma\beta)]\sigma^2 + \cdots + [\beta + (\sigma\beta) + \cdots + (\sigma^{n-2}\beta)]\sigma^{n-1}$$

It is easy to verify that $\tau - \sigma\tau = \beta(\iota + \sigma + \cdots + \sigma^{n-1})$ and so if $\operatorname{Tr}_{E/F}(\gamma) = 1$ for $\gamma \in E$ then

$$\tau\gamma - \sigma\tau\gamma = \beta\operatorname{Tr}_{E/F}(\gamma) = \beta$$

Thus, $\alpha = \tau\gamma$ is the desired element. (Since the trace map is the sum of the automorphims in the Galois group, it is not the zero map and so there is a $\gamma \in E$ for which $\operatorname{Tr}(\gamma) \neq 0$.) Proof of the converse is left to the reader.□

In this section and the previous one, we have discussed cyclic extensions of degree n where $(n, \operatorname{expchar}(F)) = 1$ or $n = p = \operatorname{char}(F) \neq 0$. A discussion of cyclic extensions of degree $n = p^k$ for $k > 1$ is quite a bit more involved and falls beyond the intended scope of this book. The interested reader may wish to consult the books by Karpilovsky (1989) and Lang (1993).

Exercises

1. Assume that F contains the nth roots of unity and suppose that $F < E$. Show that $\alpha \in E$ is a root of a binomial $x^n - u$ over F if and only if it is a root of an irreducible binomial $x^d - v$ over F, where $d \mid n$.

2. Let $F < E$ be a finite cyclic extension, with Galois group $G = \langle\sigma\rangle$. Show that if $\beta \in E$ has the form $\beta = \alpha - \sigma\alpha$ for some $\alpha \in E$, then $\mathrm{Tr}_{E/F}(\beta) = 0$.
3. Let $F < E$ be cyclic of degree p^n where p is a prime. Let $F < K < E$ with $F < K$ cyclic of degree p^d where $d < n$. Let $F < L < E$ and suppose that $E = KL$. Show that $E = L$.
4. Let $\mathrm{char}(F) = p \neq 0$ and let $F(\alpha_1) = F(\alpha_2)$ be cyclic of degree p over F, where $\min(\alpha_i, F) = x^p - x - u_i$. Show that $\alpha_2 = n\alpha_1 + b$ where $b \in F$ and $n \in \mathbb{Z}_p$.
5. Let F be a field and let E be the extension of F generated by the nth roots of unity, for all $n \geq 1$. Show that $F < E$ is abelian.
6. Let E be a field and let σ be an automorphism of E of order d. Suppose that $\beta \in E$ has the property that $\sigma\beta = \beta$ and $\beta^d = 1$. Prove that there exists an $\alpha \in E$ such that $\sigma\alpha = \alpha\beta$.
7. Let E be a field and let σ be an automorphism of E of order $d > 1$. Show that there exists an $\alpha \in E$ such that $\sigma\alpha = \alpha + 1$.
8. Let $F < E$ be finite and abelian. Show that $E = F_1 \cdots F_m$ is the composite of fields F_i such that $F < F_i$ is cyclic of prime-power degree. Thus, the study of finite abelian extensions reduces to the study of cyclic extensions of prime-power degree.
9. Let F be a field containing the nth roots of unity. We do *not* assume that $(n, \mathrm{expchar}(F)) = 1$. Let $\overline{F}$ be an algebraic closure of F. Show that if $\alpha \in \overline{F}$ is separable over F and if α is a root of the binomial $x^n - u$ with $u \in F$, then $F < F(\alpha)$ is cyclic of degree $d \mid n$.

Chapter 13
Solvable Extensions

We now turn to the question of when an arbitrary polynomial equation $p(x) = 0$ is *solvable by radicals*. Loosely speaking, this means (for $\text{char}(F) = 0$) that we can reach the roots of $p(x)$ by a finite process of adjoining nth roots of existing elements, that is, by a finite process of passing from a field K to a field $K(\alpha)$, where α is a root of a binomial $x^n - u$, with $u \in K$. We begin with some basic facts about solvable groups.

13.1 Solvable Groups

Definition *A* **normal series** *in a group G is a tower of subgroups*

$$\{1\} = G_0 < G_1 < G_2 < \cdots < G_n = G$$

where $G_i \lhd G_{i+1}$. A normal series is **abelian** *if each factor group G_{i+1}/G_i is abelian, and* **cyclic** *if each factor group is cyclic.* □

Definition *A group is* **solvable** *(or* **soluble***) if it has an abelian normal series.* □

Theorem 13.1.1 *The following are equivalent for a nontrivial finite group G.*

1) G has an abelian normal series.

2) G has a cyclic normal series.

3) G has a cyclic normal series in which each factor group G_{i+1}/G_i is cyclic of prime order.

Proof. It is clear that 3) $\Rightarrow$ 2) $\Rightarrow$ 1). Thus, we need to prove only that 1) $\Rightarrow$ 3). Let $\{G_i\}$ be an abelian normal series. We wish to refine this series by inserting subgroups until all quotients have prime order. The Correspondence Theorem says that the natural projection $\pi\colon G_{i+1} \to G_{i+1}/G_i$ is a normality-preserving bijection from the subgroups of G_{i+1} containing G_i to the subgroups of G_{i+1}/G_i. Hence, by Cauchy's Theorem, if a prime p divides $o(G_{i+1}/G_i)$ then G_{i+1}/G_i has a subgroup of order p, which must have the form H_i/G_i for $G_i < H_i < G_{i+1}$.

Since G_{i+1}/G_i is abelian, $H_i/G_i \lhd G_{i+1}/G_i$ and so $H_i \lhd G_{i+1}$. Thus, $G_i \lhd H_i \lhd G_{i+1}$. Finally, since H_i/G_i is abelian, the Third Isomorphism Theorem implies that

$$G_{i+1}/H_i \approx \frac{G_{i+1}/G_i}{H_i/G_i}$$

is also abelian.

Thus, we have refined the original abelian normal series by introducing H_i, where H_i/G_i has prime order. Since G is a finite group, we may continue the refinement process until we have an abelian normal series, each of whose quotient groups has prime order. □

The next theorem gives some basic properties of solvable groups. The proofs of these statements, with the possible exception of 2), can be found in most standard texts on group theory.

Theorem 13.1.2
1) *Any abelian group is solvable.*
2) **(Feit–Thompson)** *Any finite group of odd order is solvable.*
3) **(Subgroups)** *Any subgroup of a solvable group is solvable.*
4) **(Quotients)** *If G is solvable and $H \lhd G$, then G/H is solvable.*
5) **(Lifting property)** *If $H \lhd G$ then H and G/H solvable imply that G is solvable.*
6) **(Finite direct products)** *The direct product of a finite number of solvable groups is solvable.*
7) *The symmetric group S_n is solvable if and only if $n \le 4$.* □

13.2 Solvable Extensions

Although our results can be proved in the context of arbitrary finite extensions, we shall restrict our attention to separable extensions. As the reader knows, this produces no loss of generality for fields of characteristic 0 or finite fields. Moreover, if $p(x)$ is an inseparable polynomial, then

$$p(x) = q(x^{p^d})$$

where $q(x)$ is separable. Thus, with respect to the solution of polynomial equations, the restriction to separable extensions is not as severe as it might first appear.

Definition *A finite separable extension $F < E$ is* **solvable** *if the finite Galois extension $F < \mathrm{nc}(E/F)$ has solvable Galois group, where $\mathrm{nc}(E/F)$ is the normal closure of E over F.* □

Theorem 13.2.2 *The class of solvable extensions is distinguished.*

Proof. Speaking in general, consider a finite separable tower of the form

$$F \triangleleft A < B$$

Since the first step is normal, we have

$$\frac{G_F(B)}{G_A(B)} \approx G_F(A)$$

Now, if the Galois groups $G_F(A)$ and $G_A(B)$ of each step are solvable, then the quotient $G_F(B)/G_A(B)$ is solvable as well. Hence, by the lifting property of solvability, the Galois group $G_F(B)$ of the full extension $F < B$ is solvable.

On the other hand, if the full extension has solvable Galois group $G_F(B)$, then the Galois group $G_F(A)$ of the lower step, being isomorphic to a quotient of $G_F(B)$ is solvable and the Galois group $G_A(B)$ of the upper step, being a subgroup of $G_F(B)$ is solvable.

Thus, in such a tower, solvability of the Galois groups has the "tower property." Note also that the implication $G_F(B)$ solvable implies $G_A(B)$ solvable does not require that the lower step $F < A$ be normal.

Now we can get to the business at hand. Suppose first that

$$F < K < E$$

where the full extension $F < E$ is solvable. The lower step is finite and separable, and we have the tower

$$F \triangleleft \mathrm{nc}(K/F) < \mathrm{nc}(E/F)$$

where the full extension has solvable Galois group, and therefore so does the lower step. That is, $F < K$ is solvable.

As to the upper step $K < E$, it is finite and separable. Consider the tower

$$F < K \triangleleft \mathrm{nc}(E/K) < \mathrm{nc}(E/F)$$

Since the full extension has solvable Galois group, so does the full extension

$$K \triangleleft \mathrm{nc}(E/K) < \mathrm{nc}(E/F)$$

(which is an upper step of the previous tower). Hence, the lower step has solvable Galois group, that is, $K < E$ is solvable. We have shown that if the full extension is solvable, so are the steps.

Suppose now that each step $F < K$ and $K < E$ solvable and consider Figure 13.2.1. Since all extensions are finite and separable, we will have no trouble there.

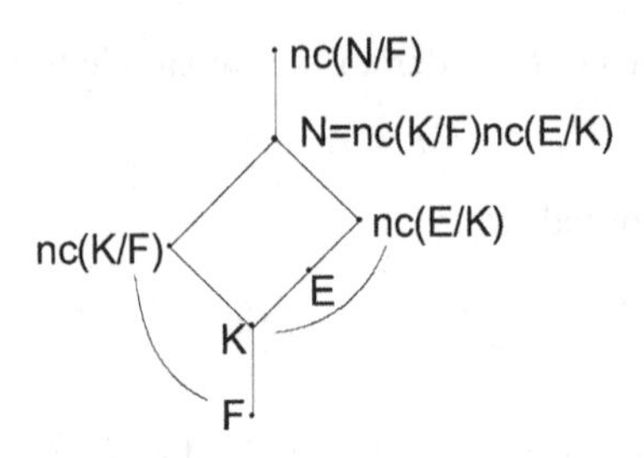

Figure 13.2.1

Since

$$F \lhd \mathrm{nc}(E/F) < \mathrm{nc}(N/F)$$

it is sufficient to show that $G_F(\mathrm{nc}(N/F))$ is solvable.

To this end, Theorem 6.5.6 implies that $G_F(\mathrm{nc}(N/F))$ is isomorphic to a subgroup of the finite direct product

$$\prod_{\sigma \in \hom_F(N,\overline{F})} \sigma G_F(N)\sigma^{-1}$$

Since this is a finite direct product and since each conjugate is isomorphic to $G_F(N)$, it suffices to show that $G_F(N)$ is solvable.

Consider the tower

$$F \lhd \mathrm{nc}(K/F) < N$$

The group $G_F(\mathrm{nc}(K/F))$ is solvable since $F < K$ is solvable. As to the Galois group $G_{\mathrm{nc}(K/F)}(N)$ of the upper step, it is a subgroup of the Galois group $G_K(N)$. Thus, it is sufficient to show that $G_F(N)$ is solvable. But $G_F(N)$ is the Galois group of a composite and is therefore isomorphic to a subgroup of the direct product of the Galois groups $G_F(\mathrm{nc}(K/F))$ and $G_K(\mathrm{nc}(E/K))$, both of which are solvable, precisely because the lower and upper steps in the tower are solvable.

For the lifting property, if $F < E$ is solvable and $F < K$ is arbitrary, then

$$F < E < \mathrm{nc}(E/F)$$

Lifting gives

$$K < KE < K \vee \mathrm{nc}(E/F)$$

and the full extension is finite and Galois, and

$$G_K(K \vee \mathrm{nc}(E/F)) \approx G_{K \cap \mathrm{nc}(E/F)}(\mathrm{nc}(E/F)) < G_F(\mathrm{nc}(E/F))$$

and since the latter is solvable, so is the former. □

13.3 Radical Extensions

Loosely speaking, when $\operatorname{char}(F) = 0$, an extension $F < E$ is *solvable by radicals* if it is possible to reach E from F by adjoining a finite sequence of nth roots of existing elements. More specifically, we have the following definitions, which also deal with the case $\operatorname{char}(F) \neq 0$.

Definition *Let* $\operatorname{expchar}(F) = p$ *and let* $F < R$*. A* **radical series** *for* $F < R$ *is a tower of fields*

$$F = R_0 < R_1 < \cdots < R_n = R$$

such that each step $R_i < R_{i+1}$ *is one of the following types:*

Pure of Class 1

$R_{i+1} = R_i(\beta_i)$ where β_i is an r_ith root of unity, where we may assume without loss of generality that $(r_i, p) = 1$.

Pure of Class 2

$R_{i+1} = R_i(\alpha_i)$ where α_i is a root of $x^{r_i} - u_i$, with $1 \neq u_i \in R_i$ and $(r_i, p) = 1$.

Pure of Class 3

(For $p > 1$ only) $R_{i+1} = R_i(\alpha_i)$ where α_i is a root of $x^p - x - u_i$, with $u_i \in R_i$.

For steps of classes 1 and 2, the number r_i is the **exponent** *(or* **type***) of the step. The* **exponent** *of a class 3 step is p. A finite separable extension $F < R$ that has a radical series is called a* **radical extension**.□

If $\operatorname{char}(F) = p \neq 0$, we may assume that the exponent in a class 1 extension is relatively prime to p, for if β is an rth root of unity where $r = mp^e$ and $(m, p) = 1$, then β is also an mth root of unity.

Note that lifting a radical series gives another radical series with the same class of steps, for if $R_{i+1} = R_i(\alpha)$, where α is a root of $f(x) \in R_i[x]$, then

$$KR_{i+1} = (KR_i)(\alpha)$$

where α is a root of $f(x) \in (KR_i)[x]$.

For convenience, we write

$$F < R/\{R_i\}$$

or

$$F < R/(R_0 < \cdots < R_n)$$

to denote the fact that $\{R_i\} = (R_0 < \cdots < R_n)$ is a radical series for the extension $F < R$.

Theorem 13.3.1 (Properties of radical extensions)

1) **(Lifting)** *If $F < R$ is a radical extension and $F < K$, then the lifting $K < RK$ is a radical extension.*

2) **(Each step implies full extension)** *If $F < R < S$, where $F < R$ and $R < S$ are radical extensions, then so is the full extension $F < S$.*

3) **(Composite)** *If $F < R$ and $F < S$ are radical extensions, then so is the composite extension $F < RS$.*

4) **(Normal closure)** *If $F < R$ is a radical extension, then so is $F < \mathrm{nc}(R/F)$.*

Proof. For 1), let $F < R/\{R_i\}$. Lifting the series $\{R_i\}$ by $F < K$ gives the radical series

$$K = R_0K < R_1K < \cdots < R_nK = RK$$

and so $K < RK$ is a radical extension.

For part 2), if $F < R/\{R_i\}$ and $R < S/\{S_i\}$, then lift the series $\{S_i\}$ by R:

$$R < SR/(R = S_0R < S_1R < \cdots < S_nR)$$

and append it to the end of $F < R/\{R_i\}$ to get

$$F < SR/(R_0 < \cdots < R_n = R = S_0R < S_1R < \cdots < S_nR)$$

and so $F < SR = S$ is a radical extension.

For part 3), if $F < R$ and $F < S$ are radical extensions, then so is the lifting $R < SR$ of $F < S$ by $F < R$ and so is the full extension $F < R < SR$.

For part 4), the normal closure is

$$\mathrm{nc}(R/F) = \bigvee_{\sigma \in \hom_F(R,\overline{R})} \sigma R$$

Since $F < R$ is a finite separable extension, $\hom_F(R,\overline{R})$ is a finite set. Hence, the composite above is a finite one. We leave it as an exercise to show that if $F < R/\{R_i\}$, then $F < \sigma R/\{\sigma R_i\}$. Hence, $F < \sigma R$ is a radical extension, and therefore so is the finite composite $F < \mathrm{nc}(R/F)$.□

13.4 Solvability by Radicals

We are interested in extensions $F < E$ where E is contained in a radical extension $F < R$.

Definition *A finite separable extension $F < E$ is* **solvable by radicals** *if $F < E < R$, where $F < R$ is a radical extension.*□

Theorem 13.4.1

1) The class of extensions that are solvable by radicals is distinguished.

2) *If $F < E$ is solvable by radicals then so is $F < \operatorname{nc}(E/F)$. In fact, if*

$$F < E < R$$

where $F < R$ is a radical extension, then

$$F < E < \operatorname{nc}(E/F) < \operatorname{nc}(R/F)$$

where $F < \operatorname{nc}(R/F)$ is a normal radical extension.

Proof. Let

$$F < K < E$$

If $F < E$ is solvable by radicals then

$$F < K < E < R$$

with $F < R$ radical. Hence, the lower step $F < K$ is solvable by radicals. For the upper step, $F < R$ radical implies $K < R$ is radical and so $K < E$ is solvable by radicals.

Now suppose the steps in the tower are radical.

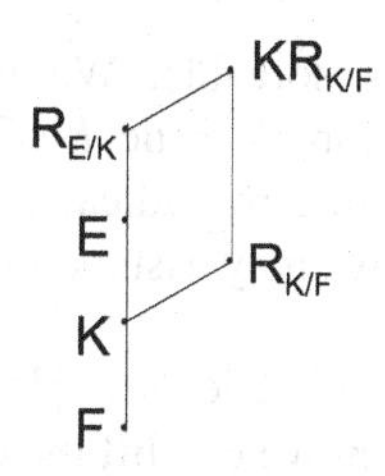

Figure 13.4.1

Referring to Figure 13.4.1, we have

$$F < K < R_{K/F}$$

and

$$K < E < R_{E/K}$$

where $F < R_{K/F}$ and $K < R_{E/K}$ are radical extensions. Lifting $K < R_{E/K}$ by $F < R_{K/F}$ gives the radical extension $R_{K/F} < R_{E/K}R_{K/F}$ and so the tower

$$F < R_{K/F} < R_{E/K}R_{K/F}$$

is radical. It follows that the full extension is radical and so $F < E$ is solvable by radicals.

As to lifting, if $F < E < R$ with $F < R$ radical, the lifting by $F < K$ gives

$$K < EK < RK$$

and since $K < RK$ is radical, $K < EK$ is solvable by radicals.

The second part of the theorem follows from the fact that if $F < E < R$ with $F < R$ is radical, then

$$F < \mathrm{nc}(E/F) < \mathrm{nc}(R/F)$$

with $F < \mathrm{nc}(R/F)$ radical.□

13.5 Solvable Equivalent to Solvable by Radicals

Now we come to the key result that links the concepts of solvable extension and solvability by radicals. Here we employ the results of Chapter 12 on cyclic extensions, taking advantage of the fact that we may assume that all appropriate roots of unity are present.

Theorem 13.5.1 *A finite separable extension $F < E$ is solvable by radicals if and only if it is solvable.*
Proof. Suppose that $F < E$ is solvable. We wish to show that $F < E$ is solvable by radicals. By definition, $F < \mathrm{nc}(E/F)$ is solvable and if we show that $F < \mathrm{nc}(E/F)$ is also solvable by radicals, then the lower step $F < E$ is also solvable by radicals. Thus, we may assume that $F < E$ is normal.

As to the presence of roots of unity, let $n = [E : F]$. If F does not contain a primitive nth root of unity ω, then we can lift the extension $F < E$ by adjoining ω to get

$$F(\omega) < E(\omega)$$

which is also solvable and normal. If we show that this extension is solvable by radicals, then so is the tower

$$F < F(\omega) < E(\omega)$$

since the lower step $F < F(\omega)$ is solvable by radicals (being pure of class 1). Hence, the lower step $F < E$ is also solvable by radicals.

Note that since $F < E$ is finite and Galois, Corollary 6.5.3 implies that

$$[E(\omega) : F(\omega)] \mid [E : F] = n$$

Hence, $F(\omega)$ contains a primitive mth root of unity, where $m = [E(\omega) : F(\omega)]$.

Thus, we may assume that $F < E$ is normal and contains a primitive nth root of unity, where $n = [E : F]$. It follows that if r is any prime dividing $[E : F]$, then F contains a primitive rth root of unity.

Since $F < E$ is finite, Galois and $G = G_F(E)$ is solvable, there is a normal series decomposition

$$\{1\} = G_0 < G_1 < G_2 < \cdots < G_n = G \tag{13.5.1}$$

where $G_i \lhd G_{i+1}$ and G_{i+1}/G_i is cyclic of prime order r_i dividing $|G| = [E : F] = n$. Taking fixed fields $F_i = \text{fix}(G_i)$ gives a tower

$$F = F_n < F_{n-1} < \cdots < F_1 < F_0 = E \tag{13.5.2}$$

Let us examine a typical step $F_{i+1} < F_i$ in this series. The relevant piece of the Galois correspondence is shown in Figure 13.5.1.

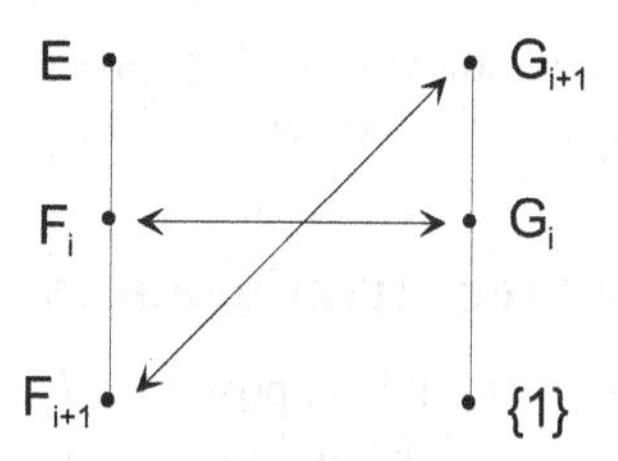

Figure 13.5.1

Since $F < E$ is finite and Galois, the Galois correspondence is completely closed, that is, all intermediate fields and subgroups are closed. Thus, since G_i is normal in G_{i+1}, it follows that $F_{i+1} < F_i$ is normal (and hence Galois) and that

$$G_{F_{i+1}}(F_i) \approx \frac{G_{i+1}}{G_i}$$

which is cyclic of degree $r_i \mid n$. Hence, $F_{i+1} < F_i$ is a cyclic extension whose base field F_{i+1} contains the r_ith roots of unity.

Now, if $r_i = \text{char}(F)$, then Theorem 12.2.1 implies that $F_{i+1} < F_i$ is pure of class 3. On the other hand, if $r_i \neq \text{char}(F)$, then $(r_i, \text{expchar}(F)) = 1$ and Theorem 12.1.1 implies that $F_{i+1} < F_i$ is pure of class 1 or class 2. Thus, $F < E$ is solvable by radicals, as desired.

For the converse, suppose that $F < E$ is solvable by radicals, with $F < E < R$ where $F < R$ is a radical extension. Then Theorem 13.4.1 implies that the full extension in the tower

$$F < E < \text{nc}(E/F) < \text{nc}(R/F)$$

has a radical series

$$F = R_0 < R_1 < \cdots < R_n = \text{nc}(R/F)$$

Let m be the product of the types of all the steps in this series. Lift the tower by adjoining a primitive mth root of unity ω, to get the radical series

$$F(\omega) = R_0(\omega) < R_1(\omega) < \cdots < R_n(\omega) = \mathrm{nc}(R/F)(\omega)$$

which contains $E(\omega)$. Note that if $R_i < R_{i+1}$ is of class 1, then the step $R_i(\omega) < R_{i+1}(\omega)$ is trivial, and we may remove it. Thus, we may assume that all steps in the lifted tower are pure of class 2 or class 3.

It follows from Theorems 12.1.1 and 12.2.1 that these pure steps are cyclic and so Theorem 6.6.2 implies that the Galois group $G_F(\mathrm{nc}(R/F)(\omega))$ is solvable. We have seen in the proof of 13.2.2 that since

$$F \lhd \mathrm{nc}(E/F) < \mathrm{nc}(R/F) < \mathrm{nc}(R/F)(\omega)$$

where the full extension has solvable Galois group, so does the lower step. Hence, $F < \mathrm{nc}(E/F)$ is a radical extension, which implies that $F < E$ is solvable by radicals.□

13.6 Natural and Accessory Irrationalities

Let us assume that $\mathrm{char}(F) = 0$ and suppose that $F < E$ is finite, normal and solvable by radicals. Let $n = [E : F]$ and assume that F contains the nth roots of unity.

Then, by definition, there is a radical series of the form

$$F = R_0 < R_1 < \cdots < R_n = R \tag{13.6.1}$$

where $F < E < R$. A typical step in this series has the form $K < K(\sqrt[k]{u})$, where $u \in K$. Elements of the form $a + b\sqrt[k]{u} \in R$, for $b \neq 0$, might reasonably be referred to as *irrationalities*, at least with respect to K (or F).

Kronecker coined the term *natural irrationalities* for those irrationalities of R that lie in E and *accessory irrationalities* for those irrationalities of R that do not lie in E.

Given a radical series (13.6.1) containing E, it is natural to wonder whether there is another radical series

$$F = S_0 < S_1 < \cdots < S_n = E$$

that contains only natural irrationalitites, that is, for which the top field S_n is E itself.

We begin by refining the steps in (13.6.1) so that each has prime degree. Consider a step $K < K(\alpha)$. Since the steps in the series are cyclic, every subgroup of the Galois group G of $K < K(\alpha)$ is normal, and so all lower steps $K < L < K(\alpha)$ are normal. If $m = pm'$ where p is prime, then G has a subgroup H of index p and so $K < \mathrm{fix}(H)$ is cyclic of degree p.

Hence, any step $K < K(\alpha)$ of (13.6.1) can be decomposed into a tower $K < L < K(\alpha)$, where the lower step $K < L$ is cyclic of prime degree. The upper step has the form $L < L(\alpha)$ and is cyclic of degree $m' < m$. We may repeat this decomposition on the upper step until each step is decomposed into a tower of cyclic extensions of prime degree.

So, let us assume that each step in (13.6.1) is cyclic of prime degree. Consider the tower obtained by intersecting each field in (13.6.1) by E

$$F = S_0 < (S_1 = R_1 \cap E) < \cdots < (S_n = R_n \cap E = E) \qquad (13.6.2)$$

We wish to show that each step in (13.6.2) is also cyclic of prime degree. This is the content of the following theorem. It will follow that $F < E$ has a radical series that starts at F and ends precisely at E.

Theorem 13.6.1 *Let* $\operatorname{char}(F) = 0$. *Let* $F < A < B$, *where* $A < B$ *is Galois of prime degree* p. *Let* $F < E$ *be finite and normal. Then* $A \cap E < B \cap E$ *is either trivial or Galois of degree* p.
Proof. Figure 13.6.1 shows the situation.

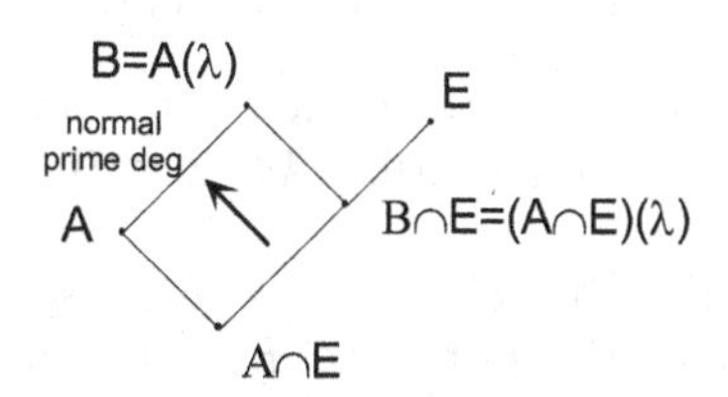

Figure 13.6.1

We first show that $A \cap E < B \cap E$ is normal by showing that it is closed in the Galois correspondence of $A \cap E < B \cap E$, that is,

$$\operatorname{fix}(G_{A\cap E}(B \cap E)) \subseteq A \cap E$$

(the reverse inclusion is clear). The plan is as follows. Let

$$\alpha \in \operatorname{fix}(G_{A\cap E}(B \cap E)) \subseteq E$$

Since $A < B$ is normal, if we show that $\sigma\alpha = \alpha$ for any $\sigma \in G_A(B)$, it will follow that $\alpha \in \operatorname{fix}(G_A(B)) = A$ and so $\alpha \in A \cap E$, as desired.

But it is sufficient to show that $\sigma \in G_A(B)$ implies $\sigma|_{B\cap E} \in G_{A\cap E}(B \cap E)$, since then

$$\sigma\alpha = (\sigma|_{B\cap E})\alpha = \alpha$$

To this end, since $(A \cap E) < (B \cap E)$ is finite and separable, it is simple, say

$$(A \cap E) < (B \cap E) = (A \cap E)(\lambda)$$

where $\lambda \in B$. If $\lambda \in A$, then the extension is trivial, so assume $\lambda \in B \setminus A$. We need to show that if $\sigma \in G_A(B)$ then $\sigma\lambda \in B \cap E$, since then $\sigma(B \cap E) = B \cap E$.

In the tower

$$A < A(\lambda) < B$$

the entire extension has prime degree p and the lower step is nontrivial. Hence, $B = A(\lambda)$.

Since $A < A(\lambda)$ is normal, the minimal polynomial $p(x) = \min(\lambda, A)$ splits over $A(\lambda)$ and so its roots $\{\lambda = \lambda_1, \ldots, \lambda_p\}$ lie in $A(\lambda) = B$. Also, since

$$\min(\lambda, A) \mid \min(\lambda, A \cap E)$$

each λ_i is a root of $\min(\lambda, A \cap E)$ and since $A \cap E < E$ is normal (being an upper step of $F \lhd E$), it follows that $\lambda_i \in E$ for all i. Hence,

$$\{\lambda = \lambda_1, \ldots, \lambda_p\} \subseteq B \cap E$$

But each $\sigma \in G_A(B)$ sends λ to a conjugate λ_i of λ, and so $\sigma\lambda \in B \cap E$. This is what we needed to prove and shows that $A \cap E < B \cap E$ is normal.

Finally, we must show that $[B \cap E : A \cap E] = p$. To see this, note that $A < B$ is the lifting of $A \cap E < B \cap E$ by $A \cap E < A$. Since $A \cap E < B \cap E$ is Galois, the Galois group of the lifting satisfies

$$G_A(B) \approx G_{E \cap A}(B \cap E)$$

and so $[B \cap E : A \cap E] = [B : A] = p$.□

Theorem 13.6.2 (The theorem on natural irrationalities) *Let* $\operatorname{char}(F) = 0$. *Let* $F < E$ *be finite and normal. Let* $n = [E : F]$ *and assume that* F *contains the* n*th roots of unity. If* E *is solvable by radicals, then there is a radical series starting with* F *and ending with* E.□

We remark that the requirement that F contain the appropriate roots of unity is necessary. An example is given by the casus irreducibilis, desscribed in the exercises.

13.7 Polynomial Equations

The initial motivating force behind Galois theory was the solution of polynomial equations $f(x) = 0$. Perhaps the crowning achievement of Galois theory is the statement, often phrased as follows: There is no formula, similar to the quadratic formula, involving only the four basic arithmetic operations and the taking of roots, for solving polynomial equations of degree 5 or greater over $\mathbb{Q}$.

However, this is not the whole story. The fact is that for some polynomial equations there is a formula and for others there is not, and, moreover, we can tell by looking at the Galois group of the polynomial whether or not there is such a formula. In fact, there are even algorithms for solving polynomial equations when they are "solvable," but these algorithms are unfortunately not practical.

Let us restrict attention to fields of characteristic 0. We refer to the four basic arithmetic operations (addition, subtraction, multiplication and division) and the taking of nth roots as the **five basic operations**.

Let F be a field of characteristic 0. We will say that an element $\alpha \in \overline{F}$ is **obtainable by formula from** F if we can obtain α by applying a finite sequence of any of the five basic operations, to a finite set of elements from F.

If $F < F(\sqrt[n]{\alpha})$ is a pure extension, it is clear that any element of $F(\sqrt[n]{\alpha})$, being a polynomial in $\sqrt[n]{\alpha}$, is obtainable by formula from F. Hence, any element of a radical extension $F < R$ is obtainable by formula from F.

Conversely, if $\alpha \in \overline{F}$ is obtainable by formula from F, then there is a finite set $S \subseteq F$ and a finite algorithm for obtaining α from S, where each step in the algorithm is the application of one of the five basic operations to elements of some extension E of F. If the operation is one of the four basic operations, then the result of the application is another element of the field E. It the operation is the taking of a root, then the result will lie in a pure extension of E. Thus, all the operations in the algorithm can be performed within a radical extension of F. Hence, α lies in a radical extension of F.

Theorem 13.7.1 *Let F be a field of characteristic 0. An element $\alpha \in \overline{F}$ can be obtained by formula from F if and only if α lies in a radical extension of F, that is, if and only if $F < F(\alpha)$ is solvable by radicals.* $\square$

Let us say that a root α of a polynomial $f(x) = a_0 + a_1x + \cdots + a_dx^d$ over F is **obtainable by formula** if we can obtain α by formula from $C = \mathbb{Q}(a_0, \ldots, a_d)$. Thus, a root α of $f(x)$ is obtainable by formula if and only if $C < C(\alpha)$ is solvable by radicals. Theorems 13.4.1 and 13.5.1 now imply the following.

Theorem 13.7.2 *Let* $\operatorname{char}(F) = 0$ *and let $f(x) = a_0 + a_1x + \cdots + a_dx^d$ be a polynomial over F. Let $C = \mathbb{Q}(a_0, \ldots, a_d)$ and let S be a splitting field for $f(x)$ over C.*

1) *The roots of $f(x)$ are obtainable by formula if and only if the extension $C < S$ is solvable.*
2) *Let $f(x)$ be irreducible over F. One root of $f(x)$ is obtainable by formula if and only if all roots of $f(x)$ are obtainable by formula.* □

According to Theorem 11.5.7, for any prime number p, there exists an irreducible polynomial $f_p(x)$ of degree p over $\mathbb{Q}$ whose Galois group is isomorphic to S_p. Since the group S_p is not solvable for $p \geq 5$, Theorem 13.7.2 implies that if $p \geq 5$, then *none* of the roots of $f_p(x)$ can be obtained by formula.. Although it is much harder to show, this also holds for any positive integer n [see Hadlock, 1987]. Thus, we have the following.

Theorem 13.7.3 *For any $n \geq 5$, there is an irreducible polynomial of degree n over $\mathbb{Q}$, none of whose roots are obtainable by formula.* □

As a consequence, for any given $n \geq 5$, there is no formula for the roots, similar to the quadratic formula, involving only the four basic operations and the taking of roots, that applies to all polynomials of degree n. More specifically, we have

Corollary 13.7.4 *Let $n \geq 5$ and consider the generic polynomial $p(x) = y_0 + y_1x + \cdots + y_nx^n$, where $y_0, \ldots, y_n$ are algebraically independent over $\mathbb{Q}$. Then there is no algebraic formula, involving only the five basic operations, the elements of $\mathbb{Q}$ and the variables $y_0, \ldots, y_n$, with the property that for any polynomial $f(x) = a_0 + a_1x + \cdots + a_nx^n$ of degree n over F, we can get a root of $f(x)$ by replacing y_i in the formula by a_i, for all $i = 0, \ldots, n$.* □

Exercises

1. Prove that if $H \lhd G$ then G is solvable if and only if H and G/H are solvable.
2. Prove that if $R_0 < R_1 < \cdots < R_n$ is a radical series, then there is a radical series that is a refinement of this series (formed by inserting additional intermediate fields) for which each extension has prime exponent.
3. Prove that if $F < E$ is solvable by radicals and $\sigma \in \hom_F(E, \overline{E})$ then $F < \sigma E$ is also solvable by radicals.
4. Calculate the Galois group of the polynomial $f(x) = x^5 - 4x + 2$. Is there a formula for the roots?
5. Prove that if $f(x)$ is a polynomial of degree n over F with Galois group isomorphic to S_n then $f(x)$ is irreducible and separable over F.
6. While the class of (finite, separable) solvable extensions is distinguished, show that the class of *Galois* solvable extensions does not have the tower property, and so is not distinguished. *Hint*: use the Feit–Thompson result (Theorem 13.1.2) and the proof of Theorem 11.5.1.
7. Prove that a finite separable extension $F < E$ of characteristic p is solvable by radicals if and only if there exists a finite extension $F < R$ with

$F < E < R$ and a radical series $\{R_i\}$ for $F < R$ in which each step $R_i < R_{i+1}$ is one of the following classes:

1) $R_{i+1} = R_i(\omega_i)$ where ω_i is an r_ith root of unity with r_i prime and $r_i \neq p$.
2) $R_{i+1} = R_i(\alpha_i)$ where α_i is a root of $x^{r_i} - u$, with $u \in R_i$, r_i prime and $r_i \neq p$.
3) $R_{i+1} = R_i(\beta_i)$ where β_i is a root of the irreducible polynomial $x^p - x - u$, with $u \in R_i$.

8. Prove Theorem 13.7.2. Hint: for part 2), consider the normal closure of $C(\alpha)$, where α is an obtainable root of $f(x)$.

Casus Irreducibilis

Cardano's formula for the cubic equation $x^3 + px + q = 0$ is

$$x = \sqrt[3]{-\frac{q}{2} + \sqrt{\left(\frac{p}{3}\right)^3 + \left(\frac{q}{2}\right)^2}} + \sqrt[3]{-\frac{q}{2} - \sqrt{\left(\frac{p}{3}\right)^3 + \left(\frac{q}{2}\right)^2}}$$

This formula does not always yield a "satisfactory" solution, especially to the interested parties of the 16th century. For instance, the equation $x^3 + x - 2 = 0$ has only one real solution $x = 1$, but Cardano's formula gives

$$x = \sqrt[3]{1 + \frac{2}{3}\sqrt{\frac{7}{3}}} + \sqrt[3]{1 - \frac{2}{3}\sqrt{\frac{7}{3}}}$$

(which must therefore equal 1, a handy formula to remember). The most serious "problem" with Cardano's formula comes when

$$d = \left(\frac{p}{3}\right)^3 + \left(\frac{q}{2}\right)^2 < 0$$

since in this case, the formula contains the square root of a negative number, something Cardano referred to as "impossible", "useless" and whose manipulation required "mental torture". For instance, the equation $x^3 - 15x - 4 = 0$ has a simple real solution $x = 4$, but Cardano's formula gives

$$x = \sqrt[3]{2 + \sqrt{-121}} + \sqrt[3]{2 - \sqrt{-121}}$$

(which is equal to 4). Cases where $d < 0$ are known as *casus irreducibilis* and were the subject of much debate in the 1500s. Efforts to modify the formula for the solution of a cubic with three real roots in order to avoid nonreal numbers were not successful, and we can now show why. (Actually, this turned out to be a good thing, since it sparked the development of the complex numbers.)

9. Let $f(x)$ be an irreducible cubic over $\mathbb{Q}$ with three real roots. This exercise shows that no root of $f(x)$ can be obtained by formula if we allow the

taking of *real* nth roots only, that is, we show that no root of $f(x)$ is contained in a radical series that is completely contained in $\mathbb{R}$.

a) Suppose that a root r of $f(x)$ is contained in a radical series

$$\mathbb{Q} < F_1 < \cdots < F_n < \mathbb{R}$$

Let Δ be the discriminant and let $\delta = \sqrt{\Delta}$. Show that there is a radical series

$$\mathbb{Q} < E_1 < \cdots < E_n < \mathbb{R}$$

containing r with $\delta \in E_i$.

b) Show that the radical series in part a) can be refined (by inserting more intermediate fields) into a radical series in which each step has prime exponent.

c) Let $m+1$ be the first index such that a root r of $f(x)$ is in E_{m+1} and consider the extension $E_m < E_m(r)$. Show that $E_m(r)$ is a splitting field for $f(x)$ over E_m.

d) Since $E_m < E_{m+1}$ is pure of prime exponent, we have $E_{m+1} = E_m(\beta)$, where β is a root of $x^p - u$, with p prime and $u \in E_m$. More generally, prove that if K is a field, $u \in K$ and p is a prime, then the polynomial $x^p - u$ is either irreducible over K or $u \in K^p$. *Hint*: Suppose that $f(x) = p(x)q(x)$ where $\deg(p) = d < p$ and $\deg(q) = e < p$. If $\alpha_1, \ldots, \alpha_d$ are the roots of $p(x)$ and $\beta_1, \ldots, \beta_e$ are the roots of $q(x)$, then

$$u = (-1)^p \prod \alpha_i \prod \beta_i$$

Take the pth power of this and use $\beta_i^p = u$. Then use the fact that $(d,p) = 1$.

e) Show that $u \in E_m^p$ is not possible. (The primitive pth roots of unity do not lie in $\mathbb{R}$.) Hence, $x^p - u$ is irreducible.

f) Show that $p = 3$ and $S = E_{m+1}$.

g) Show that $E_m < E_{m+1}$ is normal. What does that say about the roots of $x^3 - u$?

Galois' Result

Galois, in his memoir of 1831, proved the following result (Proposition VIII):

> "For an equation of prime degree, which has no commensurable divisors, to be solvable by radicals, it is necessary and sufficient that all roots be rational functions of any two of them."

In more modern language, this theorem says that if $f(x)$ is irreducible and separable of prime degree p, then the equation $f(x) = 0$ is solvable by radicals if and only if $F[\alpha, \beta]$ is a splitting field for $f(x)$, for *any* two roots α and β of

$f(x)$. To prove this theorem, we require some results concerning solvable transitive subgroups of S_p, the group of permutations of $\mathbb{Z}_p = \{0, 1, \ldots, p-1\}$.

Any map $\lambda_{a,b}: \mathbb{Z}_p \to \mathbb{Z}_p$ defined by $\lambda_{a,b}(x) = ax + b$, where $a, b \in \mathbb{Z}_p$ with $a \neq 0$ is called an **affine transformation** of $\mathbb{Z}_p$. Let $\mathrm{aff}(p)$ be the group of all affine transformations of $\mathbb{Z}_p$. Note that $\mathrm{aff}(p)$ is a subgroup of S_p. The **translations** are the affine maps $\tau_b = \lambda_{1,b}$. Let $\mathrm{trans}(p)$ be the subgroup of $\mathrm{aff}(p)$ consisting of the translations. Let $\tau = \tau_1$ be translation by 1. Two elements α and β of S_p are **conjugate** if there is a $\sigma \in S_p$ for which $\sigma^{-1}\alpha\sigma = \beta$.

10. a) Show that $\mathrm{trans}(p) = \langle \tau \rangle$ is a normal subgroup of $\mathrm{aff}(p)$.
    b) Show that τ is the p-cycle $(0\,1 \cdots p-1)$, that any nonidentity translation is a p-cycle and that an element $\sigma \in S_p$ is a p-cycle if and only if it is conjugate to τ.
    c) Within $\mathrm{aff}(p)$, the nonidentity translations are characterized as having no fixed points, whereas all elements of $\mathrm{aff}(p) \setminus \mathrm{trans}(p)$ have exactly one fixed point.
    d) Show that $\mathrm{aff}(p)$ acts transitively on $\mathbb{Z}_p$.
    e) Show that $|\mathrm{aff}(p)| = p(p-1)$ and $|\mathrm{trans}(p)| = p$. Hence, $\mathrm{trans}(p)$ is a Sylow p-subgroup of $\mathrm{aff}(p)$ and is the only subgroup of $\mathrm{aff}(p)$ of order p.
    f) Show that $\mathrm{trans}(p)$, $\mathrm{aff}(p)/\mathrm{trans}(p)$ and $\mathrm{aff}(p)$ are solvable.
11. Prove that if $\sigma \in S_p$ has the property that $\sigma\tau\sigma^{-1} \in \mathrm{aff}(p)$, then $\sigma \in \mathrm{aff}(p)$.
12. The following are equivalent for a subgroup G of S_p:
    1) G is transitive.
    2) G contains a subgroup conjugate to $\mathrm{trans}(p)$, that is, $\langle \gamma \rangle < G$, for some p-cycle γ.
13. The following are equivalent for a transitive subgroup G of S_p:
    1) The only element of G with two fixed points is the identity.
    2) G is conjugate to a subgroup H of $\mathrm{aff}(p)$.

We have proved that for a transitive subgroup G of S_p, the first two statements below are equivalent. We now add a third.
1) The only element of G with two fixed points is the identity.
2) G is conjugate to a subgroup H of $\mathrm{aff}(p)$.
3) G is solvable.

It is clear that 2) implies 3), since a conjugate of a solvable group is solvable. The next few exercises prove that 3) implies 1).

14. If G is a transitive subgroup of S_p, show that any normal subgroup $N \neq \{1\}$ of G also acts transitively on $\mathbb{Z}_p$.

15. Let G be a transitive, solvable subgroup of S_p. Then G has a normal series with prime indices

$$\{1\} \lhd G_1 \lhd \cdots \lhd G_{n-1} \lhd G_n = G$$

a) Show that $G_1 = \langle\gamma\rangle$, where γ is a p-cycle.
b) Show that the only element of G that has two fixed points is the identity.

We can now return to Galois' result concerning solvability by radicals for a prime-degree equation $f(x) = 0$.

16. Prove that if $f(x)$ is irreducible and separable of prime degree p, then the equation $f(x) = 0$ is solvable by radicals if and only if $F[\alpha, \beta]$ is a splitting field for $f(x)$, for *any* two roots α and β of $f(x)$.

Part III—The Theory of Binomials

Chapter 14
Binomials

We continue our study of binomials by determining conditions that characterize irreducibility and describing the Galois group of a binomial $x^n - u$ in terms of 2×2 matrices over $\mathbb{Z}_n$. We then consider an application of binomials to determining the irrationality of linear combinations of radicals. Specifically, we prove that if $p_1, \ldots, p_m$ are distinct prime numbers, then the degree of

$$\mathbb{Q}(\sqrt[n]{p_1}, \ldots, \sqrt[n]{p_m})$$

over $\mathbb{Q}$ is as large as possible, namely, n^m. This implies that the set of all products of the form

$$\sqrt[n]{p_1^{e(1)}} \cdots \sqrt[n]{p_m^{e(m)}}$$

where $0 \le e(i) \le n-1$, is linearly independent over $\mathbb{Q}$. For instance, the numbers

$$1, \sqrt[4]{3} = \sqrt[60]{3^{15}}, \sqrt[5]{4} = \sqrt[60]{2^{24}} \quad \text{and} \quad \sqrt[6]{72} = \sqrt[60]{2^{30}3^{20}}$$

are of this form, where $p_1 = 2$, $p_2 = 3$. Hence, any expression of the form

$$a_1\sqrt[4]{3} + a_2\sqrt[5]{4} + a_3\sqrt[6]{72}$$

where $a_i \in \mathbb{Q}$, must be irrational, unless $a_i = 0$ for all i.

First, a bit of notation. If $u \in F$, then $u^{1/n}$ stands for a particular (fixed) root of $x^n - u$. The set of primitive nth roots of unity is denoted by Ω_n and ω_k always denotes a primitive kth root of unity.

14.1 Irreducibility

Let us first recall a few facts about the norm. Let $F < E$ be finite with $\alpha \in E$. If the minimal polynomial of α

$$\min(\alpha, F) = x^d + a_{d-1}x^{d-1} + \cdots + a_0$$

has roots $r_1, \ldots, r_d$ then

$$N(\alpha) = \prod_{i=1}^{d} r_i = (-1)^d a_0$$

where $N = N_{F(\alpha)/F}$. Note that $N(\alpha) \in F$. Also, for all $\beta \in F(\alpha)$ and $a \in F$, we have

1) The norm is multiplicative, that is, for all $\beta, \gamma \in E$,
$$N(\beta\gamma) = N(\beta)N(\gamma)$$
In particular,
$$N(\beta^n) = N(\beta)^n$$
for any positive integer n. Also, $N(1) = 1$.
2) For $a \in F$,
$$N(a\beta) = a^d N(\beta)$$
and so
$$N(a) = a^d$$
3) If $F < E < L$ are finite and if $\alpha \in L$ then
$$N_{L/F}(\alpha) = N_{E/F}(N_{L/E}(\alpha))$$

Our technique for determining the irreducibility of a binomial $f(x) = x^n - u$ for $u \in F$ is an inductive one, beginning with the case $n = p$ prime.

Theorem 14.1.1 *Let p be a prime. Then the following are equivalent:*
1) $u \notin F^p$
2) $f(x) = x^p - u$ *has no roots in* F
3) $f(x) = x^p - u$ *is irreducible over* F

Proof. It is easy to see that 1) and 2) are equivalent and that 3) implies 2). To see that 1) implies 3), let α be a root of $x^p - u$ in $\overline{F}$ and assume that $[F(\alpha) : F] = d \leq p$. We wish to show that $d = p$, which implies that $f(x)$ is $\min(\alpha, F)$ and is therefore irreducible. Since $\alpha^p = u$, taking the norm $N = N_{F(\alpha)/F}$ gives

$$[N(\alpha)]^p = N(\alpha^p) = N(u) = u^d$$

where $N(\alpha) \in F$. Now, if $p < d$ then $(d, p) = 1$ and there exist integers a and b for which $ad + bp = 1$. Hence

$$u = u^{ad+bp} = u^{ad}u^{bp} = N(\alpha)^{ap}u^{bp} \in F^p$$

which is a contradiction. Thus $p = d$, as desired.$\square$

To generalize this to nonprime exponents, assume that

$$n = p_0 \cdots p_{t-1}$$

is a product of not necessarily distinct *odd* primes. (We will consider the even prime later.) Let us write u as α_0 and

$$f(x) = x^n - \alpha_0$$

where $\alpha_0 \in F$. Let β be a root of $f(x)$ in $\overline{F}$ and write $n = p_0 m_0$. Then

$$f(x) = (x^{m_0})^{p_0} - \alpha_0$$

Hence, $\alpha_1 = \beta^{m_0}$ is a root of $x^{p_0} - \alpha_0$ and β is a root of

$$g_0(x) = x^{m_0} - \alpha_1$$

and we have the tower

$$F = F(\alpha_0) < F(\alpha_1) < F(\beta)$$

Repeating the process with $g_0(x)$, if $m_0 = p_1 m_1$, then

$$g_0(x) = (x^{m_1})^{p_1} - \alpha_1$$

so $\alpha_2 = \beta^{m_1}$ is a root of $x^{p_1} - \alpha_1$ and β is a root of

$$g_1(x) = x^{m_1} - \alpha_2$$

over $F(\alpha_2)$ and we have the tower

$$F(\alpha_0) < F(\alpha_1) < F(\alpha_2) < F(\beta)$$

Clearly, we can repeat this process as desired to obtain a tower

$$F(\alpha_0) < F(\alpha_1) < F(\alpha_2) < \cdots < F(\alpha_{t-1}) < F(\alpha_t) = F(\beta)$$

where $\alpha_t = \beta$, and where each step $F(\alpha_i) < F(\alpha_{i+1})$ has the property that α_{i+1} is a root of the binomial $x^{p_i} - \alpha_i$ of prime degree over $F(\alpha_i)$.

Now, the binomial $x^n - u$ is irreducible if and only if

$$[F(\beta) : F] = n = p_0 \cdots p_{t-1}$$

and this happens if and only if each binomial $x^{p_i} - \alpha_i$ is irreducible, which according to Theorem 14.1.1, is equivalent to the conditions

$$\alpha_i \notin F(\alpha_i)^{p_i} \tag{14.1.1}$$

for all $i = 0, \ldots, t-1$. Let us improve upon these conditions.

First, note that if $\alpha_i \in F(\alpha_i)^{p_i}$, that is, if $\alpha_i = \gamma^{p_i}$ for $\gamma \in F(\alpha_i)$, then

$$N_{F(\alpha_i)/F(\alpha_0)}(\alpha_i) = N_{F(\alpha_i)/F(\alpha_0)}(\gamma^{p_i}) = N_{F(\alpha_i)/F(\alpha_0)}(\gamma)^{p_i} \in F^{p_i}$$

Hence, (14.1.1) is implied by the following conditions, which involve membership in a power of the base field only

$$N_{F(\alpha_i)/F(\alpha_0)}(\alpha_i) \notin F^{p_i} \tag{14.1.2}$$

for $i = 0, \ldots, t-1$. Thus, under these conditions, each binomial $x^{p_i} - \alpha_i$ is irreducible, with root α_{i+1}, and so

$$N_{F(\alpha_{i+1})/F(\alpha_i)}(\alpha_{i+1}) = -(-1)^{p_i}\alpha_i$$

for $i = 0, \ldots, t-1$. Assuming that all the primes p_i are odd or that $\text{char}(F) = 2$, this can be written as

$$N_{F(\alpha_{i+1})/F(\alpha_i)}(\alpha_{i+1}) = \alpha_i \tag{14.1.3}$$

for $i = 0, \ldots, t-1$. For $i = 0$, this is

$$N_{F(\alpha_1)/F(\alpha_0)}(\alpha_1) = \alpha_0$$

For $i = 1$, we get

$$N_{F(\alpha_2)/F(\alpha_1)}(\alpha_2) = \alpha_1$$

and applying the norm and using the case $i = 0$ gives

$$\begin{aligned} N_{F(\alpha_2)/F(\alpha_0)}(\alpha_2) &= N_{F(\alpha_1)/F(\alpha_0)}(N_{F(\alpha_2)/F(\alpha_1)}(\alpha_2)) \\ &= N_{F(\alpha_1)/F(\alpha_0)}(\alpha_1) \\ &= \alpha_0 \end{aligned}$$

In general, if $N_{F(\alpha_i)/F(\alpha_0)}(\alpha_i) = \alpha_0$, then applying the norm to (14.1.3) gives

$$\begin{aligned} N_{F(\alpha_{i+1})/F(\alpha_0)}(\alpha_{i+1}) &= N_{F(\alpha_i)/F(\alpha_0)}(N_{F(\alpha_{i+1})/F(\alpha_i)}(\alpha_{i+1})) \\ &= N_{F(\alpha_i)/F(\alpha_0)}(\alpha_i) \\ &= \alpha_0 \end{aligned}$$

Thus,

$$N_{F(\alpha_{i+1})/F(\alpha_0)}(\alpha_{i+1}) = \alpha_0$$

for $i = 0, \ldots, t-1$ and we can rephrase the conditions (14.1.2) as

$$\alpha_0 \notin F^{p_i}$$

for $i = 0, \ldots, t-1$.

Theorem 14.1.2 *If $n = p_0 \cdots p_{t-1}$ is a product of not necessarily distinct* odd *primes, then the binomial $x^n - u$ is irreducible if*

$$\alpha_0 \notin F^{p_i}$$

for $i = 0, \ldots, t-1.\square$

Let us now turn to the case where $n = 2^s$ and write

$$f_s(x) = x^{2^s} - u$$

for $u \in F$. As seems often the case, the even prime $p = 2$ causes additional problems. To illustrate, if $0 \neq b \in \mathbb{Q}$, then for any $s \geq 2$

$$x^{2^s} + 4b^4 = (x^{2^{s-1}} + 2bx^{2^{s-2}} + 2b^2)(x^{2^{s-1}} - 2bx^{2^{s-2}} + 2b^2)$$

and so the binomial $x^{2^s} + 4b^4$ is reducible even though $u = -4b^4 \notin \mathbb{Q}^2$. Thus, for $s > 1$, we must at least include the restriction that $u \neq -4b^4$ for any $b \in F$, that is, that $u \notin -4F^4$. It turns out that no further restrictions are needed.

Theorem 14.1.3 *Let F be a field and $0 \neq u \in F$.*
1) $x^2 - u$ is irreducible if and only if $u \notin F^2$
2) For $s > 1$, the binomial $f_s(x) = x^{2^s} - u$ is irreducible if and only if $u \notin F^2$ and $u \notin -4F^4$.

Proof. Part 1) is clear. For part 2), assume that $f_s(x)$ is irreducible, where $s > 1$. If $u = \gamma^2$ for some $\gamma \in F$, then

$$f_s(x) = x^{2^s} - u = x^{2^s} - \gamma^2 = (x^{2^{s-1}} - \gamma)(x^{2^{s-1}} + \gamma)$$

is reducible. Hence, $u \notin F^2$. Also, if $u = -4\gamma^4$ for $\gamma \in F$, then

$$f_s(x) = x^{2^s} + 4\gamma^4$$

factors as above. Hence, $u \notin F^2$ and $u \notin -4F^4$.

For the converse, we show that the conditions $u \notin F^2$ and $u \notin -4F^4$ imply that $f_s(x)$ is irreducible for all $s \geq 1$, by induction on s. We have seen that this holds for $s = 1$. Assume that it holds for all positive integers less than $s > 1$. Let β be a root of $f_s(x)$ in a splitting field and write

$$f_s(x) = (x^{2^{s-1}})^2 - u$$

Hence, $\alpha = \beta^{2^{s-1}}$ is a root of $x^2 - u$ and β is a root of

$$g(x) = x^{2^{s-1}} - \alpha$$

and we have the tower

$$F < F(\alpha) < F(\beta)$$

The lower step has degree 2 since $u \notin F^2$. As to the upper step, if $\alpha \notin F(\alpha)^2$ and $\alpha \notin -4F(\alpha)^4$, then the induction hypothesis implies that $g(x)$ is irreducible over $F(\alpha)$, in which case

$$[F(\beta):F] = 2 \cdot 2^{s-1} = 2^s$$

which implies that 2) holds. Hence, we need only consider the two cases wherein these hypotheses fail.

If $\alpha \in -4F(\alpha)^4$, that is, if $\alpha = -4\gamma^4$, for some $\gamma \in F(\alpha)$, we claim that $\alpha \in F(\alpha)^2$ as well. The problem is that -1 may not be a square in $F(\alpha)$. But taking norms $N = N_{F(\alpha)/F}$ gives

$$-\alpha^2 = N(\alpha) = N(-4\gamma^4) = 16[N(\gamma)]^4 = a^2$$

where $a \in F$. Hence, $-1 = a^2/\alpha^2 \in F(\alpha)^2$. It follows that

$$\alpha = -4\gamma^4 = 4\frac{a^2}{\alpha^2}\gamma^2 \in F(\alpha)^2$$

So, if either condition fails, then $\alpha = \gamma^2 \in F(\alpha)^2$ for some $\gamma \in F(\alpha)$.

We must show that this implies that $f_s(x) = x^{2^s} - u$ is irreducible over F. Applying the norm $N = N_{F(\alpha)/F}$ gives

$$-u = (-1)^2 N(\alpha) = N(\gamma^2) = [N(\gamma)]^2 = b^2 \in F^2$$

where $N(\gamma) = b \in F$. Since $u \notin F^2$, it follows that $-1 \notin F^2$. In other words, if i is a root of x^2+1 in $\overline{F}$, then $i \notin F$. Over $F(i)$, we have the factorization

$$f(x) = x^{2^s} - u = x^{2^s} + b^2 = (x^{2^{s-1}} + ib)(x^{2^{s-1}} - ib) \qquad (14.1.2)$$

If both of the factors on the right side are irreducible over $F(i)$, then $f(x)$ cannot factor nontrivially over F, because each irreducible factor of $f(x)$ over F, being over $F(i)$ as well, would be a multiple of one of the irreducible factors on the right of (14.1.2) and so would have degree greater than 2^{s-1}. But two such factors would then have product of degree greater than 2^s. Thus, in this case, $f(x)$ is irreducible over F.

On the other hand, if one of the factors in (14.1.2) is reducible, the induction hypothesis implies that one of ib or $-ib$ is in either $F(i)^2$ or $-4F(i)^4 = 4i^2F(i)^4 \subseteq F(i)^2$. Thus, in either case, one of ib or $-ib$ is in $F(i)^2$, say

$$\pm\, ib = (c+di)^2 = c^2 + 2cdi - d^2$$

Thus, $c^2 = d^2$ and $b^2 = 4c^2d^2 = 4c^4$. It follows that $u = -b^2 = -4c^4$, a contradiction to the hypothesis of the lemma. Hence, this case does not occur.□

Now we can prove the main result of this section.

Theorem 14.1.4 *Let $n \geq 2$ be an integer and let $0 \neq u \in F$.*

1) *If $4 \not| \, n$, then $f(x) = x^n - u$ is irreducible over F if and only if $u \notin F^p$ for all primes $p \mid n$.*
2) *If $4 \mid n$, then $f(x) = x^n - u$ is irreducible over F if and only if $u \notin F^p$ for all primes $p \mid n$ and $u \notin -4F^4$.*

Proof. Assume first that $f(x)$ is irreducible. Then for any prime $p \mid n$, the polynomial $x^p - u$ is irreducible, for if $x^p - u = a(x)b(x)$ is a nontrivial factorization, then $x^n - u = (x^{n/p})^p - u = a(x^{n/p})b(x^{n/p})$ is a nontrivial factorization of $f(x)$. Hence, by Theorem 14.1.3, $u \notin F^p$ for any $p \mid n$. Also, if $4 \mid n$ then the polynomial $x^4 - u$ is irreducible and so again by Theorem 14.1.3, $u \notin -4F^4$. Alternatively, we have a direct factoization

$$f(x) = x^{4d} + 4\gamma^4 = (x^{2d} + 2\gamma x^d + 2\gamma^2)(x^{2d} - 2\gamma x^d + 2\gamma^2)$$

For the converse, assume that $u \notin F^p$ for all primes $p \mid n$ and that when $4 \mid n$, we also have $u \notin -4F^4$. We proceed by induction on n. If $n = 2$, the result follows from Theorem 14.1.3. Assume that the theorem is true for integers greater than 1 and less than n. If $n = 2^m$, where $m > 1$, then Theorem 14.1.3 applies. Otherwise, n has an odd prime factor p. Suppose that $n = p^k m$ where $(p, m) = 1$ and $k \geq 1$.

Let β be a root of $x^n - u = (x^{p^k})^m - u$. Then $\alpha = \beta^{p^k}$ is a root of $x^m - u$ and β is a root of

$$g(x) = x^{p^k} - \alpha$$

The induction hypothesis implies that $x^m - u$ is irreducible over F and so the first step in the tower

$$F < F(\alpha) < F(\beta)$$

has degree m. If $g(x)$ is irreducible over $F(\alpha)$, then the second step will have degree p^k, whence $[F(\beta) : F] = mp^k = n$ and $f(x) = \min(\beta, F)$, which is irreducible.

We apply the inductive hypothesis to show that $g(x)$ is irreducible. Since p is odd, we need only show that $\alpha \notin F(\alpha)^p$. If $\alpha = \gamma^p$ for some $\gamma \in F(\alpha)$ then taking norms $N = N_{F(\alpha)/F}$ gives

$$-u = (-1)^m N(\alpha) = (-1)^m N(\gamma^p) = (-1)^m [N(\gamma)]^p$$

If m is odd, we get $u = [N(\gamma)]^p \in F^p$, a contradiction. If m is even then since p is odd, we have $u = [-N(\gamma)]^p \in F^p$, again a contradiction. Hence, $\alpha \notin F(\alpha)^p$, $g(x)$ is irreducible over $F(\alpha)$ and $f(x)$ is irreducible over F. $\square$

14.2 The Galois Group of a Binomial

Let us now examine the Galois group of a binomial $x^n - u$ over F, for $u \neq 0$ and n relatively prime to expchar(F). If α is a root of $x^n - u$ and $\omega \in \Omega_n$, then all the roots are given by

$$\alpha, \omega\alpha, \ldots, \omega^{n-1}\alpha$$

and so $S = F(\omega, \alpha)$ is a splitting field for $x^n - u$ over F. Moreover, in the tower

$$F < F(\omega) < F(\omega, \alpha) = S \tag{14.2.1}$$

the first step is a cyclotomic extension, which is abelian since its Galois group is isomorphic to a subgroup of $\mathbb{Z}_n^*$. The second step is pure of type n and so, according to Theorem 12.1.1, it is cyclic of degree $d \mid n$ and

$$\min(\alpha, F(\omega)) = x^d - \alpha^d$$

Despite the abelian nature of the lower step and the cyclic nature of the upper step, the full extension (14.2.1) need not be abelian.

The fact that α and ω both satisfy simple polynomials *over* F is the key to describing the Galois group $G_F(S)$. Since any $\sigma \in G_F(S)$ must permute the roots of $x^d - \alpha^d$, there exists an integer $k(\sigma) \in \mathbb{Z}_d \subseteq \mathbb{Z}_n$ for which

$$\sigma\alpha = \omega^{k(\sigma)}\alpha$$

Moreover, since $F(\omega)$ is a normal extension of F, the restriction of σ to $F(\omega)$ is in $G_F(F(\omega))$ and therefore σ sends ω to another primitive nth root of unity, that is,

$$\sigma\omega = \omega^{j(\sigma)}$$

where $j(\sigma) \in \mathbb{Z}_n^*$.

Multiplication in $G_F(S)$ has the following form. For $\sigma, \tau \in G_F(S)$,

$$\sigma\tau\alpha = \sigma(\omega^{k(\tau)}\alpha) = \omega^{j(\sigma)k(\tau)}\omega^{k(\sigma)}\alpha = \omega^{j(\sigma)k(\tau)+k(\sigma)}\alpha$$

and

$$\sigma\tau\omega = \sigma\omega^{j(\tau)} = \omega^{j(\sigma)j(\tau)}$$

There is something reminiscent of matrix multiplication in this. Indeed, let $\mathcal{M}_n$ be the set of all matrices of the form

$$\mathcal{M}_n = \left\{ \begin{bmatrix} 1 & 0 \\ k & j \end{bmatrix} \middle| \, k \in \mathbb{Z}_n, j \in \mathbb{Z}_n^* \right\}$$

Since

$$\begin{bmatrix} 1 & 0 \\ k & j \end{bmatrix} \begin{bmatrix} 1 & 0 \\ k' & j' \end{bmatrix} = \begin{bmatrix} 1 & 0 \\ k + jk' & jj' \end{bmatrix}$$

we see that $\mathcal{M}_n$ is a subgroup of the general linear group $GL_2(\mathbb{Z}_n)$ of all nonsingular 2×2 matrices over $\mathbb{Z}_n$. Comparing this product with the action of the product $\sigma\tau$ shows that the map $\phi: G_F(S) \to \mathcal{M}_n$ defined by

$$\lambda: \sigma \mapsto \begin{bmatrix} 1 & 0 \\ k(\sigma) & j(\sigma) \end{bmatrix}$$

satisfies

$$\lambda(\sigma\tau) = \lambda(\sigma)\lambda(\tau)$$

and is, in fact, a monomorphism from $G_F(S)$ into $\mathcal{M}_n$.

Since $|\mathcal{M}_n| = n\phi(n)$, where ϕ is the Euler phi function, the map λ is surjective if and only if

$$[S : F] = |G_F(S)| = n\phi(n)$$

But in the tower

$$F < F(\omega) < F(\omega, \alpha) = S$$

we always have $[F(\omega) : F] \leq \phi(n)$ and $[F(\omega, \alpha) : F(\omega)] \leq n$. Hence ϕ is surjective (and an isomorphism) if and only if equality holds in these two inequalities.

Theorem 14.2.1 *Let n be a positive integer relatively prime to* expchar(F). *Let S be the splitting field for $x^n - u$ over F, where $0 \neq u \in F$. Let α be a root of $x^n - u$ and $\omega \in \Omega_n$. In the tower*

$$F < F(\omega) < F(\omega, \alpha) = S$$

the first step is a cyclotomic extension and the second step is cyclic of degree $d \mid n$ with $\min(\alpha, F(\omega)) = x^d - \alpha^d$. *Also, $G_F(S)$ is isomorphic to a subgroup of the group $\mathcal{M}_n$ described above, via the embedding*

$$\lambda: \sigma \mapsto \begin{bmatrix} 1 & 0 \\ k(\sigma) & j(\sigma) \end{bmatrix}$$

where $\sigma\alpha = \omega^{k(\sigma)}\alpha$ and $\sigma\omega = \omega^{j(\sigma)}$. The map λ is an isomorphism and $G_F(S) \approx \mathcal{M}_n$ if and only if both steps in the tower (14.2.1) have maximum degree, that is, if and only if

1) $[F(\omega) : F] = \phi(n)$

2) $[F(\omega, \alpha) : F(\omega)] = n$, *or equivalently, $x^n - u$ is irreducible over $F(\omega)$.* $\square$

A Closer Look

There are two issues we would like to address with regard to the previous theorem. First, statement 2) is phrased in terms of $F(\omega)$ and we would prefer a statement involving only the base field F. Second, we would like to find conditions under which $G_F(S)$ is abelian.

We will see that for n an *odd* integer relatively prime to expchar(F), we can replace condition 2) with the condition that $x^n - u$ is irreducible over F. With respect to the commutativity of $G_F(S)$, we will derive a general necessary and sufficient condition. However, we will first prove a simpler result; namely, assuming that $[F(\omega):F] = \phi(n)$, then $G_F(S)$ is abelian if and only if the second step in (14.2.1) is trivial, that is, if and only if $x^n - u$ splits over $F(\omega)$.

The Prime Case

We first deal with both issues for $n = p$ prime. Recall that according to Theorem 14.1.1, the following are equivalent:
1) $u \notin F^p$
2) $f(x) = x^p - u$ has no roots in F
3) $f(x) = x^p - u$ is irreducible over F

Over the base field $F(\omega)$, which contains all the pth roots of unity, we have $[S:F(\omega)] = 1$ or p and the following are equivalent:
1) $[S:F(\omega)] = p$
2) $u \notin F(\omega)^p$
3) $x^p - u$ has no roots in $F(\omega)$
4) $x^p - u$ does not split over $F(\omega)$
5) $x^p - u$ is irreducible over $F(\omega)$

The next lemma ties these two situations together, and strengthens statement 2) of Theorem 14.2.1 for n prime.

Lemma 14.2.2 *Let p be a prime and let $\omega \in \Omega_p$. Then $x^p - u$ is irreducible over $F(\omega)$ if and only if it is irreducible over F.*
Proof. Certainly, if $x^n - u$ is irreducible over $F(\omega)$, it is also irreducible over F. For the converse, consider the tower

$$F < F(\omega) < F(\omega, \alpha)$$

Since $x^p - u$ is irreducible over F, we have

$$p \mid [F(\omega,\alpha):F]$$

On the other hand, the first step in the tower has degree at most $\phi(p) = p - 1$ and the second step is cyclic of degree $d \mid p$, whence $d = 1$ or p. Hence $[F(\omega,\alpha):F(\omega)] = p$, which implies that $x^p - u = \min(\alpha, F(\omega))$ is irreducible over $F(\omega)$. □

As to the question of when the Galois group $G_F(S)$ is abelian in the prime case, since both steps in the tower

$$F < F(\omega) < F(\omega, \alpha) = S$$

are abelian, if either step is trivial, then $G_F(S)$ is abelian. Thus, if $\omega \in F$ or if $\alpha \in F(\omega)$ then $G_F(S)$ is abelian. The converse is also true when n is prime.

Lemma 14.2.3 *Let p be a prime and let $\omega \in \Omega_p$. Let S be a splitting field for $x^p - u$ over F. Then the Galois group $G_F(S)$ is abelian if and only if at least one step in the tower (14.2.1) is trivial, that is, if and only if either $\omega \in F$ or $x^p - u$ is reducible over $F(\omega)$.*

Proof. As mentioned, if one step is trivial then $G_F(S)$ is abelian. Suppose now that $\omega \notin F$ and $x^p - u$ is irreducible over $F(\omega)$. Since $\omega \notin F$, it has a conjugate $\omega^j \neq \omega$ that is also not in F. Let $\tau \in G_F(F(\omega))$ be defined by $\tau\omega = \omega^j$. Since $x^p - u$ is irreducible over $F(\omega)$, for each $i \in \mathbb{Z}_p$, the map τ may be extended to a map $\sigma_i \in G_F(S)$ defined by

$$\sigma_i\omega = \omega^j, \sigma_i\alpha = \omega^i\alpha$$

For $i = 0$ and 1, we have

$$\sigma_1\sigma_0\alpha = \sigma_1\alpha = \omega\alpha$$

and

$$\sigma_0\sigma_1\alpha = \sigma_0(\omega\alpha) = \omega^j\alpha$$

and these are distinct since $\omega \neq \omega^j$. Hence, σ_1 and σ_0 do not commute and $G_F(S)$ is not abelian.□

The General Case

Armed with the previous results for n prime, we consider the general case. We use the following fact.

Suppose that $p(x) \in F[x]$ splits over F and has a nonabelian splitting field extension $F < S$. Then if $F < A$ is abelian, $p(x)$ cannot split in A because otherwise, there would be a splitting field T of $p(x)$ satisfying $F < T < A$. But $F < A$ abelian implies that the lower step $F < T$ is abelian and since all splitting fields for $p(x)$ over F are isomorphic, this contradicts the fact that $F < S$ is nonabelian.

Theorem 14.2.4 *Let n be an odd positive integer relatively prime to* expchar(F). *Let $\omega \in \Omega_n$ and suppose that F contains no nth roots of unity other than* 1. *Let $F < A$ be any abelian extension. Then $x^n - u$ is irreducible over F if and only if it is irreducible over A.*

Proof. Clearly, if $x^n - u$ is irreducible over A, it is also irreducible over the smaller field F. Suppose that $x^n - u$ is irreducible over F. Then for every prime $p \mid n$, the polynomial $x^p - u$ is irreducible over F and therefore also over $F(\omega_p)$, by Lemma 14.2.2. Now, if $x^p - u$ were reducible over A, then it would have a root in A and since $F < A$ is normal, $x^p - u$ would split over A.

But since F does not contain any primitive pth roots of unity, then if α_p is a root of $x^p - u$ in a splitting field, the tower

$$F < F(\omega_p) < F(\omega_p, \alpha_p)$$

has nontrivial steps and so is nonabelian by Theorem 14.2.3. It follows from previous remarks that $x^p - u$ cannot split over the abelian extension A. Hence, $x^p - u$ is irreducible over A for all primes p dividing n and so $x^n - u$ is irreducible over A.□

If $[F(\omega) : F] = \phi(n)$, then F cannot contain any primitive pth roots of unity for any $p \mid n$, and so it cannot contain any nth roots of unity other than 1. Thus, since $F < F(\omega)$ is an abelian extension, we may apply Theorem 14.2.4 to get the following strengthening of Theorem 14.2.1, for n odd.

Corollary 14.2.5 *Referring to Theorem 14.2.1, let n be an odd positive integer relatively prime to* expchar(F)*. Then $G_F(S) \approx \mathcal{M}_n$ if and only if $[F(\omega) : F] = \phi(n)$ and $x^n - u$ is irreducible over F.*□

Since $[\mathbb{Q}(\omega) : \mathbb{Q}] = \phi(n)$, we have the following corollary.

Corollary 14.2.6 *Referring to Theorem 14.2.1, if $F = \mathbb{Q}$ and n is an odd positive integer then $G_{\mathbb{Q}}(S) \approx \mathcal{M}_n$ if and only if $x^n - u$ is irreducible over $\mathbb{Q}$.*□

Thus, when $F < F(\omega)$ has the largest possible degree $\phi(n)$ (which includes the important special case $F = \mathbb{Q}$), we see that $G_F(S) \approx \mathcal{M}_n$ if and only if $x^n - u$ is irreducible over F. We show next that $G_F(S)$ is abelian if and only if $x^n - u$ splits over $F(\omega)$, or equivalently, $x^n - u$ has a root in F.

Note that for any positive integers a and b, we have

$$F(\omega_a, u^{1/a}) < F(\omega_{ab}, u^{1/ab})$$

This follows from the fact that since $(u^{1/ab})^b$ is a root of $x^a - u$ and since $\omega_{ab}^b \in \Omega_a$, all the roots of $x^a - u$ lie in $F(\omega_{ab}, u^{1/ab})$.

Theorem 14.2.7 *Let n be an odd positive integer relatively prime to* expchar(F)*. Let S be the splitting field for $x^n - u$ over F, where $0 \neq u \in F$. Suppose that $[F(\omega) : F] = \phi(n)$ where $\omega \in \Omega_n$. Then the following are equivalent.*

1) $G_F(S)$ is abelian

2) $x^n - u$ *has a root in* F
3) $x^n - u$ *has a root in* $F(\omega)$ *[and therefore splits over* $F(\omega)$*]*
Proof. Clearly, 2) $\Rightarrow$ 3) $\Rightarrow$ 1). We must show that 1) implies 2). Suppose that $G_F(S)$ is abelian and let k be the largest divisor of n for which $u \in F^k$, that is, $u = f^k$ for some $f \in F$. The proof will be complete if we show that $k = n$, since $u = f^n \in F^n$ implies that f is a root of $x^n - u$ in F.

If $k < n$, let p be a prime dividing n/k and consider the tower

$$F < F(\omega_p) < F(\omega_p, f^{1/p}) < F(\omega, u^{1/n}) = S$$

Note that $x^p - f$ is irreducible over F, for if not, then $f = g^p \in F^p$ for some $g \in F$, whence $u = f^k = g^{pk} \in F^{pk}$, in contradiction to the definition of k. Hence $[F(f^{1/p}) : F] = p$ and since $[F(\omega_p) : F] \le p - 1$, we deduce that neither of the first steps is trivial. Hence, Lemma 14.2.3 implies that the Galois group $G_F(F(\omega_p, f^{1/p}))$ is not abelian. But this is a contradiction to 1).□

In the exercises, we ask the reader to provide a simple example to show that Theorems 14.2.4 and 14.2.7 fail to hold when n is even.

More on When $G_F(S)$ Is Abelian

We conclude this section by generalizing the previous theorem, in order to characterize (for n odd), with no restriction on the lower step, precisely when $G_F(S)$ is abelian. The proof follows lines similar to that of Theorem 14.2.7, but is a bit more intricate and since it involves no new insights, the reader may wish to skip it on first reading. However, the result is of interest since it shows how the relationship between the nth roots of unity and the ground field F play a role in the commutativity of $G_F(S)$. We first need a result that is of interest in its own right. The proof is left as an exercise.

Theorem 14.2.8 *Let* $x^n - a$ *and* $x^n - b$ *be irreducible over* F *and suppose that* F *contains a primitive* n*th root of unity. Then* $x^n - a$ *and* $x^n - b$ *have the same splitting field over* F *if and only if* $b = c^n a^r$ *for some* $c \in F$ *and* r *relatively prime to* n. □

Note that if F is a field and U_n is the group of nth roots of unity over F, then $U_n \cap F^*$ is a (cyclic) subgroup of U_n and so is U_m, for some $m \mid n$.

Theorem 14.2.9 *Let* n *be an odd positive integer relatively prime to* $\operatorname{expchar}(F)$*. Let* U_n *be the group of* n*th roots of unity over* F *and let* $U_m = U_n \cap F^*$*. If* S *is the splitting field for* $x^n - u$*, where* $0 \ne u \in F$*, then* $G_F(S)$ *is abelian if and only if* $u^m \in F^n$.
Proof. Since $U_m = \langle \omega_m \rangle$ is cyclic, it follows that $\omega_i \in F$ if and only if $i \mid m$. Suppose first that $u^m = f^n$ for some $f \in F$. Then

$$u^{1/n} = \omega_{mn}^k f^{1/m}$$

for some integer k. (More precisely, given any nth root $u^{1/n}$ of n and any mth root $f^{1/m}$ of f, there exists a k such that this equation holds.) The field $F(f^{1/m})$ is cyclic over F, since the latter contains a primitive mth root of unity ω_m. Therefore, since the extensions $F < F(\omega_{mn})$ and $F < F(f^{1/m})$ are both abelian, so is the extension

$$F < F(\omega_{mn})F(f^{1/m}) = F(\omega_{mn}, f^{1/m}) = F(\omega_{mn}, u^{1/n})$$

Finally, since $F < S < F(\omega_{mn}, u^{1/n})$, it follows that $F < S$ is abelian.

For the converse, assume that $G_F(S)$ is abelian. Let k be the largest positive integer such that $m \mid k$, $k \mid n$ and $u^m \in F^k$, say $u^m = f^k$ for $f \in F$. We need to show that $k = n$. Suppose to the contrary that $k < n$ and let p be a prime number dividing n/k. Let p^s be the largest power of p such that $p^s \mid m$. (As an aside, the hypothesis that n is odd and $[F(\omega) : F] = \phi(n)$ in Theorem 14.2.7 implies that $m = 1$, whence $s = 0$.)

The first step is to show that the extension

$$F < F(\omega_{p^{s+1}}, f^{1/p^{s+1}})$$

is abelian. It is clear that the notation is a bit unwieldy, so let us set $q = p^{s+1}$ and note that $q \mid n$ since $p^s \mid m \mid k$ and $p \mid (n/k)$. To see that this extension is abelian, we embed it in an abelian extension. Since

$$(f^{1/q})^{kq} = f^k = u^m = (u^{m/kq})^{kq}$$

we have $f^{1/q} = \omega_{kq}^j u^{m/kq}$ for some j and so

$$F(\omega_q, f^{1/q}) < F(\omega_{kq}, f^{1/q}) = F(\omega_{kq}, u^{m/kq})$$

Now, since $p \mid (n/k)$, there is a positive integer a for which $apk = n$, and since $p^s \mid m$, it follows that

$$\frac{mn}{kq} = \frac{mn}{kp^{s+1}} = \frac{mapk}{kp^{s+1}} = \frac{ma}{p^s}$$

is a positive integer. Hence,

$$v = (u^{1/n})^{nm/kq}$$

is a root of $x^{kq} - u^m$ that lies in $F(\omega_{kq}, u^{1/n})$. Hence, all roots of $x^{kq} - u^m$ are contained in $F(\omega_{kq}, u^{1/n})$, that is,

$$F(\omega_{kq}, u^{m/kq}) < F(\omega_{kq}, u^{1/n})$$

Putting the pieces together gives

$$F < F(\omega_q, f^{1/q}) < F(\omega_{kq}, u^{m/kq}) < F(\omega_{kq}, u^{1/n}) < F(\omega_{kq})F(\omega_n, u^{1/n})$$

Since $F < F(\omega_{kq})$ and $F < F(\omega_n, u^{1/n})$ are abelian (the latter by assumption), the composite

$$F < F(\omega_{kq})F(\omega_n, u^{1/n})$$

is abelian and therefore so is

$$F < F(\omega_q, f^{1/q})$$

We now propose to arrive at a contradiction by considering the tower

$$F < F(\omega_q) < F(\omega_q, f^{1/q})$$

Note that $x^p - f$ is irreducible over F, since otherwise $f = g^p \in F^p$ for some $g \in F$, whence $u^m = f^k = g^{pk} \in F^{pk}$, in contradiction to the definition of k.

We first take the case $s = 0$, whence $q = p$. Since $x^p - f$ is irreducible over F, we have

$$[F(\omega_p, f^{1/p}) : F] \geq p$$

Since $p \not| m$, it follows that $\omega_p \notin F$ and so the lower step $F < F(\omega_p)$ is not trivial. However, since $[F(\omega_p) : F] \leq p - 1 < [F(\omega_p, f^{1/p}) : F]$, the upper step in the tower is also not trivial. Hence, Lemma 14.2.3 implies that the Galois group $H = G_F(F(\omega_p, f^{1/p}))$ is not abelian, the desired contradiction.

Now assume that $s > 0$. With regard to the first step in the tower, since p and $r = p^s$ both divide m but $q = p^{s+1}$ does not, it follows that $\omega_p, \omega_r \in F$ and $\omega_q \notin F$. Since $\omega_r \in F$, the binomial $x^p - \omega_r$ is either irreducible over F or splits over F. But ω_q is a root of this binomial that is not in F and so $x^p - \omega_r$ is irreducible over F.

Since the roots of $x^p - \omega_r$ are

$$\omega_q, \omega_r\omega_q, \ldots, \omega_r^{p-1}\omega_q$$

for each $j \in \mathbb{Z}_p$, there is a $\sigma_j \in G_F(F(\omega_q))$ for which $\sigma_j\omega_q = \omega_r^j\omega_q$. To show that $G_F(F(\omega_q, f^{1/q}))$ is not abelian, we shall need only $\sigma_0 = \iota$ (the identity) and $\sigma_1 \colon \omega_q \mapsto \omega_r\omega_q$.

There are two possibilities for the second step in the tower. If $x^q - f \in F[x]$ is irreducible over $F(\omega_q)$ then we can extend σ_0 and σ_1 to elements of $G_F(F(\omega_q, f^{1/q}))$ by defining

$$\sigma_{0,1} \colon \omega_q \mapsto \omega_q, \sigma_{0,1} \colon f^{1/q} \mapsto \omega_q f^{1/q}$$

and

$$\sigma_{1,0}\colon \omega_q \mapsto \omega_r\omega_q,\ \sigma_{1,0}\colon f^{1/q} \mapsto f^{1/q}$$

Then

$$\sigma_{0,1}\sigma_{1,0}f^{1/q} = \sigma_{0,1}f^{1/q} = \omega_q f^{1/q}$$

and

$$\sigma_{1,0}\sigma_{0,1}f^{1/q} = \sigma_{1,0}(\omega_q f^{1/q}) = \omega_r\omega_q f^{1/q}$$

which are distinct since $\omega_r \neq 1$. Hence, $G_F(F(\omega_q, f^{1/q}))$ is not abelian, a contradiction.

If $x^p - f$ is reducible over $F(\omega_q)$ then $f \in F(\omega_q)^p$. Thus $f = \beta^p$ for some $\beta \in F(\omega_q)$ and so $F(\beta) < F(\omega_q)$. Since $x^p - \omega_r$ and $x^p - f$ are both irreducible over F, it follows that $[F(\omega_q)\colon F] = p$ and $[F(\beta) : F] = p$, whence $F(\omega_q) = F(\beta)$. Thus, $x^p - f$ and $x^p - \omega_r$ have the same splitting field over F and Theorem 14.2.8 implies that

$$f = \omega_r^j v^p$$

for some $v \in F$. Taking kth powers gives, since $r \mid k$,

$$u^m = f^k = \omega_r^{kj} v^{kp} = v^{kp}$$

for $v \in F$, which contradicts the definition of k. Thus, $k = n$ and the theorem is proved. □

*14.3 The Independence of Irrational Numbers

A familiar argument (at least for $p = 2$) shows that if p is a prime number then $\sqrt{p} \notin \mathbb{Q}$ and so $[\mathbb{Q}(\sqrt{p}) : \mathbb{Q}] = 2$. Our plan in this section is to extend this result to more than one prime p and to nth roots for $n \geq 2$. Since the case in which n is even involves some rather intricate details that give no further insight into the issues involved, we will confine our attention to n odd. (The case $n = 2$ is straightforward and we invite the reader to supply a proof of Theorem 14.3.2 for this case.) If $\alpha > 0$ is rational, the notations $\sqrt[n]{\alpha}$ and $\alpha^{1/n}$ will denote the *real positive* nth root of α. The results of this section were first proved by Bescovitch [1940], but the method of proof we employ follows more closely that of Richards [1974].

Lemma 14.3.1 Let $u = a/b$ be a positive rational number, expressed in lowest terms, that is, where $(a, b) = 1$ and $a, b > 0$. If $n \geq 2$ is an integer then

$$\sqrt[n]{\frac{a}{b}} \in \mathbb{Q} \text{ if and only if } a = c^n \text{ and } b = d^n \text{ for positive integers } c \text{ and } d$$

In particular, if p is a prime, then $\sqrt[n]{p} \notin \mathbb{Q}$.

Proof. One direction is quite obvious. Suppose that

$$\left(\frac{c}{d}\right)^n = \frac{a}{b}$$

where c and d are positive integers and $(c,d) = 1$. Then $ad^n = bc^n$ and since $(a,b) = 1$, it follows that $a \mid c^n$, say $c^n = ax$, and $b \mid d^n$, say $d^n = by$. Hence, $ax/by = a/b$, which implies that $x = y$. It follows that $x \mid c^n$ and $x \mid d^n$, whence $x = y = 1$ and so $a = c^n$ and $b = d^n$.□

Suppose now that n is odd and p is prime. Since $p \notin \mathbb{Q}^r$ for any prime $r \mid n$, Theorem 14.1.4 (or Eisenstein's criterion) implies that $x^n - p$ is irreducible over $\mathbb{Q}$ and so $[\mathbb{Q}(p^{1/n}) : \mathbb{Q}] = n$. Let us generalize this to more than one prime.

Theorem 14.3.2 Let $n \geq 2$ be an integer and let $p_1, \ldots, p_m$ be distinct primes. Then

$$[\mathbb{Q}(\sqrt[n]{p_1}, \ldots, \sqrt[n]{p_m}) : \mathbb{Q}] = n^m$$

Proof. As mentioned earlier, we confine our proof to the case that $n \geq 3$ is odd. Let $\omega \in \Omega_n$. Since

$$[\mathbb{Q}(\omega)(\sqrt[n]{p_1}, \ldots, \sqrt[n]{p_m}) : \mathbb{Q}(\omega)] \leq [\mathbb{Q}(\sqrt[n]{p_1}, \ldots, \sqrt[n]{p_m}) : \mathbb{Q}] \leq n^m$$

it is sufficient to show that

$$[\mathbb{Q}(\omega)(\sqrt[n]{p_1}, \ldots, \sqrt[n]{p_m}) : \mathbb{Q}(\omega)] = n^m$$

which we shall do by induction on m.

Let p be a prime. Since $x^n - p$ is irreducible over $\mathbb{Q}$ and $\mathbb{Q}$ contains no nth roots of unity other than 1, Theorem 14.2.4 implies that $x^n - p$ is also irreducible over $\mathbb{Q}(\omega)$. Hence,

$$[\mathbb{Q}(\omega, \sqrt[n]{p}) : \mathbb{Q}(\omega)] = n$$

and the theorem holds for $m = 1$.

Now let us suppose that the theorem is true for the integer m and let p be a prime distinct from the distinct primes $p_1, \ldots, p_m$. Let

$$F = \mathbb{Q}(\omega) \text{ and } E = \mathbb{Q}(\omega)(\sqrt[n]{p_1}, \ldots, \sqrt[n]{p_m})$$

If $x^n - p$ is not irreducible over E then there exists a prime $r \mid n$ such that $p^{1/r} \in E$. Thus, $p^{1/r}$ is a linear combination, over $\mathbb{Q}(\omega)$, of terms of the form

$$\sqrt[n]{p_1^{e(1)}}, \ldots, \sqrt[n]{p_m^{e(m)}}$$

where $0 \leq e(i) \leq n-1$. There are two cases to consider.

Case 1: If the linear combination involves only one term, then

$$\sqrt[r]{p} = \sqrt[n]{p_1^{e(1)}} \sqrt[n]{p_2^{e(2)}} \cdots \sqrt[n]{p_m^{e(m)}}$$

where $c \in \mathbb{Q}(\omega)$ and not all $e(i)$ are 0. If $n = rd$, this can be written in the form

$$\sqrt{\frac{p^d}{p_1^{e(1)} \cdots p_m^{e(m)}}} \in \mathbb{Q}(\omega)$$

This says that the radicand; call it q, is a positive rational number and the polynomial $x^n - q$ has a root in $\mathbb{Q}(\omega)$. According to Theorem 14.2.7, $x^n - q$ must also have a root in $\mathbb{Q}$, which is not possible since q does not have the form a^n/b^n, for relatively prime integers a, b. Hence, this case cannot occur.

Case 2: At least two terms in the linear combination are nonzero. It follows that one of the primes p_i, which we may assume for convenience is p_m, appears to different powers in at least two distinct terms. Collecting terms that involve like powers of p_m gives

$$p^{1/r} = A_0 + A_1 p_m^{1/n} + A_2 p_m^{2/n} + \cdots + A_{n-1} p_m^{(n-1)/n} \tag{14.3.1}$$

where $A_i \in \mathbb{Q}(\omega)(\sqrt[n]{p_1}, \ldots, \sqrt[n]{p_m})$ and where at least two of the A_i's are nonzero. Now, since

$$\mathbb{Q}(\omega) < \mathbb{Q}(\omega)(\sqrt[n]{p_1}, \ldots, \sqrt[n]{p_m})$$

is a Galois extension (this is why we adjoined ω in the first place), the inductive hypothesis implies that its Galois group G has size n^m. Since any $\sigma \in G$ must send roots of $x^n - p_i$ to other roots, it must send $p_i^{1/n}$ to $\omega^{j_i} p_i^{1/n}$ for some choice of $j_i \in \{0, \ldots, n-1\}$. Since there are n^m such choices, all these choices must occur.

Thus, there is a $\sigma \in G$ for which σ is the identity on $\mathbb{Q}(\omega)(\sqrt[n]{p_1}, \ldots, \sqrt[n]{p_m})$ and

$$\sigma p_m^{1/n} = \omega p_m^{1/n}$$

Since $\sigma p^r = \omega^k p^r$ for some $0 \leq k \leq n-1$, applying σ to (14.3.1) gives

$$\omega^k p^r = A_0 + A_1 \omega p_m^{1/n} + A_2 \omega^2 p_m^{2/n} + \cdots + A_{n-1} \omega^{n-1} p_m^{(n-1)/n}$$

We now multiply (14.3.1) by ω^k and subtract the previous equation to get

$$0=(\omega^k-1)A_0+(\omega^k-\omega)A_1p_m^{1/n}+\cdots+(\omega^k-\omega^{n-1})p_m^{(n-1)/n}$$

where at least one of the coefficients $(\omega^k-\omega^i)A_i$ is nonzero. This is a contradiction to the inductive hypothesis. We have therefore established that x^n-p is irreducible over E and the proof is complete. □

Exercises

1. Let n be relatively prime to char(F). Show that the group

$$\mathcal{M}_n=\left\{\begin{bmatrix}1&0\\k&j\end{bmatrix}\;\middle|\;k\in\mathbb{Z}_n, j\in\mathbb{Z}_n^*\right\}$$

is generated by two elements σ and τ, where $o(\sigma)=n$, $o(\tau)=\phi(n)$ and $\sigma\tau\sigma^{-1}=\tau^r$. What is r?
2. (Van der Waerden) Let n be relatively prime to char(F). Show that the Galois group of x^n-u is isomorphic to a subgroup of the group G of linear substitutions modulo n, that is, maps on $\mathbb{Z}_n$ of the form $x\to cx+d$ where $d\in\mathbb{Z}_n$, $c\in\mathbb{Z}_n^*$.
3. Let $x^n-u\in GF(q)[x]$. Show that the following are equivalent:
   a) $r\mid n$, r prime implies $u\notin GF(q)^r$
   b) $r\mid n$, r prime implies $r\mid o(u)$ but $r\nmid (q-1)/o(u)$ where $o(u)$ is the multiplicative order of u in $GF(q)$.
4. Prove the following without using any of the results of Section 14.1. If $u\in F$ and $(m,n)=1$ then $x^{mn}-u$ is irreducible over F if and only if x^m-u and x^n-u are irreducible over F.
5. Let char$(F)=p\neq 0$ and let $F<E$ be cyclic of degree p^k, with Galois group $G=\langle\sigma\rangle$. If there exists a $\beta\in E$ with $\mathrm{Tr}_{E/F}(\beta)=1$ show that there exists an $\alpha\in E$ for which the polynomial $f(x)=x^p-x-\alpha$ is irreducible over E.
6. Let char$(F)=p>0$ and let $n=p^e m$ where $(m,p)=1$. Show that the Galois groups of

$$x^n-u \quad\text{and}\quad x^m-u^{p^{-e}}$$

are the same.
7. Let n be a positive integer relatively prime to expchar(F) and let ω be a primitive nth root of unity over F. Let $S=F(\omega,u^{1/n})$ be the splitting field for $f(x)=x^n-u$ over F, where $u\in F$, $u\neq 0$. If $4\mid n$ and if $u^{1/2}\notin F$ then $G_F(S)$ is not abelian.
8. Show that Theorem 14.2.4 and Theorem 14.2.7 fail to hold when n is even. *Hint*: $\sqrt{2}\in\mathbb{Q}(\omega)$, where ω is a primitive 8th root of unity.
9. Prove the following: Let $f(x)$ be a monic irreducible polynomial of degree m over F, with constant term $-a_0$. Let $n\geq 2$ be an integer for which $(m,n)=1$, $4\nmid n$ and $a_0\notin F^r$, for all primes $r\mid n$. Then the polynomial $f(x^n)$ is also irreducible over F.

10. Let ω be a primitive nth root of unity over F, n odd, and let α be a root of $x^n - u$ over F. Then $S = F(\omega, \alpha)$ is the splitting field for $x^n - u$. Assume that $G_F(S) \approx \mathcal{M}_n$. In this exercise, we determine the largest abelian subextension F^{ab} of S.
    a) If G is a group, the subgroup G' generated by all **commutators** $\alpha\beta\alpha^{-1}\beta^{-1}$, for $\alpha, \beta \in G$, is called the **commutator subgroup**. Show that G' is the smallest subgroup of G for which G/G' is abelian.
    b) If the commutator subgroup $G_F(E)'$ of a Galois group $G_F(E)$ is closed, that is, if $G_F(E)' = G_K(E)$ for some $F < K < E$, then K is the largest abelian extension of F contained in E.
    c) The commutator subgroup of $\mathcal{M}_n$ is

$$\mathcal{M}'_n = \left\{ \begin{bmatrix} 1 & 0 \\ k & 1 \end{bmatrix} : k \in \mathbb{Z}_n \right\}$$

    and if λ is defined as in Theorem 14.2.1, then

$$\lambda(G_{F(\omega)}(S)) = \lambda(G_F(S)) \cap \mathcal{M}'_n = \left\{ \begin{bmatrix} 1 & 0 \\ i\frac{n}{d} & 1 \end{bmatrix} : i \in \mathbb{Z}_d \right\}$$

    where $d = [F(\omega, \alpha) : F(\omega)]$.
    d) Prove that $G_F(S)' = G_{F(\omega)}(S)$, and so $F(\omega)$ is the largest abelian extension of F contained in $F(\omega, \alpha)$.
11. Prove that if $p_1, \ldots, p_m$ are distinct primes then

$$[\mathbb{Q}(\sqrt{p_1}, \ldots, \sqrt{p_m}) : \mathbb{Q}] = 2^m$$

    by induction on m.
12. Show that $\sqrt{60} \notin \mathbb{Q}(\sqrt{42}, \sqrt{10})$.
13. Let $x^n - a$ and $x^n - b$ be irreducible over F and suppose that F contains a primitive nth root of unity. Then $x^n - a$ and $x^n - b$ have the same splitting field over F if and only if $b = c^n a^r$ for some $c \in F$ and r relatively prime to n. *Hint*: if the splitting fields are the same, consider how the common Galois group acts on a root of each binomial.
14. Let $F < E$ be a finite Galois extension and let $\alpha, \beta \in E$ have degrees m and n over F, respectively. Suppose that $[F(\alpha, \beta) : F] = mn$.
    a) Show that if α_i is a conjugate of α and β_j is a conjugate of β, then there is a $\sigma \in G_F(E)$ such that $\sigma\alpha = \alpha_i$ and $\sigma\beta = \beta_j$. Hence, the conjugates of $\alpha + \beta$ are $\alpha_i + \beta_j$.
    b) Show that if the difference of two conjugates of α is never equal to the difference of two conjugates of β then $F(\alpha, \beta) = F(\alpha + \beta)$.
    c) Let r be a prime different from $\mathrm{char}(F)$. Let $f(x) = x^r - u$ and $g(x) = x^r - v$ be irreducible over F, with roots α and β, respectively. Show that if $[F(\alpha, \beta) : F] = r^2$ then $F(\alpha, \beta) = F(\alpha + \beta)$.

Chapter 15
Families of Binomials

In this chapter, we look briefly at families of binomials and their splitting fields and Galois groups. We have seen that when the base field F contains a primitive nth root of unity, cyclic extensions of degree $d \mid n$ correspond to splitting fields of a single binomial $x^n - u$. More generally, we will see that abelian extensions correspond to splitting fields of families of binomials. We will also address the issue of when two families of binomials have the same splitting field.

15.1 The Splitting Field

Let F be a field containing a primitive nth root of unity and consider a family $\mathcal{F}$ of binomials given by

$$\mathcal{F} = \{x^n - u \mid u \in U\}$$

where $U \subseteq F$ is the set of constant terms. We will refer to n as the **exponent** of the family $\mathcal{F}$.

If S_u is the splitting field for $x^n - u$, then $S = \bigvee\{S_u \mid u \in U\}$ is the splitting field for the family $\mathcal{F}$. Since each extension $F < S_u$ is Galois, so is $F < S$ and Theorem 6.5.4 implies that $G_F(S)$ is isomorphic to a subgroup of the product

$$H = \prod_{u \in U} G_F(S_u)$$

Since each $F < S_u$ is cyclic of degree dividing n, the group H is the direct product of cyclic groups of order dividing n and so $G_F(S)$ is abelian with exponent n. An abelian extension $F < S$ whose Galois group $G_F(S)$ has exponent n will be referred to as an abelian extension with **exponent** n.

Thus, if F contains a primitive nth root of unity, the splitting field of any family of binomials over F of exponent n is an abelian extension of F with exponent n. Happily, the converse is also true.

Suppose that $F < E$ is an abelian extension with exponent n. Let K be any field for which $F < K < E$ where $F < K$ is finite. Since $F < E$ is abelian, so is $F < K$. In addition, $G_F(K)$ is finite and has exponent n. Since a finite abelian group is a direct product of cyclic subgroups, we have

$$G_F(K) \approx G_1 \times \cdots \times G_m$$

where each G_i is cyclic with exponent n and hence order $n_i \mid n$. Corollary 6.5.5 implies that K is a composite $K = K_1 \cdots K_m$ where $G_F(K_i) \approx G_i$ is cyclic of order $n_i \mid n$. Since F contains the n_ith roots of unity and $F < K_i$ is cyclic, Theorem 12.1.1 implies that $K_i = F(\alpha_i)$ is the splitting field for

$$\min(\alpha_i, F) = x^{n_i} - \alpha_i^{n_i}$$

where $\alpha_i \in E$. Hence $K = F(\alpha_1, \ldots, \alpha_m)$ is the splitting field over F for the family

$$\mathcal{F}_K = \{x^{n_i} - \alpha_i \mid i = 1, \ldots, m\}$$

It follows that E is the splitting field for the union $\bigcup \mathcal{F}_K$, taken over all finite intermediate fields K.

Theorem 15.1.1 *Let F be a field containing a primitive nth root of unity. An extension $F < E$ is abelian with exponent n if and only if E is the splitting field for a family of binomials over F of exponent n.* □

Definition *Let F be a field containing a primitive nth root of unity. An extension $F < E$ is a* **Kummer extension** *of exponent n if $F < E$ is abelian and has exponent n.* □

Thus, according to Theorem 15.1.1, the Kummer extensions of F of exponent n are precisely the splitting fields over F of families of binomials of exponent n.

15.2 Dual Groups and Pairings

Before proceeding, we need a few concepts from group theory. If A and B are groups, we denote by $\hom(A, B)$ the set of all group homomorphisms from A to B. Note that $\hom(A, B)$ is a group under the product

$$(\lambda\theta)(\alpha) = (\lambda\alpha)(\theta\alpha)$$

with identity being the constant map $\lambda\alpha = 1$ for all $\alpha \in A$.

Lemma 15.2.1

1) *If A, B and C are abelian groups, then*

$$\hom(A \times B, C) \approx \hom(A, C) \times \hom(B, C)$$

2) *Let U_n be the group of all nth roots of unity over a field F. If A is a finite abelian group of exponent n, then*

$$\hom(A, U_n) \approx A$$

Proof. We leave it as an exercise to show that the map

$$\mathcal{P}: \hom(A, C) \times \hom(B, C) \to \hom(A \times B, C)$$

defined by

$$\mathcal{P}(\lambda, \theta)(\alpha, \beta) = \lambda(\alpha)\theta(\beta)$$

is an isomorphism, proving part 1). For part 2), since A can be written as the product of finite cyclic groups, part 1) implies that we need only show that $\hom(A, U_n) \approx A$ when $A = \langle \alpha \rangle$ is cyclic. Suppose that A has order $m \mid n$. If A has order $m \mid n$, then $\lambda \in \hom(A, U_n)$ maps A into U_m, since for any $\alpha \in A$, we have

$$(\lambda\alpha)^m = \lambda(\alpha^m) = 1$$

Hence, $\hom(A, U_n) = \hom(A, U_m)$. Suppose that $U_m = \langle \omega \rangle$ and define $\lambda \in \hom(A, U_m)$ by setting $\lambda(\alpha) = \omega$, which is easily seen to define a group homomorphism. Then

$$\langle \lambda \rangle = \{1, \lambda, \lambda^2, \ldots, \lambda^{m-1}\}$$

is a cyclic subgroup of $\hom(A, U_m)$ of order $m = |U_m|$. But $\hom(A, U_m)$ has size at most m and so $\hom(A, U_n) = \langle \lambda \rangle$ is cyclic of order m, whence $\hom(A, U_n) \approx A$. □

Definition *If A, B and C are abelian groups, a* **pairing** *of $A \times B$ into C is a map $\langle , \rangle: A \times B \to C$ that is a* **bihomomorphism**, *that is,*

1) *For each $\alpha \in A$, the map $\lambda_\alpha: B \to C$ defined by $\lambda_\alpha(\beta) = \langle \alpha, \beta \rangle$ is a group homomorphism.*
2) *For each $\beta \in B$, the map $\theta_\beta: A \to C$ defined by $\theta_\beta(\alpha) = \langle \alpha, \beta \rangle$ is a group homomorphism.* □

A pairing is the analogue of a bilinear map between vector spaces (and is sometimes referred to as a bilinear map). Note that $\langle 1, \beta \rangle = \langle \alpha, 1 \rangle = 1$ for all $\alpha \in A$ and $\beta \in B$ and $\langle \alpha, \beta \rangle^{-1} = \langle \alpha^{-1}, \beta \rangle = \langle \alpha, \beta^{-1} \rangle$. If $S \subseteq A$ and $T \subseteq B$, we set

$$\langle S, T \rangle = \{\langle s, t \rangle \mid s \in S, t \in T\}$$

(We will write $\langle \{\alpha\}, T \rangle$ as $\langle \alpha, T \rangle$ and $\langle S, \{\beta\} \rangle$ as $\langle S, \beta \rangle$.) The **left kernel** of a pairing is the set

$$K_L = \{\alpha \in A \mid \langle \alpha, B \rangle = \{1\}\}$$

and the **right kernel** is defined similarly:

$$K_R = \{\beta \in B \mid \langle A, \beta \rangle = \{1\}\}$$

It is easy to see that these kernels are normal subgroups of their respective parent groups.

Note that $\langle \alpha_1, \beta \rangle = \langle \alpha_2, \beta \rangle$ for all $\beta \in B$ if and only if $\langle \alpha_1 \alpha_2^{-1}, B \rangle = \{1\}$, that is, if and only if $\alpha_1 \alpha_2^{-1} \in K_L$, or equivalently, $\alpha_1 K_L = \alpha_2 K_L$. Similar statements hold for the right kernel. Thus, we may define a pairing from $A/K_L \times B/K_R$ to C by

$$\langle \alpha K_L, \beta K_R \rangle = \langle \alpha, \beta \rangle$$

and this pairing is **nonsingular**, that is, both the left and right kernels are trivial.

Theorem 15.2.2 *Let $\langle , \rangle : A \times B \to U_n$ be a nonsingular pairing from abelian groups A and B into U_n, the group of nth roots of unity over a field F. Then A and B both have exponent n and*

1) A is isomorphic to a subgroup of $\hom(B, U_n)$

2) B is isomorphic to a subgroup of $\hom(A, U_n)$

Moreover, A is finite if and only if B is finite, in which case

3) $A \approx \hom(B, U_n)$ and $B \approx \hom(A, U_n)$

4) $A \approx B$, in particular, $|A| = |B|$.

Proof. First observe that if $\alpha \in A$, then $\langle \alpha^n, \beta \rangle = \langle \alpha, \beta \rangle^n = 1$ for all $\beta \in B$, and so $\alpha^n \in K_L$, whence $\alpha^n = 1$ and A has exponent n. A similar statement holds for B. Now consider the map $A \to \hom(B, U_n)$ defined by $\alpha \mapsto \lambda_\alpha$, where $\lambda_\alpha : \beta \mapsto \langle \alpha, \beta \rangle$. Since

$$\lambda_{\alpha\alpha'}(\beta) = \langle \alpha\alpha', \beta \rangle = \langle \alpha, \beta \rangle \langle \alpha', \beta \rangle = \lambda_\alpha(\beta) \lambda_{\alpha'}(\beta)$$

the map $\alpha \mapsto \lambda_\alpha$ is a group homomorphism from A to $\hom(B, U_n)$. If $\lambda_\alpha = 1$ is the constant homomorphism then $\langle \alpha, \beta \rangle = 1$ for all $\beta \in B$, that is, $\alpha \in K_R$, whence $\alpha = 1$. Hence, the map $\alpha \mapsto \lambda_\alpha$ is injective and 1) holds. Similarly, 2) holds.

It follows from Lemma 15.2.1 that if B is finite, then

$$|A| \le |\hom(B, U_n)| = |B|$$

The dual argument shows that $|B| \le |A|$ and so $|A| = |B|$. This also implies that the monomorphism $\alpha \mapsto \lambda_\alpha$ is an isomorphism. □

We can now return to binomials.

15.3 Kummer Theory

While each family of binomials gives rise to a unique Kummer extension, different families may produce the same extension, that is, different families may have the same splitting field. We seek a collection of families of binomials such that there is a one-to-one correspondence between families in the collection and Kummer extensions.

Let us phrase the problem a little differently, for which we require some notation. Recall that if $u \in F$, then by $u^{1/n}$ we mean a particular (fixed) root of $x^n - u$. If $A \subseteq F$, we let $A^{1/n}$ denote the set of *all* nth roots of all elements of A. Also, if $A \subseteq F$ and n is a nonnegative integer then $A^n = \{a^n \mid a \in A\}$.

Let F be a field containing a primitive nth root of unity. Of course, we may identify a family $\mathcal{F} = \{x^n - b \mid b \in U\}$ of binomials of a fixed exponent n with the set $U \subseteq F^*$ of constant terms. (Since binomials with zero constant term are not very interesting, we exclude such binomials.) Moreover, the splitting field for $\mathcal{F}$ is $S = F(U^{1/n})$.

In seeking a bijective correspondence between sets $U \subseteq F^*$ of constant terms (that is, families of binomials) and splitting fields $S = F(U^{1/n})$, it is natural to restrict attention to maximal sets $U \subseteq F^*$ that generate the given splitting field. As we now show, if $S = F(U^{1/n})$ for some $U \subseteq F^*$, then

$$S = F(U')$$

where $U' = \langle U, (F^*)^n \rangle$ is the multiplicative subgroup of F^* generated by U and the nth powers $(F^*)^n$ of elements of F^*. To see this, note that any element of $\langle U, (F^*)^n \rangle$ has the form $\alpha = f^n u_1^{e_1} \cdots u_k^{e_k}$ for $u_1, \ldots, u_k \in U$ and $f \in F^*$. The nth roots of α have the form

$$\omega_n^j f (u_1^{1/n})^{e_1} \cdots (u_k^{1/n})^{e_k}$$

and since each of the factors in this product is in $F(U^{1/n})$, so is the product. Hence, nothing new is added to the splitting field by increasing the set of constants to $\langle U, (F^*)^n \rangle$, that is,

$$F([U']^{1/n}) = F(U^{1/n})$$

Thus, for sets of constant terms, we may restrict attention to the lattice $\mathcal{U}_n$ of all intermediate subgroups U satisfying

$$(F^*)^n < U < F^*$$

Indeed, we will show that the association $U \mapsto F(U^{1/n})$ is a bijection from $\mathcal{U}_n$ onto the class $\mathcal{K}_n$ of all Kummer extensions E of F with exponent n. We will also obtain a description of the Galois group G of $F < E$ in terms of U.

For $U \in \mathcal{U}_n$, let $F < F(U^{1/n}) = S$ be a Kummer extension with Galois group $G = G_F(S)$, and let $\sigma \in G$ and $u \in U$. If α is a root of $x^n - u$ then $\sigma\alpha$ is also a root of $x^n - u$ and so

$$\sigma\alpha = \omega_{\sigma,\alpha}\alpha$$

for some nth root of unity $\omega_{\sigma,\alpha}$. We claim that $\omega_{\sigma,\alpha}$ does not depend on α, that is, σ is simply multiplication by an nth root of unity. To see this, if β is another root of $x^n - u$, then $\beta = \omega^i \alpha$ where $\omega \in \Omega_n$ and so

$$\omega_{\sigma,\beta}\beta = \sigma\beta = \omega^i \sigma\alpha = \omega^i \omega_{\sigma,\alpha}\alpha = \omega_{\sigma,\alpha}\beta$$

and so $\omega_{\sigma,\beta} = \omega_{\sigma,\alpha}$; that is, $\omega_\sigma = \omega_{\sigma,\alpha}$ depends only on σ.

It follows that the map $\langle , \rangle : G \times U \to U_n$ defined by

$$\langle \sigma, u \rangle = \frac{\sigma\alpha}{\alpha} = \omega_\sigma, \text{ for any } \alpha \text{ with } \alpha^n = u$$

is well-defined (does not depend on α) and we may write

$$\langle \sigma, u \rangle = \frac{\sigma u^{1/n}}{u^{1/n}} \tag{15.2.1}$$

without ambiguity. Moreover, the map $\langle , \rangle$ is a pairing of $G \times U$ into U_n, that is, a group bihomomorphism. Specifically, we have $\langle \iota, u \rangle = 1$ and $\langle \sigma, 1 \rangle = 1$. Also,

$$\langle \sigma\tau, u \rangle = \frac{\sigma\tau\alpha}{\alpha} = \frac{\omega_\tau \omega_\sigma \alpha}{\alpha} = \omega_\tau \omega_\sigma = \frac{\sigma\alpha}{\alpha}\frac{\tau\alpha}{\alpha} = \langle \sigma, u \rangle \langle \tau, u \rangle$$

and

$$\langle \sigma, uv \rangle = \frac{\sigma(\alpha\beta)}{\alpha\beta} = \frac{\sigma\alpha}{\alpha}\frac{\sigma\beta}{\beta} = \langle \sigma, u \rangle \langle \sigma, v \rangle$$

The left kernel of this pairing is

$$K_L = \{\sigma \in G \mid \sigma u^{1/n} = u^{1/n} \text{ for all } u \in U\} = G_S(S) = \{\iota\}$$

Also, since we are assuming that $(F^*)^n < U$,

$$\begin{aligned} K_R &= \{u \in U \mid \sigma u^{1/n} = u^{1/n} \text{ for all } \sigma \in G\} \\ &= \{u \in U \mid u^{1/n} \in \mathrm{fix}(G)\} \\ &= U \cap \mathrm{fix}(G)^n \\ &= U \cap (F^*)^n \\ &= (F^*)^n \end{aligned}$$

It follows that the pairing $\langle , \rangle : G \times (U/(F^*)^n) \to U_n$ given by

$$\langle \sigma, u(F^*)^n \rangle = \frac{\sigma u^{1/n}}{u^{1/n}}$$

is nonsingular. We may thus apply Theorem 15.2.2.

Theorem 15.3.1 *Let F be a field containing a primitive nth root of unity and let U be a subset of an extension K of F. If $E = F(U^{1/n})$, where $(F^*)^n \subseteq U$, then* *the pairing*

$$\langle , \rangle : G_F(E) \times U/(F^*)^n \to U_n$$

given by

$$\langle \sigma, u(F^*)^n \rangle = \frac{\sigma u^{1/n}}{u^{1/n}}$$

is nonsingular. Also,

1) $G_F(E)$ *and* $U/(F^*)^n$ *have exponent* n
2) $|G_F(E)| = [E : F]$ *is finite if and only if* $(U : (F^*)^n)$ *is finite, in which case*

$$[E : F] = (U : (F^*)^n)$$

and

$$G_F(E) \approx \hom(U/(F^*)^n, U_n)$$

□

The previous theorem not only describes the Galois group of a Kummer extension, but allows us to show that the map $U \mapsto F(U^{1/n})$, from $\mathcal{U}_n$ to $\mathcal{K}_n$, is a bijection.

Theorem 15.3.2 *Let* F *be a field containing a primitive nth root of unity. Let* $\mathcal{K}_n$ *be the class of all Kummer extensions* $F < E$ *with exponent* n *and let* $\mathcal{U}_n$ *be the class of all subgroups of* F^* *containing* $(F^*)^n$. *Then the map* $U \mapsto F(U^{1/n})$ *is a bijection from* $\mathcal{U}_n$ *onto* $\mathcal{K}_n$, *with inverse given by* $E \mapsto E^{*n} \cap F^*$.

Proof. To show that the map in question is injective, suppose that $F(U^{1/n}) = F(V^{1/n})$, with $U, V \in \mathcal{U}_n$. If $u \in U$, then $u^{1/n} \in F(V^{1/n})$ and so there exists a finite subset V_0 of V for which $u^{1/n} \in F(V_0^{1/n})$. Let $V_1 = \langle V_0, (F^*)^n \rangle$ be the subgroup generated by V_0 and $(F^*)^n$. Then

$$V_0^{1/n} \subseteq V_1^{1/n} \subseteq V^{1/n}$$

and

$$u^{1/n} \in F(V_0^{1/n}) \subseteq F(V_1^{1/n})$$

Note that $V_1 \in \mathcal{U}_n$ is finitely generated (by V_0) over F^{*n} and hence $(V_1 : (F^*)^n)$ is finite. Theorem 15.3.1 implies that

$$[F(V_1^{1/n}) : F] = (V_1 : (F^*)^n)$$

Let us now adjoin u. Let $V_2 = \langle u, V_1 \rangle$ be the subgroup generated by u and V_1. Then $V_2 \in \mathcal{U}_n$ and

$$F(V_2^{1/n}) = F(\langle u, V_1 \rangle^{1/n}) = F(\langle \alpha, V_1^{1/n} \rangle) = F(V_1^{1/n})$$

Another application of Theorem 15.3.1 gives

$$(V_2 : (F^*)^n) = (V_1 : (F^*)^n)$$

and since $V_1 \subseteq V_2$ we get $V_1 = V_2$. It follows that $u \in V_1 \subseteq V$ and since u was arbitrary, $U \subseteq V$. A symmetric argument gives $V \subseteq U$, whence $U = V$. This proves that the map $U \mapsto F(U^{1/n})$ is injective. We have seen that any Kummer extension $F < E$ in $\mathcal{K}_n$ is a splitting field extension for a family $\mathcal{F}$ of binomials with exponent n. If C is the set of constant terms and if U is the subgroup of F^* generated by C and F^{*n} then $E = F(U^{1/n})$ and so the map is surjective.

Let $F < E$ be a Kummer extension with exponent n and let $U = E^{*n} \cap F^*$. Then U is a subgroup of F^* containing F^{*n}, that is, $U \in \mathcal{U}_n$. It is clear that $E \subseteq F(U^{1/n})$. For the reverse inclusion, let $\beta^n \in U$. Then $\beta^n = \alpha^n$ for some $\alpha \in E^*$, which implies that β is a root of $x^n - \alpha^n \in F[x]$ and so $\beta = \omega^i \alpha \in E^*$. This shows that $U^{1/n} \subseteq E^*$ and so $E = F(U^{1/n})$. Hence, $E \mapsto U = E^{*n} \cap F^*$ is the inverse map. □

Exercises

1. Referring to Lemma 15.2.1, show that the map

$$\mathcal{P}: \hom(A, C) \times \hom(B, C) \to \hom(A \times B, C)$$

defined by

$$\mathcal{P}(\lambda, \theta)(\alpha, \beta) = \lambda(\alpha)\theta(\beta)$$

is an isomorphism.

2. Let A be a finite group and let $\sigma \in \hom(G, U_n)$. Show that $\sum_{a \in A} \sigma(a) = |A|$ if $\sigma(a) = 1$ for all $a \in A$ and $\sum_{a \in A} \sigma(a) = 0$ otherwise.
3. Let A be a finite abelian group with exponent n. If $\beta \in A$ satisfies $\sigma(\beta) = 1$ for all $\sigma \in \hom(A, U_n)$ then $\beta = 1$.
4. Let B be a proper subgroup of a finite abelian group A of exponent n and let $\alpha \in A \setminus B$. Then there exists $\sigma \in \hom(A, U_n)$ such that $\sigma B = \{1\}$ but $\sigma\alpha \neq 1$.
5. Let B be a subgroup of a finite abelian group A of exponent n. Let

$$B^\perp = \{\sigma \in \hom(A, U_n) \mid \sigma B = \{1\}\}$$

Show that

$$\hom(B, U_n) \approx \hom(A, U_n)/B^\perp$$

6. Let B be a subgroup of a finite abelian group A of exponent n. Let

$$B^\perp = \{\sigma \in \hom(A, U_n) \mid \sigma B = \{1\}\}$$

Show that $\hom(A/B, U_n) \approx B^\perp$.

7. Let $\mathcal{F} = \{x^{n_i} - u_i \mid i \in I\}$ be a family of binomials over F of varying degrees. Suppose that $n_i \mid n$ for all $i \in I$ and that F contains a primitive nth root of unity. Show that there is a family $\mathcal{G}$ of binomials over F, each of which has degree n, with the same splitting field as $\mathcal{F}$.
8. In this exercise, we develop the analogous theory for families of polynomials of the form $\mathcal{F} = \{x^p - x - u_i\}$ where $p = \text{char}(F) \neq 0$.
   a) Prove that $F < E$ is abelian with exponent p if and only if E is the splitting field of a family of the form $\mathcal{F}$.
   b) Let $\mathcal{P}: \overline{F} \to \overline{F}$ be the map $\mathcal{P}\alpha = \alpha^p - \alpha$. Let $\mathcal{P}^{-1}U = \{\alpha \in \overline{F}$ such that $\mathcal{P}\alpha \in U\}$. Let $\mathcal{U}$ be the class of all additive subgroups of F with $\mathcal{P}^{-1}F \subseteq U$. Let $\mathcal{E}_p$ be the class of all abelian extensions $F < E$ of F with exponent p. Prove the following theorem: The map $U \mapsto F(\mathcal{P}^{-1}U)$ is a bijection between $\mathcal{U}$ and $\mathcal{E}_p$. If $F < E = F(\mathcal{P}^{-1}U)$ is in $\mathcal{E}_p$ and has Galois group G then there is a well-defined pairing $\langle , \rangle: G \times U \to U_p$ given by $\langle \sigma, \alpha \rangle = \sigma\beta - \beta$ for any $\beta \in \mathcal{P}^{-1}\alpha$. The left kernel is $\{1\}$ and the right kernel is $\mathcal{P}F$. The extension $F < E$ is finite if and only if $(U : \mathcal{P}F)$ is finite, in which case $[E : F] = (U : \mathcal{P}F)$ and $G \approx (U/\mathcal{P}F)^{\wedge}$.

Appendix
Möbius Inversion

Möbius inversion is a method for inverting certain types of sums. The classical form of Möbius inversion was originally developed independently by P. Hall and L. Weisner, in 1935. However, in 1964, Gian-Carlo Rota generalized the classical form to apply to a much wider range of situations. To describe the concept in its fullest generality, we require some facts about partially ordered sets.

Partially Ordered Sets

Definition *A partial order on a nonempty set P is a binary relation, denoted by $\leq$ and read "less than or equal to," with the following properties:*
1) **(reflexivity)** *For all $a \in P$,*

$$a \leq a$$

2) **(antisymmetry)** *For all $a, b \in P$,*

$$a \leq b \textit{ and } b \leq a \textit{ implies } a = b$$

3) **(transitivity)** *For all $a, b, c \in P$,*

$$a \leq b \textit{ and } b \leq c \textit{ implies } a \leq c$$

□

Definition A **partially ordered set** is a nonempty set P, together with a partial order $\leq$ defined on P. The expression $a \leq b$ is read "a is less than or equal to b." If $a, b \in P$, we denote the fact that a is *not* less than or equal to b by $a \not\leq b$. Also, we denote the fact that $a \leq b$, but $a \neq b$, by $a < b$.□

If there exists an element $z \in P$ for which $z \leq x$ for all $x \in P$, we call z a **zero** element and denote it by 0. Similarly, if there exists an element $y \in P$ for which $x \leq y$ for all $x \in P$, then we call y a **one** and denote it by 1.

As is customary, when the partial order $\leq$ is understood, we will use the phrase "let P be a partially ordered set."

Note that in a partially ordered set, it is possible that not all elements are comparable. In other words, it is possible to have $x, y \in P$ with the property that $x \not\leq y$ and $y \not\leq x$. Thus, in general, $x \not\leq y$ is *not* equivalent to $y \leq x$. A partially ordered set in which every pair of elements is comparable is called a **totally ordered set** or a **linearly ordered set**.

Example A.2.1

1) The set $\mathbb{R}$ of real numbers, with the usual binary relation $\leq$, is a partially ordered set. It is also a totally ordered set.
2) The set $\mathbb{N}$ of natural numbers, together with the binary relation of divides, is a partially ordered set. It is customary to write $n \mid m$ (rather than $n \leq m$) to indicate that n divides m.
3) Let S be any set, and let $\mathcal{P}(S)$ be the power set of S, that is, the set of all subsets of S. Then $\mathcal{P}(S)$, together with the subset relation $\subseteq$, is a partially ordered set. z

Definition *Let P be a partially ordered set. For $a, b \in P$, the* (*closed*) *interval $[a, b]$ is the set*

$$[a, b] = \{x \in P \mid a \leq x \leq b\}$$

We say that the partially ordered set P is **locally finite** *if every closed interval is a finite set.*□

Notice that if P is locally finite and contains a zero element 0, then the set $\{x \in P \mid x \leq a\}$ is finite for all $a \in P$, for it is the same as the interval $[0, a]$.

The Incidence Algebra of a Partially Ordered Set

Now let P be a locally finite partially ordered set, and let F be a field. We set

$$\mathcal{A}(P) = \{f : P \times P \to F \mid f(x, y) = 0 \text{ if } x \not\leq y\}$$

Addition and scalar multiplication are defined on $\mathcal{A}(P)$ by

$$(f + g)(x, y) = f(x, y) + g(x, y)$$

and

$$(kf)(x, y) = k[f(x, y)]$$

We also define multiplication by

$$(f * g)(x, y) = \sum_{x \leq z \leq y} f(x, z)g(z, y)$$

the sum being finite, since P is assumed to be locally finite. Using these definitions, it is not hard to show that $\mathcal{A}(P)$ is a noncommutative algebra, called the **incidence algebra** of P. The identity in this algebra is

$$\delta(x,y) = \begin{cases} 1 & \text{if } x = y \\ 0 & \text{if } x \neq y \end{cases}$$

The next theorem characterizes those elements of $\mathcal{A}(P)$ that have multiplicative inverses.

Theorem A.2.1 *An element $f \in \mathcal{A}(P)$ is invertible if and only if $f(x,x) \neq 0$ for all $x \in P$.*
Proof. A right inverse g_R of f must satisfy

$$\sum_{x \leq z \leq y} f(x,z) g_R(z,y) = \delta(x,y) \tag{A.2.1}$$

In particular, for $x = y$, we get

$$f(x,x) g_R(x,x) = 1$$

This shows the necessity and also that $g_R(x,x)$ must satisfy

$$g_R(x,x) = \frac{1}{f(x,x)} \tag{A.2.2}$$

Equation (A.2.2) defines $g_R(x,y)$ when the interval $[x,y]$ has cardinality 1, that is, when $x = y$. We can use (A.2.1) to define $g_R(x,y)$ for intervals $[x,y]$ of all cardinalities.

Suppose that $g_R(x,y)$ has been defined for all intervals with cardinality at most n, and let $[x,y]$ have cardinality $n+1$. Then, by (A.2.1), since $x \neq y$, we get

$$f(x,x) g_R(x,y) = -\sum_{x < z \leq y} f(x,z) g_R(z,y)$$

But $g_R(z,y)$ is defined for $z > x$ since $[z,y]$ has cardinality at most n, and so we can use this to define $g_R(x,y)$.

Similarly, we can define a left inverse g_L using the analogous process. But

$$g_L = g_L * (f * g_R) = (g_L * f) * g_R = g_R$$

and so $g_R = g_L$ is an inverse for f.□

Definition *The function $\zeta \in \mathcal{A}(P)$, defined by*

$$\zeta(x,y) = \begin{cases} 1 & \text{if } x \leq y \\ 0 & \text{if } x \not\leq y \end{cases}$$

is called the **zeta function**. *Its inverse $\mu(x,y)$ is called the* **Möbius function**.□

The next result follows from the appropriate definitions.

Theorem A.2.2 *The Möbius function is uniquely determined by either of the following conditions:*
1) $\mu(x,x) = 1$ *and for* $x < y$,

$$\sum_{x \le z \le y} \mu(z,y) = 0$$

2) $\mu(x,x) = 1$ *and for* $x < y$,

$$\sum_{x \le z \le y} \mu(x,z) = 0$$

□

Now we come to the main result.

Theorem A.2.3 (Möbius Inversion) *Let* P *be a locally finite partially ordered set with zero element* 0. *If* f *and* g *are functions from* P *to the field* F, *then*

$$g(x) = \sum_{y \le x} f(y) \Rightarrow f(x) = \sum_{y \le x} g(y)\mu(y,x) \tag{A.2.4}$$

If P *is a locally finite partially ordered set with* 1, *then*

$$g(x) = \sum_{x \le y} f(y) \Rightarrow f(x) = \sum_{x \le y} \mu(x,y)g(y) \tag{A.2.5}$$

Proof. Since all sums are finite, we have, for any x,

$$\begin{aligned}
\sum_{y \le x} g(y)\mu(y,x) &= \sum_{y \le x} \left[\sum_{z \le y} f(z) \right] \mu(y,x) \\
&= \sum_{z \le x} \sum_{z \le y \le x} f(z)\mu(y,x) \\
&= \sum_{z \le x} f(z) \sum_{z \le y \le x} \mu(y,x) \\
&= \sum_{z \le x} f(z)\delta(z,x) \\
&= f(x)
\end{aligned}$$

The rest of the theorem is proved similarly.□

The formulas (A.2.4) and (A.2.5) are called **Möbius inversion formulas**.

Example A.2.2 (Subsets) Let $P = \mathcal{P}(S)$ be the set of all subsets of a finite set S, partially ordered by set inclusion. We will use the notation $\subseteq$ for subset and $\subset$ for *proper* subset. The zeta function is

$$\zeta(A, B) = \begin{cases} 1 & \text{if } A \subseteq B \\ 0 & \text{otherwise} \end{cases}$$

The Möbius function μ is computed as follows. From Theorem A.2.2, we have

$$\mu(A, A) = 1$$

and

$$\mu(A, B) = -\sum_{A \subseteq X \subset B} \mu(A, X)$$

So, for $x, y, z \notin A$, we have

$$\begin{aligned} \mu(A, A \cup \{x\}) &= -\mu(A, A) = -1 \\ \mu(A, A \cup \{x, y\}) &= -\mu(A, A) - \mu(A, A \cup \{x\}) - \mu(A, A \cup \{y\}) \\ &= -1 + 1 + 1 = 1 \\ \mu(A, A \cup \{x, y, z\}) &= -\mu(A, A) - \mu(A, A \cup \{x\}) - \mu(A, A \cup \{y\}) \\ &= -\mu(A, A \cup \{x, y\}) - \mu(A, A \cup \{x, z\}) \\ &\quad - \mu(A, A \cup \{y, z\}) \\ &= -1 + 1 + 1 + 1 - 1 - 1 - 1 \\ &= -1 \end{aligned}$$

It begins to appear that the values of μ alternate between +1 and -1 and that

$$\mu(A, A \cup B) = (-1)^{|B|}$$

Asume this is true for $|B| < n$ and let $|B| = n$. Then

$$\sum_{A \subseteq A \cup X \subseteq A \cup B} (-1)^{|X|} = \sum_{X \subseteq B} (-1)^{|X|} = \sum_{k=0}^{|B|} \binom{|B|}{k} (-1)^k = 0$$

Now let $P_1, \ldots, P_n$ be "properties" that the elements of a set S may or may not possess, that is, $P_i \subseteq S$. For $K \subseteq \{1, \ldots, k\}$, let $E(K)$ be the number of elements of S that have properties P_i for $i \in K$, *and no others*. Let $F(K)$ be the number of elements of S that have *at least* properties P_i, for $i \in K$. Thus, for $K \neq \emptyset$,

$$F(K) = \left| \bigcap_{k \in K} P_k \right|$$

where an empty intersection is defined to be S, and

$$E(K) = \left| \bigcap_{k \in K} P_k \cap \bigcap_{j \notin K} P_j^c \right|$$

Then

$$F(K) = \sum_{K \subseteq L} E(L)$$

Hence, by Möbius inversion,

$$E(K) = \sum_{K \subseteq L} (-1)^{|L \setminus K|} F(L)$$

that is,

$$E(K) = \sum_{K \subseteq L} (-1)^{|L \setminus K|} \left| \bigcap_{j \in L} P_j \right|$$

In particular, if $K = \emptyset$ is the empty set, then

$$\begin{aligned} E(\emptyset) &= \sum_{L \subseteq S} (-1)^{L} \left| \bigcap_{j \in L} P_j \right| \\ &= \sum_{k \geq 0} \sum_{i_1 < \cdots < i_k} (-1)^k |P_{i_1} \cap \cdots \cap P_{i_k}| \end{aligned}$$

where $E(\emptyset)$ is the number of elements of S that have *none* of the properties. Since

$$E(\emptyset) = |S| - |P_1 \cup \cdots \cup P_k|$$

and since the first term in the previous expression for $E(\emptyset)$ is $|S|$, we get

$$|P_1 \cup \cdots \cup P_k| = \sum_{k \geq 1} \sum_{i_1 < \cdots < i_k} (-1)^{k+1} |P_{i_1} \cap \cdots \cap P_{i_k}|$$

For example, if $k = 2$, then

$$|P_1 \cup P_2| = |P_1| + |P_2| - |P_1 \cap P_2|$$

This formula is the well-known *Principle of Inclusion–Exclusion*, which we now see is just a special case of Möbius inversion.□

Classical Möbius Inversion

Consider the partially ordered set of positive natural numbers, ordered by division. That is, x is less than or equal to y if and only if x divides y, which we will denote by $x \mid y$. The zero element is 1.

In this case, the Möbius function $\mu(x, y)$ depends only on the ratio y/x, and is given by

$$\mu(x,y)=\mu\left(\frac{y}{x}\right)=\begin{cases}1 & \text{if } \frac{y}{x}=1\\ (-1)^k & \text{if } \frac{y}{x}=p_1\cdots p_k \text{ for } \textit{distinct} \text{ primes } p_i\\ 0 & \text{otherwise}\end{cases}$$

Notice that the "otherwise" case can occur if either $x \nmid y$ (x does not divide y) or if $p^2 \mid (y/x)$ for some prime p.

To verify that this is indeed the Möbius function, we first observe that $\mu(x,x)=\mu(1)=1$. Now let $x \mid y,\ x \neq y$ and

$$\frac{y}{x}=p_1^{e_1}p_2^{e_2}\cdots p_n^{e_n}$$

where the p_i are *distinct* primes. Then

$$\sum_{x|z|y}\mu\left(\frac{z}{x}\right)=\sum_{1|\frac{z}{x}|\frac{y}{x}}\mu\left(\frac{z}{x}\right)=\sum_{1|k|\frac{y}{x}}\mu(k)=1+\sum_{1\le j\le n}\binom{n}{j}(-1)^j=0$$

Now, in the present context, the Möbius inversion formula becomes

$$g(n)=\sum_{k|n}f(k)\Rightarrow f(n)=\sum_{k|n}g(k)\mu\left(\frac{n}{k}\right)$$

This is the important classical formula, which often goes by the name Möbius inversion formula.

Multiplicative Version of Möbius Inversion

We now present a multiplicative version of the Möbius inversion formula.

Theorem A.2.4 Let P be a locally finite partially ordered set with zero element 0. If f and g are functions from P to F, then

$$g(x)=\prod_{y\le x}f(y)\Rightarrow f(x)=\prod_{y\le x}[g(y)]^{\mu(y,x)}$$

Proof. Since all products are finite, we have, for any x,

$$\begin{aligned}
\prod_{y \le x} [g(y)]^{\mu(y,x)} &= \prod_{y \le x} \left[\prod_{z \le y} f(z) \right]^{\mu(y,x)} \\
&= \prod_{z \le x} \prod_{z \le y \le x} [f(z)]^{\mu(y,x)} \\
&= \prod_{z \le x} f(z)^{\sum\limits_{z \le y \le x} \mu(y,x)} \\
&= \prod_{z \le x} f(z)^{\delta(z,x)} \\
&= f(x)
\end{aligned}$$

□

Example A.2.3 Let $P = \mathbb{N}$, and let F be the field of rational functions in x. Consider the formula

$$x^n - 1 = \prod_{k|n} Q_k(x)$$

Then, if we let $f(k) = Q_k(x)$ and $g(n) = x^n - 1$, Theorem A.2.4 gives

$$Q_n(x) = \prod_{k|n} (x^k - 1)^{\mu(n/k)} = \prod_{k|n} (x^{n/k} - 1)^{\mu(k)}$$

□

References

Artin, E., *Galois Theory*, 2nd ed., Notre Dame Press, 1959.

Bescovitch, A.S., On the linear independence of fractional powers of integers, *J. London Math. Soc.* 15 (1940) 3–6.

Brawley, J. and Schnibben, G., *Infinite Algebraic Extensions of Finite Fields*, AMS, 1989.

Edwards, H., *Galois Theory*, Springer-Verlag, 1984.

Gaal, L., *Classical Galois Theory*, 4th ed., Chelsea, 1988.

Jacobson, N., *Basic Algebra II*, Freeman, 1989.

Lidl, R. and Niederreiter, H., *Introduction to Finite Fields and Their Applications*, Cambridge University Press, 1986.

Richards, I., An application of Galois theory to elementary arithmetic, *Advances in Mathematics* 13 (1974) 268-273.

Roman, S., *Advanced Linear Algebra*, second edition, Springer-Verlag, 2005.

Schinzel, A., Abelian binomials, power residues, and exponential congruences, *Acta. Arith.* 32 (1977) 245-274.

Index

(continued from page ii)

64 EDWARDS. Fourier Series. Vol. I. 2nd ed.
65 WELLS. Differential Analysis on Complex Manifolds. 2nd ed.
66 WATERHOUSE. Introduction to Affine Group Schemes.
67 SERRE. Local Fields.
68 WEIDMANN. Linear Operators in Hilbert Spaces.
69 LANG. Cyclotomic Fields II.
70 MASSEY. Singular Homology Theory.
71 FARKAS/KRA. Riemann Surfaces. 2nd ed.
72 STILLWELL. Classical Topology and Combinatorial Group Theory. 2nd ed.
73 HUNGERFORD. Algebra.
74 DAVENPORT. Multiplicative Number Theory. 3rd ed.
75 HOCHSCHILD. Basic Theory of Algebraic Groups and Lie Algebras.
76 IITAKA. Algebraic Geometry.
77 HECKE. Lectures on the Theory of Algebraic Numbers.
78 BURRIS/SANKAPPANAVAR. A Course in Universal Algebra.
79 WALTERS. An Introduction to Ergodic Theory.
80 ROBINSON. A Course in the Theory of Groups. 2nd ed.
81 FORSTER. Lectures on Riemann Surfaces.
82 BOTT/TU. Differential Forms in Algebraic Topology.
83 WASHINGTON. Introduction to Cyclotomic Fields. 2nd ed.
84 IRELAND/ROSEN. A Classical Introduction to Modern Number Theory. 2nd ed.
85 EDWARDS. Fourier Series. Vol. II. 2nd ed.
86 VAN LINT. Introduction to Coding Theory. 2nd ed.
87 BROWN. Cohomology of Groups.
88 PIERCE. Associative Algebras.
89 LANG. Introduction to Algebraic and Abelian Functions. 2nd ed.
90 BRØNDSTED. An Introduction to Convex Polytopes.
91 BEARDON. On the Geometry of Discrete Groups.
92 DIESTEL. Sequences and Series in Banach Spaces.
93 DUBROVIN/FOMENKO/NOVIKOV. Modern Geometry—Methods and Applications. Part I. 2nd ed.
94 WARNER. Foundations of Differentiable Manifolds and Lie Groups.
95 SHIRYAEV. Probability. 2nd ed.
96 CONWAY. A Course in Functional Analysis. 2nd ed.
97 KOBLITZ. Introduction to Elliptic Curves and Modular Forms. 2nd ed.
98 BRÖCKER/TOM DIECK. Representations of Compact Lie Groups.
99 GROVE/BENSON. Finite Reflection Groups. 2nd ed.
100 BERG/CHRISTENSEN/RESSEL. Harmonic Analysis on Semigroups: Theory of Positive Definite and Related Functions.
101 EDWARDS. Galois Theory.
102 VARADARAJAN. Lie Groups, Lie Algebras and Their Representations.
103 LANG. Complex Analysis. 3rd ed.
104 DUBROVIN/FOMENKO/NOVIKOV. Modern Geometry—Methods and Applications. Part II.
105 LANG. $SL_2(\mathbf{R})$.
106 SILVERMAN. The Arithmetic of Elliptic Curves.
107 OLVER. Applications of Lie Groups to Differential Equations. 2nd ed.
108 RANGE. Holomorphic Functions and Integral Representations in Several Complex Variables.
109 LEHTO. Univalent Functions and Teichmüller Spaces.
110 LANG. Algebraic Number Theory.
111 HUSEMÖLLER. Elliptic Curves. 2nd ed.
112 LANG. Elliptic Functions.
113 KARATZAS/SHREVE. Brownian Motion and Stochastic Calculus. 2nd ed.
114 KOBLITZ. A Course in Number Theory and Cryptography. 2nd ed.
115 BERGER/GOSTIAUX. Differential Geometry: Manifolds, Curves, and Surfaces.
116 KELLEY/SRINIVASAN. Measure and Integral. Vol. I.
117 J.-P. SERRE. Algebraic Groups and Class Fields.
118 PEDERSEN. Analysis Now.
119 ROTMAN. An Introduction to Algebraic Topology.
120 ZIEMER. Weakly Differentiable Functions: Sobolev Spaces and Functions of Bounded Variation.
121 LANG. Cyclotomic Fields I and II. Combined 2nd ed.
122 REMMERT. Theory of Complex Functions. *Readings in Mathematics*
123 EBBINGHAUS/HERMES et al. Numbers. *Readings in Mathematics*

124 DUBROVIN/FOMENKO/NOVIKOV. Modern Geometry—Methods and Applications. Part III
125 BERENSTEIN/GAY. Complex Variables: An Introduction.
126 BOREL. Linear Algebraic Groups. 2nd ed.
127 MASSEY. A Basic Course in Algebraic Topology.
128 RAUCH. Partial Differential Equations.
129 FULTON/HARRIS. Representation Theory: A First Course. *Readings in Mathematics*
130 DODSON/POSTON. Tensor Geometry.
131 LAM. A First Course in Noncommutative Rings. 2nd ed.
132 BEARDON. Iteration of Rational Functions.
133 HARRIS. Algebraic Geometry: A First Course.
134 ROMAN. Coding and Information Theory.
135 ROMAN. Advanced Linear Algebra. 2nd ed.
136 ADKINS/WEINTRAUB. Algebra: An Approach via Module Theory.
137 AXLER/BOURDON/RAMEY. Harmonic Function Theory. 2nd ed.
138 COHEN. A Course in Computational Algebraic Number Theory.
139 BREDON. Topology and Geometry.
140 AUBIN. Optima and Equilibria. An Introduction to Nonlinear Analysis.
141 BECKER/WEISPFENNING/KREDEL. Grübner Bases. A Computational Approach to Commutative Algebra.
142 LANG. Real and Functional Analysis. 3rd ed.
143 DOOB. Measure Theory.
144 DENNIS/FARB. Noncommutative Algebra.
145 VICK. Homology Theory. An Introduction to Algebraic Topology. 2nd ed.
146 BRIDGES. Computability: A Mathematical Sketchbook.
147 ROSENBERG. Algebraic *K*-Theory and Its Applications.
148 ROTMAN. An Introduction to the Theory of Groups. 4th ed.
149 RATCLIFFE. Foundations of Hyperbolic Manifolds.
150 EISENBUD. Commutative Algebra with a View Toward Algebraic Geometry.
151 SILVERMAN. Advanced Topics in the Arithmetic of Elliptic Curves.
152 ZIEGLER. Lectures on Polytopes.
153 FULTON. Algebraic Topology: A First Course.
154 BROWN/PEARCY. An Introduction to Analysis.
155 KASSEL. Quantum Groups.
156 KECHRIS. Classical Descriptive Set Theory.
157 MALLIAVIN. Integration and Probability.
158 ROMAN. Field Theory. 2nd ed.
159 CONWAY. Functions of One Complex Variable II.
160 LANG. Differential and Riemannian Manifolds.
161 BORWEIN/ERDÉLYI. Polynomials and Polynomial Inequalities.
162 ALPERIN/BELL. Groups and Representations.
163 DIXON/MORTIMER. Permutation Groups.
164 NATHANSON. Additive Number Theory: The Classical Bases.
165 NATHANSON. Additive Number Theory: Inverse Problems and the Geometry of Sumsets.
166 SHARPE. Differential Geometry: Cartan's Generalization of Klein's Erlangen Program.
167 MORANDI. Field and Galois Theory.
168 EWALD. Combinatorial Convexity and Algebraic Geometry.
169 BHATIA. Matrix Analysis.
170 BREDON. Sheaf Theory. 2nd ed.
171 PETERSEN. Riemannian Geometry.
172 REMMERT. Classical Topics in Complex Function Theory.
173 DIESTEL. Graph Theory. 2nd ed.
174 BRIDGES. Foundations of Real and Abstract Analysis.
175 LICKORISH. An Introduction to Knot Theory.
176 LEE. Riemannian Manifolds.
177 NEWMAN. Analytic Number Theory.
178 CLARKE/LEDYAEV/ STERN/WOLENSKI. Nonsmooth Analysis and Control Theory.
179 DOUGLAS. Banach Algebra Techniques in Operator Theory. 2nd ed.
180 SRIVASTAVA. A Course on Borel Sets.
181 KRESS. Numerical Analysis.
182 WALTER. Ordinary Differential Equations.

183 MEGGINSON. An Introduction to Banach Space Theory.
184 BOLLOBAS. Modern Graph Theory.
185 COX/LITTLE/O'SHEA. Using Algebraic Geometry. 2nd ed.
186 RAMAKRISHNAN/VALENZA. Fourier Analysis on Number Fields.
187 HARRIS/MORRISON. Moduli of Curves.
188 GOLDBLATT. Lectures on the Hyperreals: An Introduction to Nonstandard Analysis.
189 LAM. Lectures on Modules and Rings.
190 ESMONDE/MURTY. Problems in Algebraic Number Theory. 2nd ed.
191 LANG. Fundamentals of Differential Geometry.
192 HIRSCH/LACOMBE. Elements of Functional Analysis.
193 COHEN. Advanced Topics in Computational Number Theory.
194 ENGEL/NAGEL. One-Parameter Semigroups for Linear Evolution Equations.
195 NATHANSON. Elementary Methods in Number Theory.
196 OSBORNE. Basic Homological Algebra.
197 EISENBUD/HARRIS. The Geometry of Schemes.
198 ROBERT. A Course in p-adic Analysis.
199 HEDENMALM/KORENBLUM/ZHU. Theory of Bergman Spaces.
200 BAO/CHERN/SHEN. An Introduction to Riemann–Finsler Geometry.
201 HINDRY/SILVERMAN. Diophantine Geometry: An Introduction.
202 LEE. Introduction to Topological Manifolds.
203 SAGAN. The Symmetric Group: Representations, Combinatorial Algorithms, and Symmetric Functions.
204 ESCOFIER. Galois Theory.
205 FÉLIX/HALPERIN/THOMAS. Rational Homotopy Theory. 2nd ed.
206 MURTY. Problems in Analytic Number Theory. *Readings in Mathematics*
207 GODSIL/ROYLE. Algebraic Graph Theory.
208 CHENEY. Analysis for Applied Mathematics.
209 ARVESON. A Short Course on Spectral Theory.
210 ROSEN. Number Theory in Function Fields.
211 LANG. Algebra. Revised 3rd ed.
212 MATOUek. LECTURES ON DISCRETE GEOMETRY.
213 FRITZSCHE/GRAUERT. From Holomorphic Functions to Complex Manifolds.
214 JOST. Partial Differential Equations.
215 GOLDSCHMIDT. Algebraic Functions and Projective Curves.
216 D. SERRE. Matrices: Theory and Applications.
217 MARKER. Model Theory: An Introduction.
218 LEE. Introduction to Smooth Manifolds.
219 MACLACHLAN/REID. The Arithmetic of Hyperbolic 3-Manifolds.
220 NESTRUEV. Smooth Manifolds and Observables.
221 GRÜNBAUM. Convex Polytopes. 2nd ed.
222 HALL. Lie Groups, Lie Algebras, and Representations: An Elementary Introduction.
223 VRETBLAD. Fourier Analysis and Its Applications.
224 WALSCHAP. Metric Structures in Differential Geometry.
225 BUMP: Lie Groups
226 ZHU. Spaces of Holomorphic Functions in the Unit Ball.
227 MILLER/STURMFELS. Combinatorial Commutative Algebra.
228 DIAMOND/SHURMAN. A First Course in Modular Forms.
229 EISENBUD. The Geometry of Syzygies.
230 STROOCK. An Introduction to Markov Processes.